Engineering Hydrology

Xi Feng
Editor

Engineering Hydrology

中国水利水电出版社
www.waterpub.com.cn

Springer

Editor
Xi Feng
College of Harbour, Coastal and Offshore Engineering
Hohai University
Nanjing, China

ISBN 978-981-95-9088-9 ISBN 978-981-95-9089-6 (eBook)
https://doi.org/10.1007/978-981-95-9089-6

Jointly published with China Water & Power Press.
The print edition is not for sale in China mainland. Customers from China mainland please order the print book from: China Water & Power Press.
ISBN of the Co-Publisher's edition: 978-7-5226-3339-8

This Springer imprint is published by the registered company Springer Nature Singapore Pte Ltd.
The registered company address is: 152 Beach Road, #21-01/04 Gateway East, Singapore 189721, Singapore

The Painting of Yu the Great Controlling the Waters, a 12th-century masterpiece attributed to Zhao Boju, a renowned Southern Song Dynasty painter and royal descendant, is presented here in a detailed partial view from the original scroll. Executed in an exquisite azure-green and gold palette, the scroll vividly depicts Yu the Great—a legendary ruler and hydraulic engineer—directing laborers to channel floodwaters through mountain clefts and riverbanks. As a cultural symbol, this artwork immortalizes China's earliest recorded large-scale water management efforts, reflecting the ancient philosophy of harmonizing human ingenuity with natural forces.

Yu's legacy, rooted in the mytho-historical narrative of taming the Yellow River's deluges, marks a pivotal shift from futile containment to systematic channeling—a principle foundational to modern hydrology. His 13-year effort, involving sustained perseverance and engineering innovation, corresponds to the topics in Chapter 1 concerning hydrological cycles, river systems, and the utilization of aquatic resources. By bridging antiquity's wisdom with contemporary hydrological science, we honor a tradition where water stewardship shaped civilizations, echoing across millennia from Yu's era to today's global challenges.

▲ *The Painting of Yu the Great Controlling the Waters* (《禹王治水图》) — Collection of the National Palace Museum of Taipei, Taiwan Province, PRC.

The Painting of Competing Streams in Myriad Valleys, a Ming-Dynasty landscape masterpiece by Wen Zhengming, one of the "Four Masters of the Wu School". Rendered in dynamic ink washes and delicate brushwork, the painting captures the vitality of mountain streams cascading through rugged gorges, converging into powerful rivers. The artwork embodies the Chinese philosophical concept of shan shui ("mountain-water")—a harmonious interplay of geological forces and hydrological energy—mirroring the interconnected systems of watersheds and river networks explored in Chapter 2.

Historically, Chinese scholars-artists like Wen Zhengming observed nature not merely aesthetically but as a dynamic hydrological laboratory. The painting's intricate depiction of tributaries carving valleys, seasonal runoff patterns, and erosional processes resonates with modern principles of fluvial hydrology. Chapter 2 systematically examines the mechanisms of runoff generation, drainage basin morphology, and hydrological regimes that govern the formation and functioning of riverine ecosystems.

By framing these scientific concepts through Wen's 16th-century lens, we honor an enduring tradition where art and hydrology converged to decode Earth's aqueous rhythms—a legacy now amplified by contemporary engineering to sustain our planet's arterial waterways.

▲ *The Painting of Competing Streams in Myriad Valleys* (《万壑争流图》) — Collection of the Nanjing Museum, Jiangsu Province, PRC.

The picture is sourced from Zhang Tuantuan, Liang Zhongmin, Li Wentao, et al., 2023. Statistical post-processing of precipitation forecasts using circulation classifications and spatiotemporal deep neural networks[J]. Hydrology and earth system sciences, 27(10): 1945-1960.

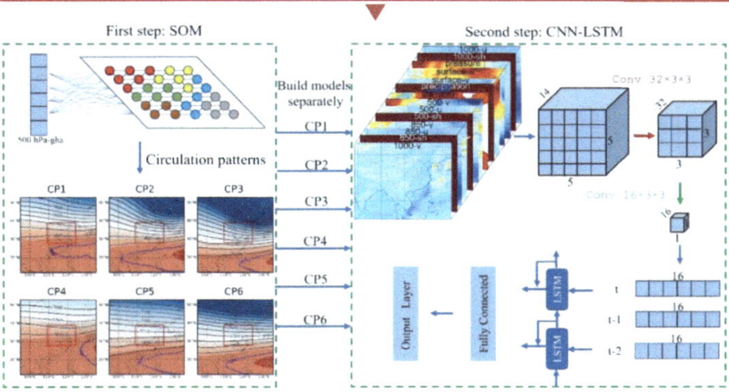

Chinese calligraphy "Li Chun", meaning the first of the 24 solar terms, marking the beginning of spring in China (Picture is generated by OpenAI, Feb, 24, 2026, chatgpt.com).

Along the River During the Qingming Festival is the 12th- century masterpiece by Zhang Zeduan. The selected vignette captures a cargo ship lowering its mast to navigate beneath Bianliang's iconic Hongqiao—a wooden arch structure spanning the Bian River. This meticulous detail reveals the Song Dynasty's advanced understanding of navigational hydrology: the bridge's clearance height was calibrated to seasonal water levels, ensuring safe passage for vessels during peak flows.

The Rainbow Bridge, constructed without piers to accommodate turbulent flow conditions, reflects an early empirical understanding of fluvial hydraulics. By integrating structural stability with hydrological constraints, its design anticipated modern principles of channel conveyance and flood-stage control. Accordingly, Chapter 4 examines contemporary methodologies for determining design flood levels and discharge rates, emphasizing the trade-offs between hydraulic safety and infrastructural feasibility.

By anchoring these calculations in Zhang Zeduan's vivid tableau—where geometry, hydraulics, and human ingenuity converge—we honor a legacy where art and science once charted the pulse of rivers, now amplified by stochastic modeling to safeguard communities against nature's caprices.

▲*Along the River During the Qingming Festival* (《清明上河图》) — Collection of the Palace Museum, Beijing, PRC.

The Picture of Fuxi and Nüwa was created during the Jin and Tang Dynasties. In 1965, this silk scroll, which is two meters long, was unearthed in Astana, Xinjiang, PRC. It is now housed in the Xinjiang Uygur Autonomous Region Museum.

The main characters in the picture are the legendary Fuxi and Nüwa. They have human faces and snake bodies. On the right side is Fuxi, who is known as the prime ancestor of Chinese culture. On the left side is Nüwa, with her hair piled up in a high bun and holding a compass in her right hand. She is the legendary goddess who created humans of the ancient China.

According to the legend, after the great flood, the two intermarried and reproduced humanity. In the picture, Fuxi and Nüwa embrace each other with their upper bodies, and their lower bodies in the shape of snake tails are intertwined in a spiral structure, symbolizing the endless reproduction of humanity. This coincides remarkably with the double helix structure of DNA discovered more than 1,000 years later by Nobel laureates James Watson and Francis Crick.

The historical record of the highest flood level observed at the Nanjing section of the Yangtze River is documented here, with the peak value reaching 10.22 meters in 1954. The most recent notable flood event occurred in 1998, marked by a water level of 10.14 meters at the junction of the iron and glass components depicted in the diagram.

The Baopingkou of Dujiangyan. The waterway shown in the photo is the inner branch of the Minjiang River, which is an artificial waterway excavated by Li Bing and his son more than 2,000 years ago through cutting through Yulei Mountain. The water of this waterway irrigates the present Chengdu Plain, PRC.
Photo by Feng Xi, Dujiangyan Scenic Area, October 2024.

The beautiful night view of the Tianjin-Haihe River section of the Beijing-Hangzhou Grand Canal at present. Cruise ships and cargo ships on the canal are still coming and going continuously.
Photo by Feng Xi, Tianjin, PRC, May 2024.

The unique water towns in the Jiangnan region of China. Small bridges and flowing water are the standard features of Chinese ancient towns. In the past, these waterways and black-awning boats supported people's livelihoods. Nowadays, tourists glide through water towns on black-awning boats, soaking in their charm and beauty.

Photo by Feng Xi, Wuzhen, PRC, October 2024.

The story of "Ne Zha Making Havoc in the Sea" originates from Buddhist tradition and can be traced back to the Song and Liao Dynasties. Rooted deeply in traditional Chinese mythology, it gained widespread popularity through its vivid depiction in the Ming Dynasty classic, Investiture of the Gods. Over the centuries, this beloved tale has inspired numerous adaptations, including animations, television series, and films. For instance, the animated movie *Ne Zha II: the Demon Child Making Havoc in the Sea,* directed by Jiaozi and released in early 2025, achieved remarkable success, surpassing 12 billion yuan in box office revenue by the time this textbook was compiled, and secured a position among the top ten highest- grossing films globally.

The image on the top vividly captures the dramatic moment when Ne Zha unleashes chaos upon the sea. The surging, towering waves depicted in this scene effectively convey the immense power and over- whelming force of nature, amplifying the story's tension and heightening the conflict within the narrative.

▲Image created by artificial intelligence with reference to film *Ne Zha* II: *the Demon Child Making Havoc in the Sea* released in 2025.

The Painting of Moonlit Night Watching the Tide is a 12th-century masterpiece by Southern Song Dynasty artist Li Song, housed in Taipei's National Palace Museum. The painting captures the ethereal grandeur of the Qiantang River tidal bore under a full moon, where literati gather in riverside pavilions to witness nature's rhythmic spectacle—a testament to China's ancient fascination with tidal phenomena.

Li Song's meticulous brushwork immortalizes the interplay of lunar cycles and hydrodynamic forces: the moon's gravitational pull, symbolized by its luminous arc, orchestrates the surging "silver dragons" of the Qiantang River tide. As Song scholars decoded tidal patterns for flood control and navigation, Chapter 6 explores modern tidal prediction techniques, including harmonic analysis of lunar-solar cycles and probabilistic modeling of extreme water levels. The painting's spectral human figures symbolize humanity's ongoing challenge with tidal uncertainty, now managed through satellite altimetry and real-time gauging.

The Painting of Moonlit Night Watching the Tide (《月夜观潮图》) — Collection of the National Palace Museum, Taipei.

The photo taken by the author on the expressway in Hainan after Typhoon Yagi passed shows that the woods, which are 30 meters high, were knocked down to the ground.

Zheng He's Seven Voyages, undertaken during the 15th century of the Ming Dynasty, navigated the monsoon-driven currents of the Indian Ocean, relying on sophisticated maritime knowledge, and established a lasting maritime legacy. Between 1405 and 1433, Zheng He's armada, consisting of over 200 ships and 27,000 crew members, utilized seasonal winds and coastal currents to navigate from China to East Africa. Their voyages epitomized ancient mastery of nearshore hydrodynamics: by timing departures with the northeast monsoon in winter, the fleets harnessed southward coastal currents, while the southwest monsoon in summer propelled their return—a harmonious interplay of wind and water now commemorated by UNESCO's maritime heritage.

Zheng He's navigators, attuned to tidal rips and estuarine eddies, prefigured modern coastal current analysis. Their logs, detailing the Bianliang Rainbow Bridge's hydraulic lessons (as immortalized in *Along the River During the Qingming Festival*), reveal how Song-Ming engineers optimized hull designs for variable flow regimes—anticipating today's computational fluid dynamics. Just as Ming admirals mapped the Malacca Strait's countercurrents to avoid (upstream struggle), Chapter 7 deciphers nearshore systems: from wind-driven littoral drift to tidal asymmetry's role in sediment transport.

This photo was taken by the author at Baochuan Park in Nanjing, which features large-scale models of Zheng He's treasure ships and serves as a significant site for the appreciation and preservation of ancient Chinese maritime history.

The Kuroshio Current, also known as the "Black Tide", is one of the strongest western boundary currents in the world's oceans, comparable to the Gulf Stream. It originates from a branch of the North Equatorial Current in the tropical Pacific, flowing northward before turning eastward and merging into the North Pacific Current.

The Kuroshio is characterized by deep clear blue water, high temperature, high salinity, fast velocity, and large volume transport. The main stream is approximately 100-200 km wide, with an influence depth of several hundred meters. As a major conveyor of heat and moisture in the North Pacific, it significantly regulates the climate of coastal East Asia, and represents a key component in the study of western Pacific circulation and climate systems.

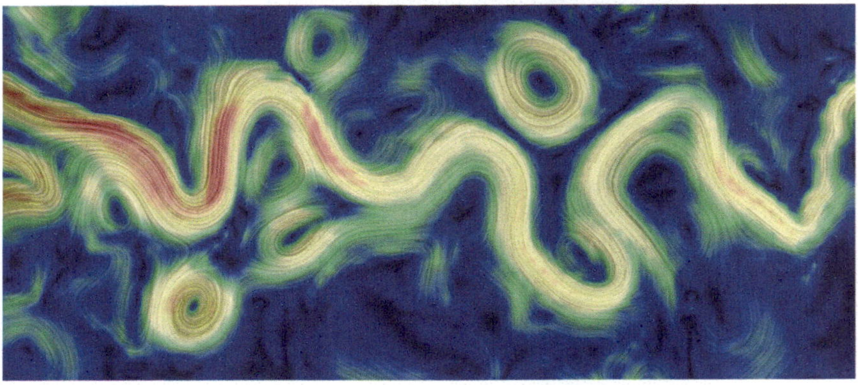

The image depicts the Kuroshio Current merging into the North Pacific Current, as simulated by the Global Ocean Physics Analysis and Forecast System (CMEMS).

Preface

Currently, there is no English textbook on Engineering Hydrology in the market that specifically caters to the major of Harbour, Waterway, and Coastal Engineering, as well as Coastal Resources. Most of the existing English textbooks are suitable for majors such as Hydrology and Water Resources, which are more land-oriented. However, for majors oriented towards engineering services, including coastal, ocean, and waterway engineering, the engineering hydrology knowledge delivered by these textbooks is far from sufficient. The fourth edition of *Engineering Hydrology*, published by Academician Qiu Dahong from Dalian University of Technology, is regarded as a classic in Engineering Hydrology for students and peers in the field of Harbour, Coastal, and Offshore Engineering in mainland China. Unfortunately, this textbook is only available in Chinese.

In the past decade, technology in the engineering field has been constantly innovating. For example, nature-based coastal protection methods and deep-learning tools have achieved rapid development from theory to practical applications. However, most of these advanced theories and methods are only found in journals and magazines, and very few have been incorporated into textbooks. The compilation of this textbook, *Engineering Hydrology*, not only focuses on knowledge dissemination but also hopes to serve as a bridge connecting industry applications and the transformation of professional technologies in the academic community.

In the process of compilation, this textbook mainly targets third-year undergraduate students majoring in Harbor, Waterway, and Coastal Engineering, and is also applicable to students and practitioners of other water-related engineering majors, including Hydraulic Engineering, Civil Engineering, and Ocean Engineering. This book adheres to the professional characteristics of equal emphasis on ports and waterways, and consideration of both rivers and seas, systematically explains the relationship between abstract hydrological knowledge and concrete engineering practices, aims to serve sea-related, navigation-related, and port-related projects, deeply analyzes engineering design concepts, engineering design parameter standards and definitions, and is committed to spreading the manufacturing technologies and design concepts of major waterway, coastal and ocean construction projects with international leading levels.

The chief editor of the textbook has been engaged in the curriculum reform of Engineering Hydrology since 2018. In 2019, newly recorded high-quality teaching videos were launched and successfully selected for the Construction Plan of Online Open Courses in Jiangsu Provincial Universities. Textbooks are indispensable teaching resources in curriculum construction, and the compilation of a new textbook is undoubtedly a crucial part of curriculum reform.

This textbook has the following main features and innovations:

(1) This is the first English textbook on Engineering Hydrology specifically designed for the major of Harbour, Waterway and Coastal Engineering, comprehensively covering the major of River Engineering Hydrology and Ocean Engineering Hydrology. Different from the *Engineering Hydrology* textbooks for majors in Hydrology and Water Resources or Water Conservancy and Hydropower Engineering, this textbook has irreplaceable value in determining sea-related and navigation-related engineering parameters.
(2) This is a characteristic textbook that organically integrates hydrology and engineering, with a particular focus on the cross-integration of interdisciplinary knowledge. At the same time, it incorporates multiple concepts such as green, low-carbon, and digital twin systems. It is suitable for both in-school students who will be engineers in the future and professional practitioners, providing them with professional and technical support.
(3) This is an English textbook that showcases the characteristics of Chinese water conservancy. While spreading hydrological knowledge, it skillfully integrates the unique water culture of the Chinese nation, adding fun points to reading. It fully demonstrates the contributions and leading roles of Chinese scientists and engineers across knowledge and technology in both engineering practice and theory.

Nanjing, China
February 2025

Xi Feng

Acknowledgements This book is humbly presented in tribute to Academician Qiu Dahong.

The first author, Xi Feng, extends her deepest gratitude to all contributors and translators of this textbook, including:

- The entire Engineering Hydrology teaching team at Hohai University's College of Harbour, Coastal and Offshore Engineering: Prof. Gong Zheng (Dean), Associate Professor Tan Ya, Associate Professor Ni Xingye, and Associate Professor Zhang Yu.
- Professor Kong Jun, Associate Professor Yang Jie, Associate Professor Zhang Qian, and Postdoctoral researcher Gu Haoyang (College of Harbour, Coastal and Offshore Engineering, Hohai University)
- Associate Professor Qin Youwei (College of Hydrology and Water Resources, Hohai University)
- Associate Professor Cao Haijin (College of Oceanography, Hohai University)

Special thanks to Prof. Feng Weibing (Hohai University) and Prof. Chen Yuanfang (College of Hydrology and Water Resources), whose pioneering works in *Marine Engineering Hydrology* and *Engineering Hydrology* (*River Hydrology*) profoundly inspired this edition.

The authors extend their sincere gratitude to their research students for their dedicated support in preparing this manuscript, including formula editing, figure organization, and illustration work. Dr. Li Huichao spearheaded the manuscript coordination. Liu Maoyuan, Gao Sheng, Xie Xiaolei, Yan Yi, Gao Fangfang, Liu Suning, Hu Zhengyuan, Lin Jie, Shu Ying, He Ruihang, Mao Yashi, Wu Haoyu, and Wen Shiwei contributed to the preparation of figures. Xu Hang prepared all the mind maps. Lin Jie, Cen Lin, Wu Junxiang, and Ding Jiayao provided valuable assistance in refining the text for clarity and readability. The authors also thank Zhang Di, Li Zheng, Fan Hongfei, Yan Wenju, and Mao Wei for their careful proofreading of the manuscript.

The authors wish to express gratitude for the support from the Construction Project of the Quality Engineering for the Cultivation of International Students in 2023 of Hohai University (Document No. 16 [2023] issued by the College of International Education, Hohai University), as well as the Brand Professional Project for the Cultivation of International Talents in Jiangsu Provincial Universities of the College of Harbour, Coastal and Offshore Engineering, Hohai University (Project No. B2203205). The authors want to express gratitude to Prof. Zhang Wei, the Vice Dean in charge of teaching affairs and personnel, for his comprehensive support for this book from its inception to its publication.

Competing Interests The author has no competing interests to declare that are relevant to the content of this manuscript.

Editorial Board

Introduction to Editor-in-Chief

This textbook was compiled and edited by Prof. Xi Feng at Hohai University.

Professor Xi Feng has long been engaged in the teaching and construction of Engineering Hydrology in the field of harbour, coastal and offshore engineering.

She has presided over key courses, textbooks, and education reform projects. Meanwhile, she has led over 10 key state research foundations, including the National Natural Science Foundation, the Foundation from the Ministry of Education, PRC, etc. She was awarded the "Young Scientist Award" by the China Association for Disaster Prevention and the first prize in science and technology.

Since 2017, she has taught core courses, including Engineering Hydrology and Coastal Dynamics, with a total teaching load of over 600 class hours and more than 2000 students. She has served as the principal investigator of five teaching reform projects.

She received the First Prize in the National Teaching Competition for Higher Education Institutions organized by the Ministry of Education of the People's Republic of China. She has also supervised students who have been awarded the Best Academic Thesis Awards (Ministry of Education, PRC), the National Scholarship, and the title of Outstanding Graduates of Jiangsu Province. In recognition of her contributions to teaching and research, she was awarded the Outstanding Teacher under the Qing-lan Project of Jiangsu Province in 2022.

Resource List

Resource number	Resource name	Resource type	Section number
ESM 1.1	Global water balance relationship	PDF	1.3.2
ESM 1.2	Units of runoff measurement	PDF	1.3.3
ESM 3.1	Example	PDF	3.2.2
ESM 3.2	Example	PDF	3.2.2, 3.4.1
ESM 3.3	Distribution parameters of random variables	PDF	3.2.3
ESM 3.4	Example	PDF	3.2.4 (3.2.4.2), 3.4.1
ESM 3.5	Statistical parameter estimation and Sampling error	PDF	3.3.2, 3.4.2 (3.4.2.3), 3.4.3 (3.4.3.3)
ESM 3.6	Example	PDF	3.4.2 (3.4.2.3)
ESM 3.7	Example	PDF	3.4.3 (3.4.3.3)
ESM 3.8	Related analysis	PDF	3.4.3 (3.4.3.4) 4.3.1 (4.3.1.1)
ESM 4.1	Compilation of daily mean water level tables, and plotting of daily mean hydrographs and duration curves	PDF	4.2.2 (4.2.2.3)
ESM 4.2	Method for extending the high water level portion of the stage-discharge relationship curve	PDF	4.3.2 (4.3.2.1), 4.5.4 (4.5.4.2)
ESM 4.3	Method for extending the low water level portion of the stage-discharge relationship curve	PDF	4.3.2 (4.3.2.2)

(continued)

(continued)

Resource number	Resource name	Resource type	Section number
ESM 4.4	Data compilation on flow rates	PDF	4.3.3
ESM 4.5	Example	PDF	4.5.2 (4.5.2.1)
ESM 4.6	Example	PDF	4.5.2(4.5.2.2)
ESM 5.1	Four factors that control the horizontal movement of air near the ground (wind)	PDF	5.3.1 (5.3.1.1)
ESM 5.2	Wind power rating table	PDF	5.3.2
ESM 5.3	Wind speed conversion table	PDF	5.3.3 (5.3.3.2)
ESM 5.4	Table of sea condition level	PDF	5.4.1
ESM 5.5	Wave measurement techniques	PDF	5.4.1
ESM 5.6	Example	PDF	5.4.3 (5.4.3.2), 5.6.2 (5.6.2.3)
ESM 5.7	The relationship between different cumulative wave heights and average wave heights	PDF	5.4.3 (5.4.3.3)
ESM 5.8	Common wind wave spectra	PDF	5.5.2
ESM 5.9	Example	PDF	5.6.2 (5.6.2.3)
ESM 5.10	Example	PDF	5.6.3
ESM 5.11	Fetch division and wind wave development stages	PDF	5.7.1 (5.7.1.2)
ESM 5.12	Introduction to commonly used methods for wind wave calculation at home and abroad	PDF	5.7.3
ESM 5.13	Example	PDF	5.8.2 (5.8.2.1)
ESM 5.14	Introduction to numerical methods for plotting refraction	PDF	5.8.2 (5.8.2.2)
ESM 5.15	Example	PDF	5.8.3 (5.8.3.3)
ESM 5.16	Wave deformation under current action	PDF	5.8.4
ESM 5.17	Illustration of the calculation of the diffraction coefficient of an irregular wave	PDF	5.8.5 (5.8.5.1)
ESM 6.1	Astronomical knowledge related to tidal phenomena	PDF	6.2.1
ESM 6.2	Example	PDF	6.4.2 (6.4.2.1)
ESM 6.3	The formation and propagation of storm surges	PDF	6.5.1

(continued)

(continued)

Resource number	Resource name	Resource type	Section number
ESM 6.4	Storm surges and their hazards	PDF	6.5.2
ESM 6.5	Forecast of storm surges	PDF	6.5.3
ESM 7.1	Example	PDF	7.3.2

Contents

Chapter 1
Introduction

Yu's Temple in the deserted mountains, autumn wind, the setting sun slants down.
禹庙空山里, 秋风落日斜。
A weed-grown courtyard, hung with tangerines, ancient chambers, painted with dragons.
荒庭垂橘柚, 古屋画龙蛇。
Cloudy vapors puff slowly over green cliffs, river sounds speed across white sands.
云气嘘青壁, 江声走白沙。
Long I've known that he rode on four vehicles, his dredging and cutting brought the Three Ba regions under control.
早知乘四载, 疏凿控三巴。
Yu Temple, Du Fu, Tang Dynasty
—《禹庙》, 杜甫, 唐代.
▲
This poem depicts the desolate scene of Yu Temple, employing imagery of autumn wind, setting sun, tangerines, and dragon motifs to convey reverence for Yu the Great's monumental achievements in flood control.

Abstract This chapter introduces engineering hydrology, a geophysical branch supporting projects for river and ocean resource utilization, widely used in water conservancy, transportation, etc. Critical for port and waterway engineers, it covers river/coastal hydrology, data observation/analysis. This textbook has seven chapters, excluding some content taught in other courses, with research methods combining causal and statistical approaches, plus emerging tools.

For a visual overview of the chapter's key concepts, refer to the knowledge map below.

Supplementary Information The online version contains Electronic Supplementary Material (ESM) available at https://doi.org/10.1007/978-981-95-9089-6_1.

X. Feng (ed.), *Engineering Hydrology*, https://doi.org/10.1007/978-981-95-9089-6_1

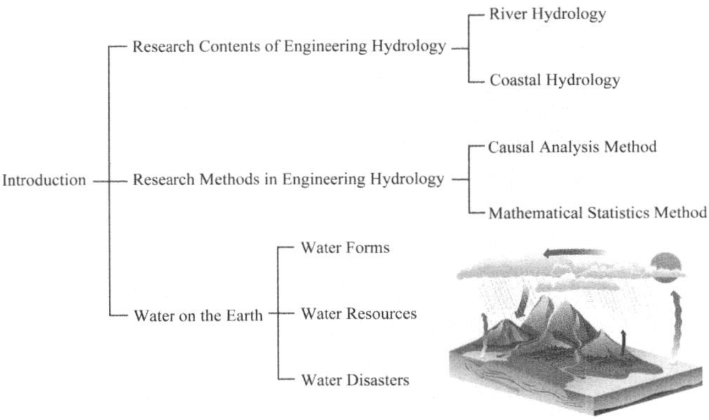

1.1 Engineering Hydrology

Hydrology deals with the existence, circulation, and distribution of the various bodies of water (rivers, streams, lakes, and seas) on the Earth, as well as their chemical and physical properties and their impact on the environment. Hydrology is a branch of geophysics.

Hydrology as applied to practical engineering is called engineering hydrology and includes the knowledge of hydrology required for the planning, design, construction, and operation management of construction projects that control or utilize river and ocean resources.

Engineering hydrology has a wide range of applications: water conservancy and hydropower, transportation, agricultural irrigation, urban construction, nuclear energy use, offshore oil and gas exploration and development, fishery production and environmental protection, etc., all of which need various hydrological data and information as the basis for engineering planning, design, construction, and operation management.

As far as engineers and technicians in port and waterway engineering are concerned, they must understand and master all kinds of hydrological and natural raw information required for the construction of ports and waterways, among which hydrological information and data are very crucial.

In order to build ports and waterways on rivers, it is necessary to understand and master the knowledge of river hydrology, including: according to the river hydrological characteristics of the river section where the project is to be constructed to provide the maximum flood level, design minimum navigable water level, design flood flow, etc., which can be used as the basis for the design of the elevation of wharves, locks, approach channels, buildings crossing waterways (such as bridges), waterway improvement, dredging, etc. and the design of the size of overflow buildings of canalized hubs. In the comprehensive use of water resources, we need to understand annual runoff and its distribution throughout the year, so as to coordinate the requirements of various water-using sectors and ensure navigation needs. When it comes to the construction and operation management of the project, it is also necessary to know the flood and dry water forecasts in order to organize construction and formulate operation plans, of which the hydrological forecasts themselves are generally issued by specialized hydrological agencies.

To build a coastal port, it is necessary to understand and master coastal hydrology, which mainly includes the historical conditions and future development trends of the coastal area and adjacent waters where the port is to be built, covering waves, inshore currents, and sediment movement. It is also necessary to provide hydrological data and information such as wave parameters for engineering purposes, water levels under various conditions, current velocity and direction for design, coastal evolution, and the trends and intensity of sediment transport. These serve as the basis for coastal port site selection, port general layout, channel selection and design, and the structural design of wharves, breakwaters, and other coastal works.

The main task of this course is to observe, collect, master, and analyze the hydrological data of rivers and coasts involved in this major. The course covers: ① the study of the laws of hydrological phenomena; ② the methods of testing and collecting hydrological data; and ③ the methods of collating, analyzing, counting, and calculating hydrological data.

The textbook for the course Engineering Hydrology is divided into seven chapters: Chap. 1 (Introduction), Chaps. 2, 3, 4 and 5 (River Hydrology), and Chaps. 6 and 7 (Coastal Hydrology). The contents of the textbook are selected according to the teaching requirements and teaching hours. It should be noted that sediment movement in rivers, estuaries, and coastal areas is a very important issue to be studied for the construction of ports and waterways, and it is also an important hydrological phenomenon. However, according to the Steering Committee for Teaching Port and Waterway Engineering in Higher Education of the Ministry of Communications, knowledge about this part will be taught in two other courses of the subject, River Dynamics and Coastal Dynamics, and therefore, this aspect is not covered in this textbook. In addition, the integrated use of water resources and their corresponding hydrology will also be taught in other similar disciplines and is not included in this textbook.

Chapter 1 offers an overview of hydrology, covering the distribution of Earth's water resources, the causes of hydrological disasters, and their impact on the environment. Chapter 2 delves deeper into the hydrological cycle, the processes of runoff formation, and the key influencing factors, while analyzing the structure of rivers and watersheds and examining hydrological patterns. Chapter 3 focuses on statistical methods commonly used in hydrology, including random variables, probability distributions, and parameter estimation, providing essential tools for frequency analysis and statistical inference. Chapter 4 emphasizes river hydrological observations and flow measurement techniques, detailing how to use stage-discharge relationships for design flood calculations and the derivation of navigable water flow. Chapter 5 explores the impact of ocean waves and wind, introducing the basic elements and classification of waves, as well as the role of wind conditions in wave generation, propagation, and decay, with a focus on wave data statistical patterns and their applications in design. Chapter 6 discusses tidal phenomena, tidal forces, tidal forecasting, tidal level calculations, and the generation mechanisms and hazards of storm surges. Finally, Chap. 7 examines nearshore flow systems, providing methods for observing and analyzing nearshore currents. By blending theory with practical application, this book offers a thorough exploration of the relationship between hydrological phenomena and engineering applications, providing a rich knowledge framework and practical guidance that enables readers to understand the significance and application of hydrology in water resources engineering.

1.2 Research Methods in Engineering Hydrology

Hydrological phenomena are physical processes with deterministic relationships to influencing factors, studied via the causal analysis method. However, their stochastic nature under combined factors makes deterministic results hard to obtain through causal analysis alone, as seen in complex cases like river flood levels or wind-induced waves.

Probabilistic and statistical methods are essential to grasp stochastic patterns and predict future trends, complementing rather than conflicting with causal analysis. Engineering Hydrology emphasizes field surveys, data collection, and applying the "practice-knowledge" cycle to refine estimates. Additionally, big data, AI, and digital twins are emerging tools for hydrological forecasting and smart water conservancy.

1.3 Water on the Earth

1.3.1 Water Forms

Earth holds about 1.38 billion km^3 of water, mostly ocean water (96.7%), with 3.6% as land water (glaciers accounting for 52.8% of land water) and 0.001% as atmospheric water, dynamically cycled by solar radiation and gravity.

1.3.2 Water Resources

Usable water resources mainly refer to renewable land freshwaters (e.g., river water, shallow groundwater), but global distribution is uneven in time and space, serving multiple functions like drinking, irrigation, power generation, and shipping.

Please refer to Electronic Supplementary Material (ESM) 1.1 for more detailed information.

1.3.3 Water Disasters

Natural factors (e.g., heavy rain, drought) and human activities (e.g., pollution, over-exploitation) cause floods (rapid water level rise) and droughts (supply–demand imbalance), aggravating water ecological risks.

Please refer to ESM 1.2 for more detailed information.

Chapter 2
Basics of River Hydrology

The Peaks of Qin are lush and green.
秦峰苍翠,
The Ye River flows with carefree grace.
耶溪潇洒,
Thousands of cliffs and countless valleys vie in their streams.
千岩万壑争流。
Watching the Tide: The Verdant Peaks of Qin, Qin Guan, Northern Song Dynasty
—《望海潮·秦峰苍翠》, 秦观, 北宋.
▲
This poem portrays the majestic scenery of verdant peaks, the graceful flow of the Ye River, and the competing streams of countless cliffs and valleys, highlighting the dynamic hydrology and geomorphology of a natural river system.

Abstract This chapter starts from the concept of constant dripping wears away stone to explore the origins of rivers, the mechanisms of runoff generation, the storage and regulation capacity of catchments, and the morphological evolution of river systems under the combined influence of natural factors and human activities. Additionally, we will examine the basic characteristics of hydrological regimes in nature. This fundamental knowledge lays the groundwork for the subsequent study of hydrological statistics and the derivation of engineering design parameters.

For a visual overview of this chapter's key concepts, refer to the knowledge map below.

X. Feng (ed.), *Engineering Hydrology*, https://doi.org/10.1007/978-981-95-9089-6_2

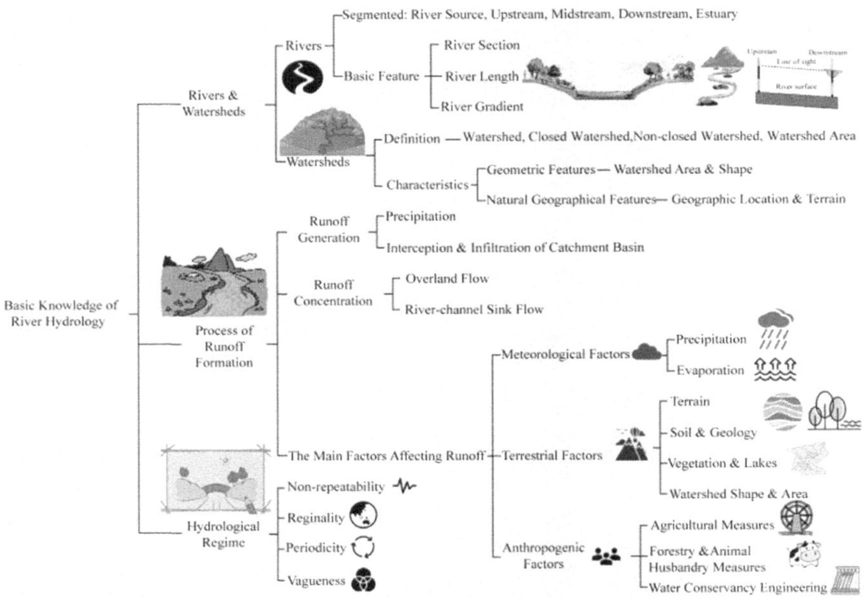

This chapter starts from the concept of constant dripping wears away stone to explore the origins of rivers, the mechanisms of runoff generation, the storage and regulation capacity of catchments, and the morphological evolution of river systems under the combined influence of natural factors and human activities. Additionally, we will examine the basic characteristics of hydrological regimes in nature. This fundamental knowledge lays the groundwork for the subsequent study of hydrological statistics and the derivation of engineering design parameters.

2.1 Introduction

In ancient legends, the story of the Great Yu harnessing the floods suggests that early civilizations had already started to observe and manage hydrological phenomena. The Great Yu successfully controlled the flood disasters by dredging, embodying a preliminary understanding and utilization of the laws of flow dynamics.

During the Pre-Qin period, a book titled *Shangshu* emerged. In this book, a chapter entitled "Yu Gong" records the mountains, rivers, and soils across ancient China. It is one of the earliest works on hydro-geography in China. During this period, attention shifted to issues such as river distribution and the replenishment of water sources, accompanied by specific practices in the construction of hydraulic engineering projects. For example, the Quebei Water Conservancy Project in the State of Chu demonstrated the early development and utilization of water resources.

During the Qin and Han dynasties, the Chapter of "Records of Rivers and Canals" in the book of *Shi Ji* (*The Grand Historian*) written by Sima Qian constituted the first general history of water conservancy in China. It recorded in detail the construction of hydraulic projects and flood control activities at that time, marking the beginning of the development of ancient hydrology in China from practice to theory. Simultaneously, hydraulic technology advanced significantly. The construction of large-scale water conservancy projects such as the Dujiangyan Irrigation System and the Lingqu Canal reflected an in-depth understanding and ingenious utilization of hydrological laws. Specifically, in river management, artificial water conservancy facilities such as dams, sluices, and canals are designed according to river discharge variations and the topography. During the flood season, flow discharge and direction were regulated via dams and sluices to prevent inundation. During the dry season, river water was diverted into farmlands for irrigation through canals.

During the Tang and Song dynasties, the theory of hydrology developed further, and many works dedicated to hydrological phenomena emerged. For example, the book named *Shui Bu Shi* (*Regulations of the Water Bureau*) of the Tang Dynasty was the first hydraulic code for water conservancy in China, establishing detailed regulations on the management and utilization of water resources. Shen Kuo in the Song Dynasty recorded and studied diverse hydrological phenomena in many aspects in his book *Meng Xi Bi Tan* (*Dream Pool Essays*), including river erosion and sedimentation, reflecting the scientific understanding level of hydrology at that time.

During the Yuan, Ming, and Qing dynasties, hydrology matured both in theory and practice. Regarding hydraulic engineering construction, the recurrent dredging and regulation of the Beijing-Hangzhou Grand Canal, as well as the training projects of major rivers such as the Yellow River and the Huai River, all reflected the profound understanding and comprehensive application of river dynamics. Simultaneously, numerous hydraulic experts and treatises emerged. For example, the theory of constricting flow to scour sediment proposed by Pan Jixun in the Ming Dynasty had a profound impact on the management of the Yellow River, signifying that ancient Chinese hydrology had reached a sophisticated level.

By comprehensively examining the general history of hydrology represented by ancient China, it is evident that early civilizations had already mastered methods of harnessing, utilizing, and benefiting from nature-based solutions. Only through careful observation of hydrological characteristics and the summarization of natural laws could humanity align with nature to create enduring hydraulic engineering marvels that have stood the test of time for millennia.

The following chapters will start from the concept of constant dripping wears away stone to explore the origins of rivers, the mechanisms of runoff generation, the storage and regulation capacity of catchments, and the morphological evolution of river systems under the combined influence of natural factors and human activities. Additionally, we will examine the basic characteristics of hydrological regimes in nature. This fundamental knowledge lays the groundwork for the subsequent study of hydrological statistics and the derivation of engineering design parameters.

2.2 Hydrological Cycle

The Earth is enveloped in water, yet estimating its precise volume is challenging, particularly regarding deep ocean reserves and subsurface water.

According to data from the United Nations Intergovernmental Panel on Climate Change (IPCC) Sixth Assessment Report 2021, there is a total of about 1.380 billion km^3 of stored water on Earth, of which about 1.335 billion km^3 (96.7%) is oceanic water, about 49.2 million km^3 (3.6%) is terrestrial water, and about 13,000 km^3 (just about 0.001%) is atmospheric water. The largest amount of terrestrial water is stored in ice caps and glaciers, with 26 million km^3, accounting for 52.8% of the water on land.

As illustrated in Fig. 2.1, oceanic water evaporates into the atmosphere under solar radiation to form water vapor. Driven by the movement of air masses, this vapor condenses under suitable conditions to form precipitation (rain, snow, or hail), which subsequently falls to the ground or the sea. Upon reaching the land surface, water follows two primary paths under the influence of gravity. A significant portion flows over the surface into rivers, forming surface runoff. Another portion infiltrates the soil. While a small amount is retained temporarily in the topsoil and returns to the atmosphere via evaporation and plant transpiration, the remainder percolates downward to the saturated zone, forming groundwater flow (or subsurface runoff). The sun serves as the primary energy source for this cycle, while solar radiation and gravity act as the driving forces. The ocean functions as both the source and the sink of the cycle.

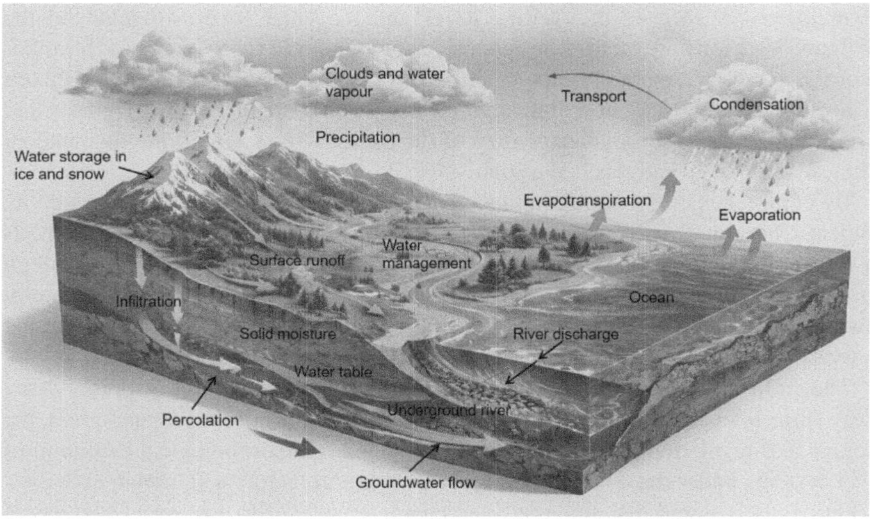

Fig. 2.1 Schematic diagram of the hydrological cycle

In the context of the global water cycle, evaporation from the oceans exceeds oceanic precipitation. Conversely, terrestrial evaporation is less than terrestrial precipitation, generating runoff that flows toward the sea, primarily via rivers. On a multi-year average, this global system maintains equilibrium. In the case of a closed basin, precipitation, runoff, and evaporation also achieve a balance over the long term, maintaining relatively stable storage levels. These dynamic and static terrestrial water stores constitute vital resources upon which humanity depends.

2.3 Runoff Formation Processes and Their Main Impact Factors

2.3.1 *Runoff Formation Processes*

Precipitation falling on a watershed, after deducting losses, partly forms surface runoff and partly infiltrates into the soil, where the latter forms groundwater flow in the aquifer. Both components converge in the river network, flow along the channel, and finally discharge through the watershed outlet; this combined flow is termed runoff. The comprehensive physical sequence from the onset of precipitation to the flow passing through the outlet cross-section is known as the runoff formation process. It results from the combined influence of atmospheric precipitation, evaporation, and the physiographic conditions of the watershed. Due to the variability of atmospheric precipitation and evaporation, coupled with the diversity of physiographic conditions, the runoff formation process is highly complex. To conceptualize the phenomenon of runoff formation and facilitate its analysis, the physical processes involved can be generalized into four stages: the precipitation process, the watershed storage and infiltration process, the overland flow process, and the channel flow process. The watershed storage and infiltration process is commonly known as runoff generation; the overland flow and channel flow processes are collectively known as flow concentration. Figures 2.2 and 2.3 illustrate the runoff formation process and the flow concentration process.

2.3.1.1 Precipitation Processes

Precipitation is the main factor in the formation of runoff. Precipitation is the collective term for rainfall, snowfall, and hail. Some rivers are primarily fed by rainfall, while others receive contributions from both rainfall and snowfall, and still others are predominantly sustained by snowmelt and glacial meltwater induced by rising temperatures. When discussing the concept of runoff formation, precipitation is often narrowly defined as rainfall alone.

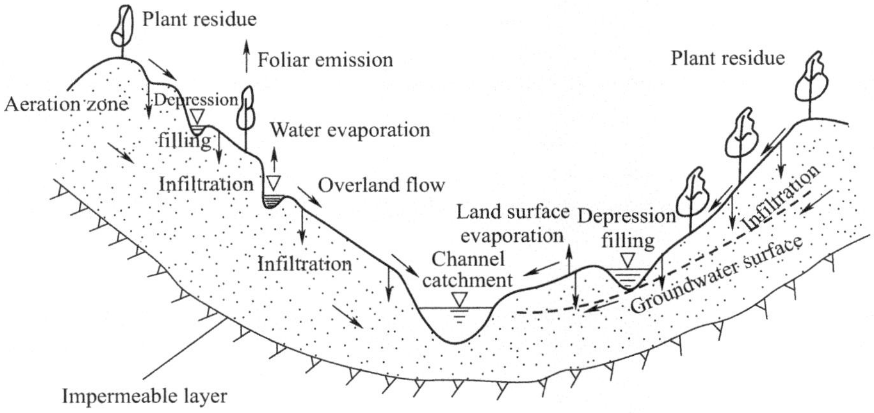

Fig. 2.2 Schematic diagram of the runoff formation processes

Fig. 2.3 Schematic diagram of the flow concentration process

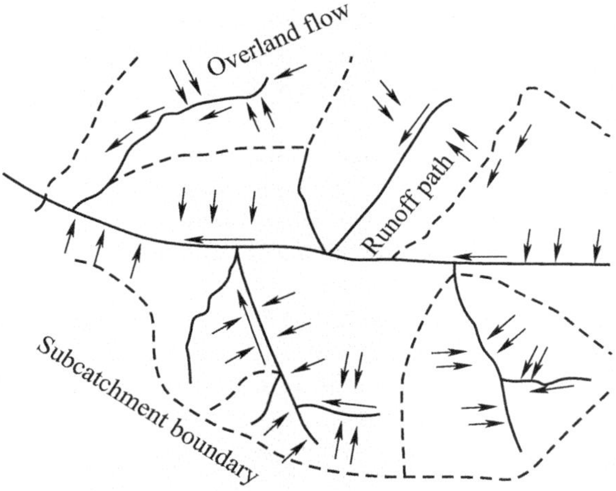

The rainfall process is characterized by physical quantities, including rainfall amount, intensity, temporal distribution, rain pattern, spatial distribution, areal extent, location of the storm center and storm track.

In fact, the distinct variations in the physical parameters characterizing rainfall processes lead to differences in the characteristics of the runoff formation process. Consequently, the rainfall process constitutes a critical component of the runoff formation process.

2.3.1.2 Watershed Storage and Infiltration Processes

At the onset of rainfall, the majority of rainwater falls onto the catchment surface and does not immediately generate runoff, with the exception of precipitation falling directly onto the water surface of river channels. Initially, a portion of the rainwater is intercepted by vegetation, including trees, grasses, and crop foliage; this phenomenon is known as vegetation interception. Subsequently, rainwater reaching the ground partially infiltrates into the soil, a process termed infiltration. The rate of infiltration per unit time is referred to as the infiltration rate. Infiltration is rapid at the beginning of rainfall. If the rainfall intensity exceeds the infiltration rate, the rainfall excess is called excess rainfall, which generates overland flow. As rainfall continues and soil moisture approaches saturation, the infiltration rate decreases to a stable value, termed steady infiltration. Additionally, a portion of rainwater is retained in surface depressions, a process known as depression storage.

The combined processes of vegetation interception, infiltration, and depression storage are known as the watershed storage and infiltration processes. Since this portion of rainwater does not produce surface runoff, it is classified as a loss. The rainfall remaining after deducting these losses is referred to as net rainfall. Essentially, the runoff generation process is the process of determining net rainfall by deducting losses from total rainfall.

2.3.1.3 Slope Surface Overland Flow Processes

After the completion of the watershed storage and infiltration process, the remaining rainwater flows along the slope, a phenomenon known as the overland flow process. The start time of overland flow is not uniform across the watershed. Overland flow from localized areas gradually expands to form watershed-wide overland flow. If the rainfall duration is short or occurs as a localized rainstorm, it is difficult to form watershed-wide overland flow. Objectively, the overland flow process is a complex process, as it is also accompanied by infiltration, rainfall, and evaporation processes.

2.3.1.4 River Channel Concentration Processes

After rainwater from overland flow enters the channel, it flows along the river course from small gullies to tributaries, then to the main stream, and finally reaches the watershed outlet cross-section; this process is called the river channel concentration process. The flow entering the channel, on one hand, continues to flow rapidly downstream along the channel; on the other hand, it increases the water volume within the channel, causing the water level to rise accordingly. The water volume stored in the channel flows slowly downstream only after the rainfall ceases and overland flow terminates, making the discharge at the watershed outlet cross-section smoother and the duration extended; this is known as the regulation effect of the river channel on floods. In addition, the groundwater flow formed by infiltration into the saturated soil layer takes a longer duration to collect into the river channel, as it is first regulated by the groundwater reservoir; it usually overlaps with the runoff formed by the next rainfall event.

It is evident that the flow concentration process is essentially a process of temporal redistribution of net rainfall through the regulation (or storage) of the slope surface, groundwater reservoirs, and river channels.

It should also be noted that rainfall, watershed storage and infiltration, overland flow, and river channel concentration constitute the entire process from the onset of rainfall to the generation of runoff at the outlet cross-section. These processes have no distinct boundaries in time but proceed in an interlaced manner.

2.3.2 Main Factors Affecting Runoff

The main factors affecting river runoff include meteorological and climatic variables, characteristics of the underlying surface, and human activities.

2.3.2.1 Meteorological and Climatic Factors

The meteorological and climatic factors of precipitation (rain and snow) and evaporation are the most important factors influencing runoff.

(1) Rainfall. Water vapor in the air rises with the airflow, undergoing adiabatic expansion, cooling, and condensing into water droplets before finally descending to the ground, thus forming rainfall. The formation of rainfall is generally divided into four categories according to the reasons for the rise of water vapor: ① cyclonic rain, resulting from the passage of cyclones (i.e., low-pressure systems); ② convective rain, formed by the ascent, cooling, and condensation of moist air near the surface due to ground heating; ③ orographic rain, generated when warm, moist air encounters high mountains, causing uplift and condensation; ④ typhoon rain, precipitated by the passage or peripheral influence of a typhoon. Rainfall factors are not only an important influence on runoff formation, but also on the full process of runoff formation.

There are differences in the rainfall characteristics of different rainfall types; the differences in rainfall amount, rainfall intensity, the temporal-spatial distribution of rainfall, and rainfall duration will affect river runoff formation to different degrees. For example, typhoon rain brings a large rainfall amount, and it is intense and widespread, it produces flood process with a high peak and large flood volume. But for orographic rain, the rainfall amount usually increases gradually from the foot of the hill to the top of the mountain, so the flood process has a high peak but a later timing of the flood peak. Another example is convective rain, it is mostly thunderstorm weather, although the intensity of rainfall is strong, but with short duration and small coverage, so the formed flood has a high peak and small volume, and the water level rises and falls sharply. For example, cyclonic rain generally has a long duration and widespread coverage. If the rainfall belt moves from the lower reaches of the watershed to its upper reaches, the flood process formed is of the early rise, low peak, and long duration type; if the rainfall belt moves from the upper reaches of the watershed to its lower reaches, the flood process formed is of the late rise and high peak type. Sometimes rainfall events with longer duration are a combination of two weather types, which result in a flood process with large flood volume, long-lasting, and bimodal shape.

(2) Evaporation. It is the physical process by which water changes from a liquid state to a gaseous state. Total evaporation from a watershed is made up of evaporation from the water surface (lakes, depression ponds, river channels, and reservoir surfaces), evaporation from the land surface, and plant transpiration. The amount of antecedent rainfall resulting from vegetation interception, depression storage, and infiltration into the soil of the vadose zone is called rainfall losses, and is consumed through evaporation and transpiration. If this loss amount is consumed less by evaporation, then the loss during the next rainfall will be smaller, and the corresponding runoff generation will be larger; on the contrary, the loss during the next rainfall will be larger, and the runoff will be smaller when the loss amount by evaporation is larger. Annual evaporation in wet areas of China accounts for 30%–50% of the annual precipitation, and this ratio is 80%–95% in arid areas, where the runoff is only formed by residual rainfall. It can be seen that evaporation is another important factor affecting the formation of runoff, and it mainly affects the process of runoff generation.

2.3.2.2 Underlying Surface Factors

The underlying surface factors mainly include topography, soil, geology, vegetation, lakes, marshes, and watershed morphology. These factors, on the one hand, directly influence the size and course of river runoff; on the other hand, they indirectly influence runoff through meteorological and climatic factors.

(1) Topography. The topographic features of the basin, such as ground elevation, slope, and degree of incision, directly influence the concentration process. The steeper the terrain slope and the deeper the cutting, will lead to higher the slope overland flow and channel collection velocity, as well as shorter concentration time, at the same time, the rainfall loss is small, and the shape of the discharge hydrograph is relatively sharp and thin; and vice versa. In addition, as the ground elevation increases, the average annual rainfall also increases; however, the temperature decreases, and therefore evaporation decreases. The combined effects enhance the runoff with increasing elevation.
(2) Soil and geology. The physical properties of the soil, moisture content, stratification and slope of the rock layers, thickness of permeable rock layers, and the geological conditions for groundwater storage significantly influence the infiltration process of rainfall in the watershed. On the surface of deep weathered

crusts and Quaternary loose deposits, due to their strong permeability, a portion of rainwater infiltrates to recharge groundwater, which is then continuously replenished to the river channel through underground reservoirs, thereby slowing down the changes in river runoff. The Loess Plateau with deep soil layers has almost no infiltration of rainwater to form groundwater or recharge river channels. The distribution area of karst landforms can cause a large amount of rainfall and surface runoff to flow underground, often resulting in non-overlapping of ground and underground water distribution lines.

(3) Vegetation, lakes, and marshes. The interception of rainwater by vegetation reduces runoff. However, it can also increase ground roughness, change soil structure, and give it high water holding capacity, which facilitates infiltration and slows down slope surface overland flow speed, and provides good regulation for river runoff. On the one hand, lakes and marshes have a regulating effect on the water storage capacity of the watershed, which affects runoff variation. On the other hand, due to the increase in the water storage area of the watershed, the evaporation is increased and the total runoff is reduced.
(4) Watershed shape and area. These are two important factors that affect the process of runoff formation. If the watershed area is extensive, the total runoff will be substantial, and the subsurface runoff will correspondingly increase, making the runoff process more stable. If the shape of the watershed is feathered or narrow, the duration of the watershed concentration is long, and the flood hydrograph is relatively gentle; if it is a fan-shaped watershed, all tributaries almost simultaneously converge into the outlet section, so the flood hydrograph is steep and the concentration time is short.

Some of the influencing factors on the underlying surface mentioned above affect the entire process of runoff formation, while others only affect one or two stages.

2.3.2.3 Human Activities

The impacts of human activities on runoff can be mainly classified into three categories: agriculture, forestry and animal husbandry, and water conservancy measures.

(1) Measures for agriculture. The relevant measures mainly include converting sloping land into terraced fields, converting dry land into paddy fields, and deep cultivation and dense planting. These measures serve to intercept runoff and sediment, increase infiltration and evaporation, and reduce runoff.
(2) Measures for forestry and animal husbandry. The relevant measures mainly include planting grasses and trees to increase vegetation coverage, thereby reducing sediment loss, increasing infiltration and evaporation, slowing down the speed of overland flow on the slope, slowing down the concentration process of the watershed, and increasing the regulation capacity.
(3) Measures for water conservancy. The relevant measures include valleys, irrigation channels, ponds, and reservoirs of large, medium, and small sizes, as well

as inter-basin water diversion projects, which mainly regulate and store water during the runoff period and change the concentration process.

In addition, with the growth of the national economy and population, the level of urbanization in larger watersheds also falls within the scope of human activities. The increase in hard coverage and drainage is also changing the characteristics of the generation and concentration processes of river runoff.

The above-mentioned factors associated with human activities primarily affect one or two stages of the runoff formation process.

2.4 Rivers and Watersheds

2.4.1 Rivers

2.4.1.1 Formation and Segmentation of Rivers

After excluding losses of precipitation, the water forms that flow in a specific direction and pathway over and underground, under the influence of gravity, initially eroding the ground and washing into gullies, thereby forming small streams. The valley through which the water flows is called a river valley, and the part of the valley where water flows at the bottom is called the riverbed. As water flows downstream, it continuously cuts and scours the riverbed and erodes to the sides, gradually enlarging the riverbed. A number of small rivers merge to form larger rivers, eventually flowing into the ocean or inland lakes.

Based on the valley and bed conditions, the degree of siltation and changes in water conditions, natural rivers are generally divided into five sections, namely headwaters, upper reaches, middle reaches, lower reaches, and estuary.

The headwater of a river is the place where it first has a flow of water, which may be a stream, spring, glacier, lake, or swamp.

The upper reaches are the upper sections of the rivers immediately adjacent to the headwaters, mostly located in deep mountain valleys, where steep slopes and rushing flows result in low discharge, rapid rises and falls of the water level, and strong valley undercutting. There are often sharp shoals or waterfalls, and the longitudinal section of the river bottom is mostly stepped.

The middle reaches are the middle sections of rivers. The banks are hilly, often with mudflats, and the riverbed slopes gently, so that siltation is not obvious.

The lower reaches of rivers are generally in the plain area. Here, the river channels are broad, with high discharge rates, low flow velocities, gentle slope bottoms, significant siltation, and shallower banks and bays.

The estuary is the stretch of a river where it flows into the ocean or a lake. Some rivers vanish into the desert and are called blind-ended rivers, meaning there is no estuary. When the flow paths of the estuary suddenly widen, the flow velocity decreases sharply, and abundant sediment deposits form sandbars or barrage sand shoals.

2.4.1.2 Basic Characteristics of Rivers

The basic characteristics of a river are generally described in terms of river cross-section, river length, and river gradient.

(1) River cross-section. There are longitudinal and cross-sectional. The section perpendicular to the direction of flow is called the river cross section, its general shape is presented in Fig. 2.4. Within the cross-section, the free water surface, which is referenced to the elevation of a specific level datum, is called the water level. When the water level changes, the wetted cross-sectional area changes with it, and usually the wetted cross section reflects the size of its flow discharge, the wetted cross section, when encountering the design standard flood (or the maximum possible flood), is called the large cross section. The riverbed below the high water level is divided into two parts: the river channel and the riverbank. The river channel is the main pathway for flooding and sand transport; the higher, movable sediment accumulation along the two banks in the channel is called the riverbank or side bank, and the central part, where water often flows, is called the main channel. The riverbank only has flows during the flood season, and there is no obvious movement of the bottom sand. A section with only a channel and no bank is called a single section profile, while one with both a channel and banks is called a compound section profile.

The line connecting the locations of maximum water depth at each cross-section along the flow direction of the river is called the thalweg, and the section along the thalweg is called the longitudinal section of the river. It describes the variation of the riverbed along its course (Fig. 2.5). Both the longitudinal and cross sections of a river change over time due to the interaction between the water flow and the

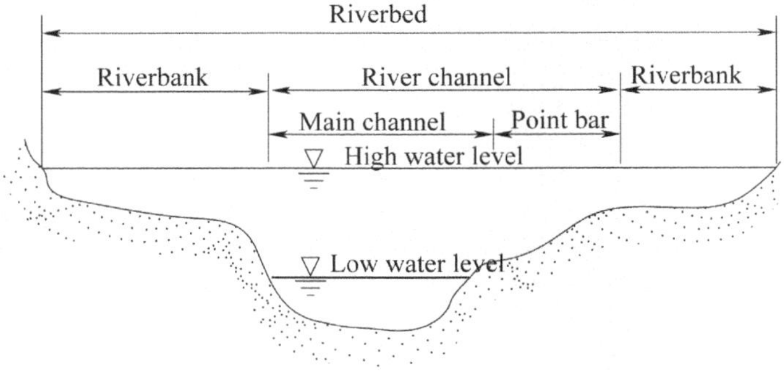

Fig. 2.4 General shape of a river along a cross-section

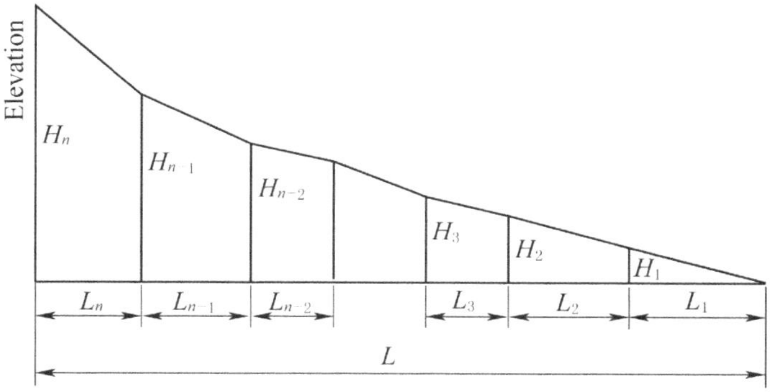

Fig. 2.5 River longitudinal section

riverbed. Usually, the longitudinal section experiences continuous sediment deposition downstream and erosion upstream, while the cross sections are often in a process of erosion and deposition on both sides of the river. The development of the river section is largely determined by the geology, topography, channel composition, and water flow in the area where the channel is located.

(2) River length. The distance from the headwater to the estuary of a river is called the river length. Approximate river lengths can be directly estimated from a topographic map of the actual river, along the dynamic axis of the channel, using a divider or other methods. However, the river length obtained using the above method usually underestimates the actual length, as the topographic map cannot reflect the detailed bend changes in the river, so corrections are necessary.
(3) River gradient. The elevation difference per unit length of the river is called the river gradient. There are surface and bottom river gradients. The gradient of a river section can be calculated via the following formula:

$$i = \left(H_{\text{upper}} - H_{\text{lower}}\right)/l = \Delta H/l \tag{2.1}$$

where i is the specific gradient of river bottom or water surface, expressed as a percentage (%) or thousandths (‰); H_{lower}, H_{upper} are the elevations of the water surface or river bottom at the downstream and upstream ends of the river section, m; l is length of river section, m; ΔH is the elevation difference in the water surface or the river bottom, m.

The river gradient is influenced by numerous factors and varies greatly. The river gradient near the estuary is influenced by sediment deposition, tidal backflow, or the topping up of large rivers and engineering structures, it has even larger variations in gradients and sometimes has negative values. The river channel is deepened by river erosion, and the gradient becomes smaller and smaller, but eventually its bottom elevation will not be lower than the sea level. The bottom gradient of the river is more stable than that of the surface, and the surface gradient will also vary with the water level variation. The river gradient generally decreases from the headwater to the estuary, and varies along the course of the river (Fig. 2.5).

2.4.1.3 General Characteristics of Mountainous Rivers and Plain Rivers

Rivers are generally divided into two categories, mountainous rivers and plain rivers, depending on the topographic features through which they flow. For larger rivers, the upper reaches are mostly mountainous, and the lower reaches are plain, while the middle reaches often have characteristics of both types of rivers and are transitional, known as a foreland river and a mid-mountainous river. For smaller rivers, all sections may be mountainous or plain rivers.

(1) Mountainous rivers. Mountainous rivers flow through mountainous areas with high terrain and complex topography. Due to the constant longitudinal cutting and lateral widening of the water flow, the section of the river valley is formed in an incomplete V or U shape (Fig. 2.6), which has narrow section shapes with straight or curved slope banks.

The rivers in the mountainous region are mostly gorges and open sections, with very complex planar patterns. There are often boulders on both banks and in the center of the river, and sharp bends and rapids can be found everywhere. The longitudinal section is steep, with sharp shoals and deep pools, often in the form of steps.

Due to the steep slopes and rocky outcrops of mountainous rivers, floods have a brief confluence period and exhibit rapid fluctuations. The water level varies greatly, often rising by as much as 10 m overnight; the discharge ratio between flood season and dry season is also very substantial.

The gradient of rivers in mountainous areas is significant, often exceeding 1%, and it varies greatly along the river course. The drops mostly appear in localized river sections, some of which may have flow velocities up to 6–8 m/s. The flow

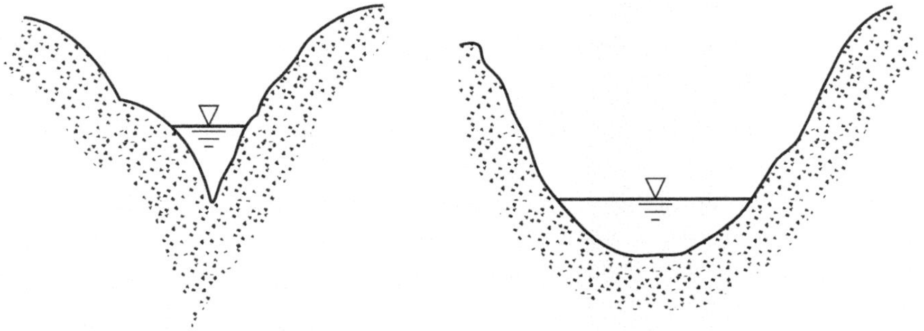

(a) V-shaped

(b) U-shaped

Fig. 2.6 Schematic diagram of a mountainous river valley pattern is chaotic, with alternate backwash, eddies, waterfalls, and jumps, which is very unfavorable to navigation.

The riverbeds of mountainous rivers are mostly composed of bedrock, boulders, and cobbles, and undergo slow scour and deformation. Although the riverbeds are relatively stable, they are susceptible to the impacts of earthquakes, rockfalls, landslides, and debris flows, and river channels may be suddenly blocked.

(2) Plain rivers. Plain rivers flow through flat plain areas. The process of river formation primarily involves the accumulation of material carried by the flow, resulting in the deposition of layers of alluvium, which can be tens to hundreds of meters thick. The layers from the bottom up are pebbles, sandy pebbles, coarse sand, medium sand, and fine sand, while the surface layer of the riverbank above the dry-season water level is clay and clay loam. The deepest layer of gravel is mostly accumulations from the glacial period, when the river runoff was higher, and the sea level was lower, while the sand layer is a more recent accumulation from a time when the river runoff was lower, and the sea level was higher.

The valleys of the plain rivers are mostly in the form of fully developed river floodplains (Fig. 2.7), with gentle valley slopes and wide river floodplains. During floods, floodplains are submerged by the flow, while during dry periods they are exposed above the water surface. Due to the soft nature of the river floodplain, one floodplain is gradually washed away by the flow, and a new floodplain is created in another location, as the main channel swings or the river is diverted.

In the main channel of a plain river, various siltation bodies are formed due to the interaction between the flow and the riverbed (Fig. 2.8). The exposed shoal immediately adjacent to the riverbank in the dry season is called a riverbank; the underwater sandbanks connecting the upper and lower point bars are called shoals; the higher and lower sandbanks located in the center of the river are called river islands and midstream bar, respectively; the narrower sandbanks that are obliquely crossing the water flow are called spits. These silt deposits are collectively called sand dunes and are in a constant state of movement, change, and development under

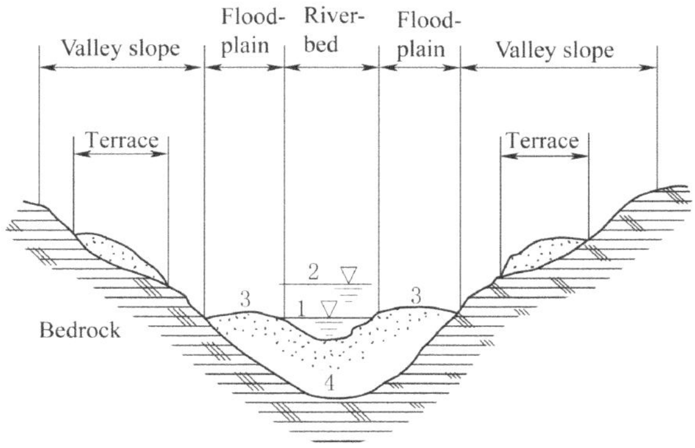

Fig. 2.7 Schematic diagram of a valley floodplain in a plain river. 1—Low water level; 2—Flood level; 3—Shoal; 4—Alluvial layer

the action of the flow, resulting in continuous changes in the morphology of the riverbed.

The cross-sectional shape of a plain river varies depending on its location and includes several types such as parabolic, asymmetrical triangular, and w-shaped (Fig. 2.9). The riverbed is generally wide and shallow, with a certain pattern. The channel planes can be classified into types of straight, slightly curved, meandering, branched, and wandering.

Because of the gentle longitudinal gradients of the riverbed and smooth flow conditions, plain rivers generally have small water surface gradients, typically below 1%. Owing to the wide channel and low flow velocity (generally less than 3 m/s), the flow is relatively smooth.

Plain rivers flow through gentle plain areas, with longer confluence times, gentler rise and fall water levels, longer duration, and smaller discharge ratios between flood

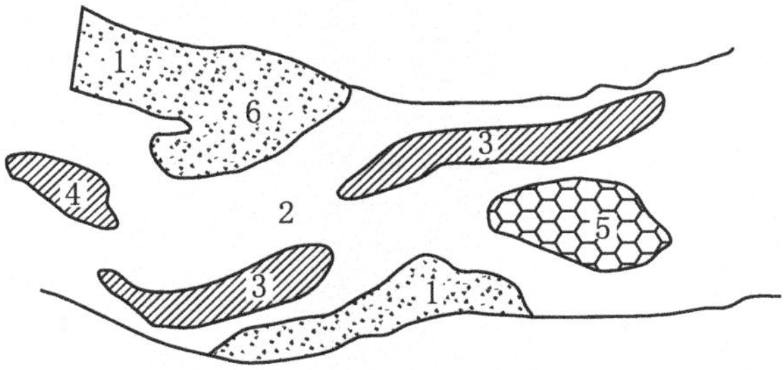

Fig. 2.8 Various-shaped silt deposits in the main channel of the river. 1—Riverbank; 2—Shallow shoals; 3—Deep channel; 4—River island; 5—Midstream bar; 6—Spit

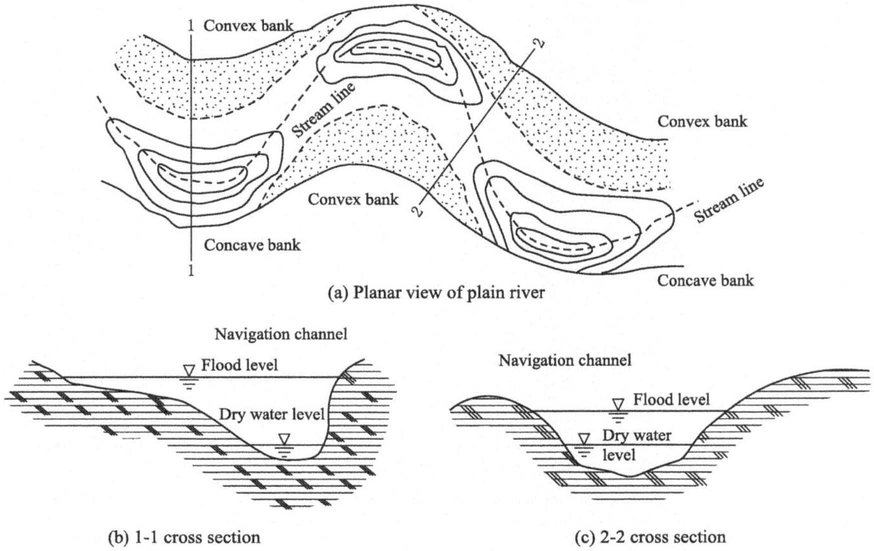

Fig. 2.9 Planar pattern and cross-sections of the meandering channel

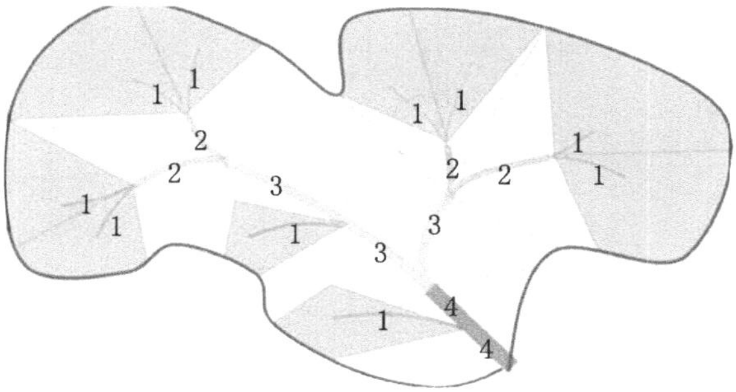

Fig. 2.10 Schematic plot of watershed and water system. 1,2,3,4—Order of rivers

season and dry season. The variation magnitude of the water level is also small, generally less than 10 m.

2.4.2 *Water System*

The system of these interconnected rivers, large and small, is called a water system (or river system). The part of the system that flows directly into the ocean or a lake is called the main stream. The rivers that feed into the main stream are called primary tributaries; the rivers that feed into primary tributaries are called secondary tributaries of the main stream, and so on. A water system is usually referred to by the name of its main stream, such as the Yangtze River, Yellow River, and Pearl River systems.

Stream orders based on Strahler's classification can be referred to Fig. 2.10. A small stream originating from the river headwaters is defined as the first order. A river formed by the confluence of two rivers of the same order is defined as one order higher than the original. The order of a river formed by the confluence of rivers of different orders is the higher of the two rivers. Accordingly, the main stream is typically the channel with the highest stream order in the river system.

2.4.3 *Watersheds*

2.4.3.1 Watershed, Divide, and Area

Precipitation falls to the ground and forms runoff that is separated by uplands and mountains and drains into different rivers. The same area where the water flows is collected is called the watershed. In summary, a watershed is the drainage basin of a river.

The ridgeline that separates the uplands and mountains where water flows is the dividing line between adjacent watersheds, is called a watershed divide (or watershed boundary). Surface water and groundwater that feed into a river often have different

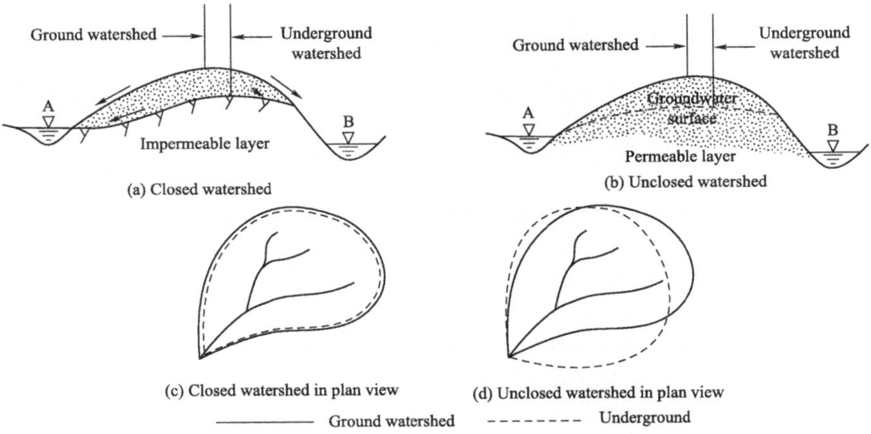

Fig. 2.11 Schematic diagram of the closed and unclosed watershed

watershed divides. A watershed where the surface and subsurface water divides coincide is called a closed watershed, while a watershed where the surface and subsurface water divides do not coincide is called an unclosed watershed (Fig. 2.11).

The watershed and outlet correspond with each other. The area enclosed by the watershed and the outlet section is called the watershed area, or catchment area, measured in units of km^2. The area of small and medium-sized watersheds is usually estimated on the basis of topographic maps by means of a planimeter or by counting squares, while the river length and watershed area of large river basins can be deduced by computer processing based on high-resolution satellite images.

2.4.3.2 Watershed Characteristics

A watershed is the source of river replenishment, and its characteristics directly affect the formation and change of river runoff. Watershed characteristics are generally divided into two categories, namely geometric and physiographic features.

(1) Geometric features. This refers mainly to the shape and size of the watershed. The size of the watershed directly influences the amount of water collected and the process of forming river runoff. If the physical and geographical conditions are the same, the larger the watershed area, the larger the runoff volume and the greater the regulation of runoff changes, resulting in gentler rise and fall of floods, and vice versa. The watershed shape mainly affects the length of the runoff collection time and the process state of the runoff. If the watershed is long and narrow and feather-shaped (Fig. 2.12a), the flood peak at the outlet section is small and the runoff process has a long duration with a small variation of discharge magnitude. If the watershed is wide and fan-shaped (Fig. 2.12b),

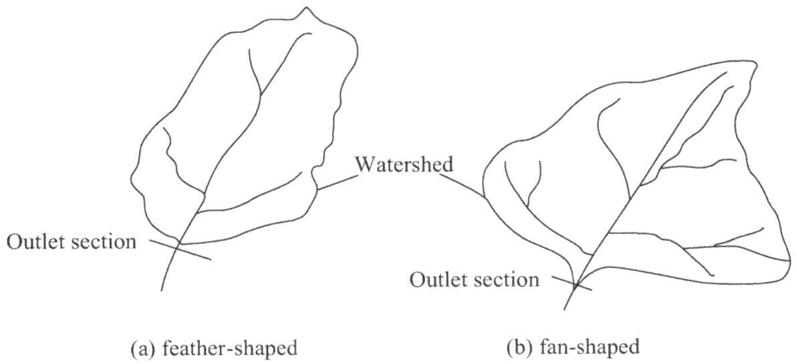

Fig. 2.12 Schematic diagram of the shape of the watershed

the flood peak at the outlet section is large, and the runoff process has a short duration with large discharge variation.

(2) Physiographic features. Physiographic features mainly include the geographical location and topography of the watershed. The geographical location of a watershed is generally expressed in terms of the latitude and longitude of the center and perimeter of the watershed, as all hydrological and meteorological features, such as rainfall, evaporation, and runoff depth, are closely related to the geographical location. The topography of the watershed is generally expressed in terms of its mean height, mean width, and mean slope, since the mean height of the watershed directly influences the precipitation, temperature, and evaporation within the watershed; the mean width and mean slope of the watershed influence the yield of runoff, infiltration, the timing and course of confluence and soil loss, etc.

The geology, soils, forest coverage rate, and percentage of lakes in the watershed are also physical geography features that are relevant to the process of runoff formation.

2.5 River Hydrological Regimes

The general changes over many years in hydrological elements of rivers, such as water level, flow rate, sediment, and ice conditions, are called river hydrological regimes. Such a regime is the result of the integrated impact of factors such as watershed climate, natural geographical conditions, underlying surface characteristics, and human activities. It can be seen that the main characteristics of river hydrological conditions are non-repeatability, regionalization, periodicity, and ambiguity by analyzing a large amount of river runoff data.

2.5.1 Non-repeatability

The flow, water level, or sediment processes of rivers are constantly changing at any time and place. The processes between years are different for the same gauge station, and vary even more among different gauge stations. The flow process of upstream stations does not overlap with that of downstream stations. This is because there are many factors that affect climate and hydrological factors, and the phenomenon is very complex. As a result of the comprehensive interplay of multiple factors, the annual runoff situation cannot be exactly the same. This property is called non-repeatability and can be sought through a multi-factor comprehensive analysis method. In addition, some years of the same gauge station have an early flood season, high flow rate, and substantial sediment content, while in other years, the flood season comes late, the flow is small, and there is also less sediment. That is to say, when the flood season of a certain year arrives, how large the flood peak is, how much sediment is contained, and how high the maximum water level can reach are all unknown in advance. This non-repeatability is also known as randomness, and its characteristics are usually analyzed using mathematical statistical methods.

2.5.2 Regionalization

Due to the climatic conditions, natural geography, and underlying surface characteristics unique to each region, there exists a distinct regional characteristic reflected in the hydrological regime. For example, in southern China, it is rainy, rivers have early flood seasons, high groundwater levels, and high vegetation coverage, the flood volume is usually large, the rise is slow, and the recession duration is long. In northern China, there is less rainfall, the flood season is relatively late, the groundwater is buried deep, and the vegetation coverage is low. Therefore, the flood volume is usually small, and the water rises and recedes quickly. Some rivers in the northwest region that rely on melting snow for replenishment have a close relationship between runoff changes and temperature. Small rivers in mountainous areas often experience sudden fluctuations, while large rivers in plains experience relatively gentle fluctuations. Rivers with similar hydrological, meteorological, and natural geographical characteristics in the same region or watershed, or upstream and downstream stations of the same river, typically exhibit analogous hydrological regimes. Due to the regional nature of hydrological regimes, when there are watersheds with similar climatic conditions and characteristics in the same region, it is possible to utilize hydrological analogy methods or regional empirical formulas and charts to address hydrological regime issues at stations lacking data.

2.5.3 *Periodicity*

The so-called "periodicity" refers to the characteristic of the changes in river runoff hydrological regimes recurring over specific time intervals. Usually, hydrological conditions have a clear cycle of one year, because among the factors affecting hydrology, climate factors change on an annual basis. Moreover, the Earth rotates around the Sun for one year, resulting in the four seasons of spring, summer, autumn, and winter, each characterized by its unique climate conditions. For example, the Yangtze River Basin in China has high temperatures and abundant rainfall in summer and low temperatures and scarce rainfall in winter, which is the case every year. In the Songhua River Basin, a spring flood is formed by snow melting in spring, and a summer flood is formed by rainstorms in summer. There has been no essential change in this cycle for a long time. Therefore, this property can be used to infer and estimate the general and possible changes in hydrological regimes within one year.

2.5.4 *Ambiguity*

There is still a clear intermediate transition in the changes of river hydrological regimes characterized by the transition from one hydrological regime to another, also known as ambiguity. If there is a transitional period between non-flood season and flood season, any day during this period is considered to belong to both non-flood season and flood season, albeit with varying degrees of affiliation. In the northern rivers of China, the closer to July and August, the greater the degree of belonging to the flood season, while the degree of belonging to the non-flood season is smaller. On the contrary, the closer it is to January and February, the greater the degree of non-flood season, while the degree of flood season is very small. Usually referred to as flood season or non-flood season, it belongs to the fuzzy concept and can be described using fuzzy set theory. A wet year or a dry year is also a vague concept.

Chapter 3
Basics of Hydrological Statistics

Hang a feather and a piece of charcoal to discern the moisture in the air.
悬羽与炭，而知燥湿之气。
From small signs, understand the greater picture. See a single leaf fall and know the year is nearing its end.
以小明大，见一叶落而知岁之将暮。
Observe the ice in a bottle and know the cold of winter approaches...
睹瓶中之冰，而知天之将寒……
Yang energy is fire. Yin energy is water.
阳气为火，阴气为水。
When water dominates, the summer solstice brings dampness. When fire dominates, the winter solstice brings dryness.
水胜，故夏至湿，火胜，故冬至燥。
Dryness makes charcoal light. Dampness makes charcoal heavy.
燥故炭轻，湿故炭重。
Huainanzi: Discourse on Mountains, Liu An, Western Han Dynasty
—《淮南子·说山训》，刘安，西汉
▲
This passage uses observations of natural phenomena (e.g., feathers, charcoal, falling leaves, ice) to infer climatic and seasonal changes, reflecting an ancient understanding of hydrological statistics and the relationship between Yin-Yang principles and weather patterns.

Abstract This chapter details the method of moments curve-fitting technique, which iteratively adjusts theoretical distribution parameters to align with empirical data, optimizing goodness-of-fit. Finally, it discusses correlation analysis to assess variable dependencies and compound extreme event risks.

Supplementary Information The online version contains Electronic Supplementary Material (ESM) available at https://doi.org/10.1007/978-981-95-9089-6_3.

X. Feng (ed.), *Engineering Hydrology*, https://doi.org/10.1007/978-981-95-9089-6_3

For a visual overview of the chapter's key concepts, refer to the knowledge map below.

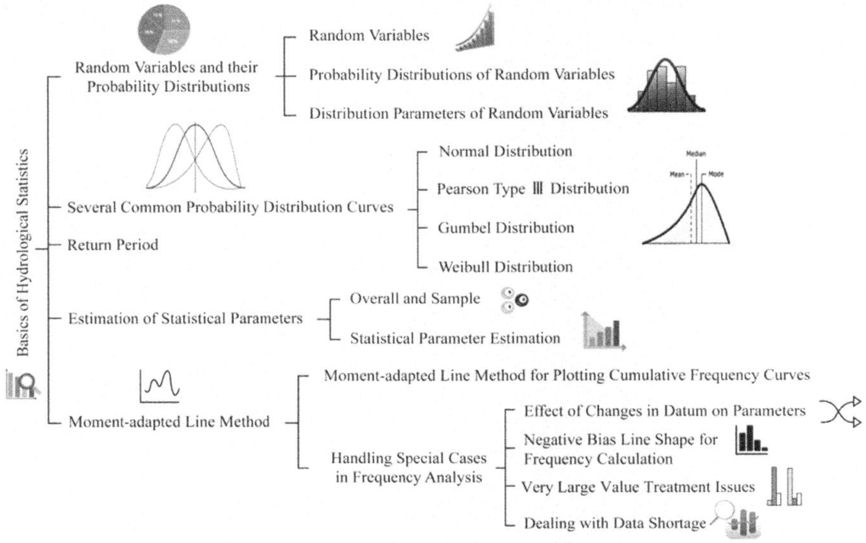

3.1 Introduction

Before diving into hydrological statistics, we acknowledge British biologist Karl Pearson (K. Pearson), a pioneer whose foundational statistical methods, including correlation coefficients and regression analysis, significantly shaped modern statistics. Among his contributions, the Pearson Type III distribution is particularly noteworthy, widely employed for hydrological data analysis due to its ability to accurately model skewed distributions.

Hydrological statistics emerged alongside advancements in water resource management, responding to the inherent randomness and uncertainty of hydrological phenomena like river flow and floods. Initially focusing on basic statistical descriptions, it evolved rapidly through the twentieth century, benefiting from advancements in probability theory and computational technologies. Modern hydrological statistics integrates with meteorology, oceanography, and remote sensing, enhancing its role in water resource assessments, flood forecasting, and hydraulic engineering. Today, artificial intelligence and big data further revolutionize hydrological statistics, enabling sophisticated analysis of extensive datasets and precise predictions of extreme hydrological events.

The chapter then details the method of moments curve-fitting technique, which iteratively adjusts theoretical distribution parameters to align with empirical data, optimizing goodness-of-fit. Finally, it discusses correlation analysis to assess variable dependencies and compound extreme event risks.

3.2 Random Variables and Their Probability Distributions

3.2.1 Random Variables

Each time a random phenomenon is tested, the result can be represented by a value x. The value of x is uncertain and often different for each test, but the occurrence of a value x_i often has a corresponding probability, indicating that this variable x is random, and is therefore called a random variable, or random variables.

The hydrological characteristics of a hydrological phenomenon are often random variables, such as the annual precipitation at a certain place, the annual maximum water level at a station, the maximum flood flow, etc. A series composed of random variables, such as $x_1, x_2, \ldots, x_n$, is called a series of random variables, which can be represented by a capital X. The range of the series can be finite or infinite.

Random variables can be classified as either discrete or continuous.

If there is no intermediate value between two adjacent random variables, such random variables are called discrete random variables. For example, the total number of days of annual rainfall at a station has only 1 to 365 (366) possible values, and no intermediate value can be taken.

A random variable is said to be continuous if any value can be obtained within a finite interval. For example, the average annual flow at any section of a river can vary between a certain flow and a limiting flow by taking any real value of it.

3.2.2 Probability Distributions of Random Variables

A random variable has a certain probability of achieving a certain value. This one-to-one relationship between a random variable and its probability is called the law of probability distribution of a random variable, or probability distribution for short. It reflects the pattern of variation of a random phenomenon.

A discrete random variable X can only take on a finite number of values or a sequence of values. Let all possible values of X be $x_1, x_2, \ldots, x_n$, and the corresponding probabilities be $p_1, p_2, \ldots, p_n$, i.e.,

$$P(X = x_1) = p_1, P(X = x_2) = p_2, \ldots, P(X = x_n) = p_n \tag{3.1}$$

Or a table of possible values of X and their corresponding probabilities, called a probability distribution table for the random variable X. There is a calculation example available via the ESM 3.1.

For a continuous random variable, possible values differ from the probability of obtaining individual values, as in the case of a discrete variable, but fill an interval such that the probability of obtaining any individual value tends to zero. Continuous random variables can therefore be studied only in terms of probabilities over finite intervals. For convenience, the probability of $X \geq x$ is studied and denoted as $P(X \geq x)$; similarly, the probability of $X < x$ may also be studied. There is a reciprocal relationship between the two, i.e.,

$$P(X > x) = 1 - P(X < x) \tag{3.2}$$

Therefore, only one probability needs to be studied.

Clearly, $P(X \geq x)$ is the distribution function of the random variable X, denoted as

$$F(x) = P(X \geq x) \tag{3.3}$$

It represents the probability that X is greater than a given value of x. The geometric curve is called the probability distribution curve; when plotted using observed data points, the hydrographic curve is called the cumulative frequency curve. Another calculation example is shown in the ESM 3.2.

The probability that the random variable X falls in the interval $(x, x + \Delta x)$ is

$$P(x + \Delta x > X \geq x) = F(x) - F(x + \Delta x) \tag{3.4}$$

The corresponding frequency value $P(x + \Delta x > X \geq x)$ is related to the group spacing Δx. To eliminate the effect of the group spacing Δx, the ratio of the probability of the random variable X falling into the interval $(x, x + \Delta x)$ to the length of the interval (Δx) can be expressed as $\frac{F(x)-F(x+\Delta x)}{\Delta x}$, which represents the average probability (density) of X falling into the interval $(x, x + \Delta x)$.

$$\overline{P} = \frac{F(x) - F(x + \Delta x)}{\Delta x} \tag{3.5}$$

Taking the limit as $\Delta x \to 0$, the resulting value is called the probability density of the random variable X at the value x:

$$\lim_{\Delta x \to 0} \frac{F(x) - F(x + \Delta x)}{\Delta x} = -\lim \frac{F(x + \Delta x) - F(x)}{\Delta x} = -\frac{\mathrm{d}F(x)}{\mathrm{d}x} = -F'(x) \tag{3.6}$$

where $F'(x)$ is the first-order derivative of the probability distribution function $F(x)$, such that

$$f(x) = -F'(x) = -\frac{\mathrm{d}F(x)}{\mathrm{d}x} \tag{3.7}$$

The function $f(x)$ possesses the property of a density and is referred to as the probability density function, or simply the density function. The density function of the geometric trilinear is called the density curve, as shown in Fig. 3.1. $f(x)\mathrm{d}x$ is called the probability element and has the geometric meaning as shown in the area shaded in Fig. 3.1. The probability distribution function $F(x)$ can be derived from the density function, i.e.,

$$F(x) = P(X \geq x) = \int_x^{\infty} f(x)dx \tag{3.8}$$

The relationship between the two is shown in Fig. 3.2. In most cases, the upper limit of the random variable X is generally taken to be infinity. In summary, the probability distribution of a continuous random variable can be better described by the density function and the distribution function.

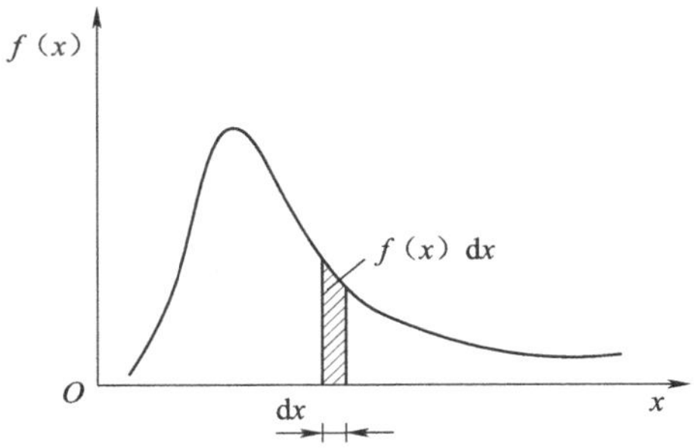

Fig. 3.1 Density curve

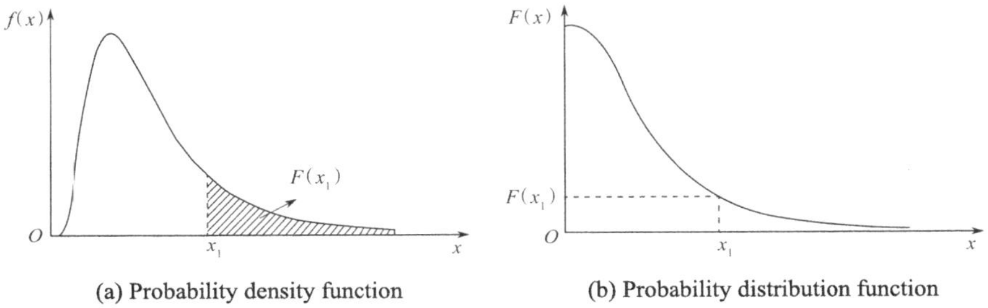

Fig. 3.2 Relationship between the probability distribution function $F(x)$ and the density function $f(x)$

3.2.3 Distribution Parameters of Random Variables

As mentioned above, from a mathematical point of view, a probability distribution curve can describe the statistical law of a random variable in its entirety, but from an engineering practical point of view, it is sometimes not always necessary to describe a random variable in its entirety in functional form, as long as some representative values of the parameters are known to illustrate the main characteristics of the distribution of the random variable. This can illustrate the statistical laws of random variables of certain characteristics of the value, known as statistical parameters, or characteristic parameters. If necessary, these parameters can be used to determine the probability distribution function of the random variable, so they are also known as distribution parameters.

In hydrological statistical calculations, commonly used parameters include: position characteristic parameters, discrete degree characteristic parameters, and skewness coefficient. Their definitions and calculation methods are shown in ESM 3.3.

3.2.4 Several Common Probability Distribution Curves

The following types of probability distributions are commonly used in current hydrological analysis calculations.

3.2.4.1 Normal Distribution

Hydrological observations and sampling errors conform to a normal distribution. The normal distribution has a probability density function of the following form:

$$f(x) = \frac{1}{\sigma\sqrt{2\pi}}e^{-\frac{(x-\overline{x})^2}{2\sigma^2}} \quad (-\infty < x < +\infty) \tag{3.9}$$

where e is the base of the natural logarithm; other symbols have the same meaning as before.

The density curve of the normal distribution has the following characteristics: ① single-peaked shape, the density function has only one extreme value ($\hat{x}$ plural); ② axis symmetry with the mean, $C_s = 0$, so $\overline{x} = \hat{x} = \breve{x}$; ③ the ends of the curve converge to $\pm\infty$ and are asymptotic to the x-axis.

The normal distribution density function contains only two parameters if a random variable is normally distributed. If a random variable is normally distributed, the distribution is completely determined by asking for the values of its $\overline{x}$ and σ parameters.

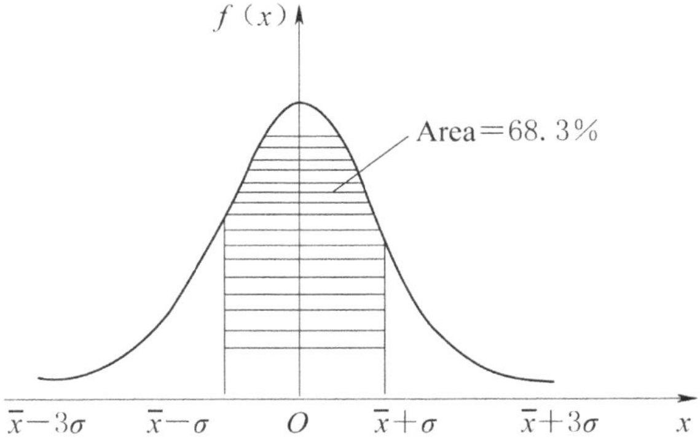

Fig. 3.3 Density profile of a normal distribution

It can be shown that the normal distribution curve has an inflection point at $\overline{x} \pm \sigma$ and that

$$\frac{1}{\sigma\sqrt{2\pi}}\int_{\overline{x}-\sigma}^{\overline{x}+\sigma} e^{-\frac{(x-\overline{x})^2}{2\sigma^2}}\,dx = 0.683 \tag{3.10}$$

$$\frac{1}{\sigma\sqrt{2\pi}}\int_{\overline{x}-3\sigma}^{\overline{x}+3\sigma} e^{-\frac{(x-\overline{x})^2}{2\sigma^2}}\,dx = 0.997 \tag{3.11}$$

This means that the probability of the random variable occurring in the interval $\overline{x} \pm \sigma$ is 68. 3% (i.e., the area shaded in Fig. 3.3); the probability of it occurring in the interval $\overline{x} \pm 3\sigma$ is 99.7%, as shown in Fig. 3.3.

3.2.4.2 Pearson Type III Distribution

(1) The origin of the density curve function. After conducting a statistical analysis of a large body of biological data, Pearson found that the probability density curve is typically unimodal, meaning that it has only one mode. The probability density decreases gradually on either side of the peak as the variable moves away from the mode, producing an approximately bell-shaped curve. Moreover, the two ends of the curve (or one end) approach the horizontal axis asymptotically, as shown in Fig. 3.4. Pearson then proposed two basic conditions for establishing a family of curves: ① if the vertical axis is at the mean, when $x = -d$, then $dy/dx = 0$, $\overline{x} = \hat{x}$; ② When $y = 0$, you should make $dy/dx = 0$, i.e., the curve is tangent to the horizontal coordinate.

From these two simple basic concepts, Pearson established a general differential equation for the probability density curve:

$$\frac{dy}{dx} = \frac{(x+d)y}{b_0 + b_1 x + b_2 x^2} \tag{3.12}$$

Fig. 3.4 Pearson probability density curve

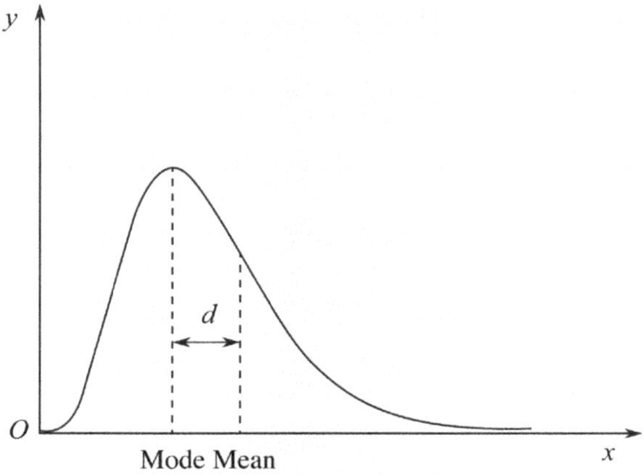

The parameter b_0, b_1, b_2 can be integrated to give different forms of probability density curves if different values are taken.

The family of Pearson curves derived from Eq. (3.12) is divided into 13 types. The curve obtained at $b_2 = 0$ is called the Pearson Type III curve. The differential equation for its probability density curve is

$$\frac{dy}{dx} = \frac{(x + d)y}{b_0 + b_1 x} \tag{3.13}$$

If the origin of the coordinates is shifted to the actual zero point of the hydrological series variable, the common form of a Pearson Type III curve is obtained:

$$y = f(x) = \frac{\beta^a}{\Gamma(\alpha)}(x - a_0)^{\alpha-1} e^{-\beta(x-a_0)} \tag{3.14}$$

where $\Gamma(\alpha) = \int_0^\infty t^{\alpha-1} e^{-t} dt$, called the gamma function of α; a_0 is the value of the horizontal coordinate of the starting point of the curve, *a* parameter to be determined; α, β are two parameters to be determined.

The Pearson Type III curve, determined by these three parameters, takes the form of an asymmetric single-peaked probability density curve with one end finite and the other end infinite, as shown in Fig. 3.5.

(2) Relationships between curve parameters and statistical parameters. The Pearson Type III curve equation contains three parameters: (α, β, a_0), which can be expressed in terms of the mean $\overline{x}$, the coefficient of variation $C_{v,}$ and the skewness coefficient C_s. The following relationships are obtained through statistical analysis.

$$\alpha = \frac{4}{C_s^2} \tag{3.15}$$

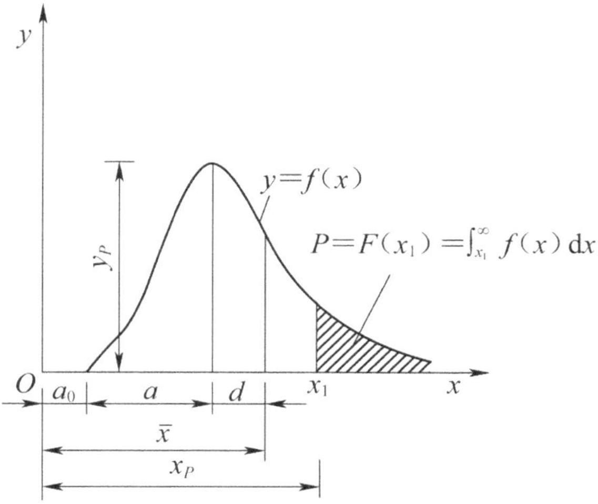

Fig. 3.5 Pearson Type III curve

$$\beta = \frac{2}{\bar{x}C_{\mathrm{v}}C_{\mathrm{s}}} \tag{3.16}$$

$$a_0 = \bar{x}\left(1 - \frac{2C_{\mathrm{v}}}{C_{\mathrm{s}}}\right) \tag{3.17}$$

(3) Method of plotting Pearson Type III probability distribution curves. It can be seen from Eqs. (3.14–3.17) that once the three parameters $\bar{x}$, C_{v}, C_{s} are determined, the probability density function of the Pearson Type III curve is completely specified. By further integrating this function, the corresponding probability distribution curve can be obtained.

$$P(X \geq x) = F(x) = \int_x^a f(x)\mathrm{d}x = P \tag{3.18}$$

or

$$P = \frac{\left[\frac{2}{\bar{x}C_{\mathrm{v}}C_{\mathrm{s}}}\right]^{\frac{4}{C_{\mathrm{s}}^2}}}{\Gamma\left(\frac{4}{C_{\mathrm{s}}^2}\right)} \int_x^{\infty} \left(x - \bar{x} + \frac{2C_{\mathrm{v}}}{C_{\mathrm{s}}}\bar{x}\right)^{\frac{4}{C_{\mathrm{s}}^2}-1} \mathrm{e}^{-\frac{2}{\bar{x}C_{\mathrm{v}}C_{\mathrm{s}}}\left(x-\bar{x}+\frac{2C_{\mathrm{v}}}{C_{\mathrm{s}}}\bar{x}\right)}\mathrm{d}x \tag{3.19}$$

Make $\frac{4}{C_{\mathrm{s}}^2} = t$, then $\Gamma(t) = \int_0^{\infty} x^{x-1}\mathrm{e}^{-x}\mathrm{d}x$.

According to Eq. (3.19), once the parameters $\bar{x}$, C_{v}, C_{s} are determined, the value of x depends only on P; therefore, x can be computed from P. In engineering hydrological analysis and calculations, it is often necessary to determine the random variable (e.g., the design water level or design discharge) corresponding to a specified cumulative frequency (such as the design cumulative frequency). This can be obtained directly by evaluating the integral in Eq. (3.19), completed by the computer. Of course, the computational work can be greatly simplified by consulting specially prepared tables.

The variable is standardized as follows:

$$\Phi = \frac{x - \overline{x}}{\overline{x}C_{\mathrm{v}}} \tag{3.20}$$

Φ is referred to as the standardized (deviation) coefficient. It has a mean of zero and a standard deviation of 1. To facilitate the preparation of a dedicated table, substitute Eq. (3.20) into Eq. (3.19) to obtain

$$P(\Phi > \Phi_P) = \int_{\Phi_P}^{\infty} f(\Phi, C_s)\mathrm{d}\Phi \tag{3.21}$$

Given C_{s} and the design cumulative frequency P, Φ_P can be obtained from schedule 1, and x_P can then be computed using Eq. (3.22).

$$x_P = (\Phi_P C_v + 1)\overline{x} \tag{3.22}$$

If you substitute the modulus ratio coefficient $k_P = x_P/\overline{x}$ into Eq. (3.22), you get

$$k_P = \Phi_P C_v + 1 \tag{3.23}$$

Based on the ratio of $C_{\mathrm{s}}/C_{\mathrm{v}}$, check the table of your values of k_P for Pearson Type III cumulative frequency curves, i.e., Schedule 2, so

$$x_P = \overline{x}k_P \tag{3.24}$$

Therefore, plotting a probability distribution curve or a cumulative frequency curve reduces to determining the values corresponding to various probabilities (or cumulative frequencies) P from the known parameters $\overline{x}$, C_{v} and C_{s} and then plotting the resulting relationship curve. There is an example in the ESM 3.4.

3.2.4.3 Gumbel Distribution

The extreme-value distribution was first applied to hydrological analysis by E.J. Gumbel in 1941, and it's commonly referred to as Gumbel distribution.

A series of random variables $x_1, x_2, \ldots, x_k$ has k variables, and its distribution function $G(x)$ is called the original distribution. If n trials are made under the same conditions, n groups of random variables can be obtained, each group of random variables has a maximum term and a minimum term, the maximum term of the n

groups form a distribution and the minimum term of the groups also form a distribution, because the maximum term or the minimum term is extreme value, so its distribution is called the extreme value distribution, expressed by $F(x)$.

The relationship between the extreme value distribution function $F(x)$ of the largest term and the original distribution function is

$$P(X \geq x) = F(x) = 1 - [1 - G(x)]^k \tag{3.25}$$

The original distribution $G(x)$ is an exponential distribution, approximated by writing

$$G(x) = \frac{1}{k}\mathrm{e}^{-y} \tag{3.26}$$

When the number of variables is sufficiently large, i.e., $k \rightarrow \infty$, Eq. (3.25) becomes

$$P(X_\alpha \geq x) = F(x) = 1 - [1 - G(x)]^k = 1 - [1 - \frac{1}{k}\mathrm{e}^{-y}]^k = 1 - \mathrm{e}^{-\mathrm{e}^{-y}} \tag{3.27}$$

where $y = \alpha(x - u)$, u is the plurality of the extreme value distribution $\hat{x}$, α is the plurality density function times, i.e.,

$$\alpha = kg(\hat{x}) \tag{3.28}$$

u and α are the two parameters of the curve. Gumbel used the least squares method to obtain

$$\left.\begin{aligned} \alpha &= \frac{\sigma_n}{S_n} \\ u &= \overline{x} - \frac{\overline{y_n}}{\alpha} \\ S_n &= \sqrt{\frac{1}{n}\sum x_i^2 - (\frac{1}{n}\sum x_i)^2} = \sqrt{\frac{1}{n}\sum (x_i - \overline{x})^2} \end{aligned}\right\} \tag{3.29}$$

Of which

$$\left.\begin{aligned} \sigma_n &= \sqrt{\frac{1}{n}\sum y_i^2 - (\frac{1}{n}\sum y_i)^2} = \sqrt{\frac{1}{n}\sum (y_i - \overline{y_n})^2} \\ \overline{x} &= \frac{1}{n}\sum x_i \\ \overline{y_n} &= \frac{1}{n}\sum y_i \end{aligned}\right\} \tag{3.30}$$

σ_n, $\overline{y_n}$ are only related to the cumulative frequency P, which is determined by the sample size n. For a given n, P can be computed from the Weibull plotting position:

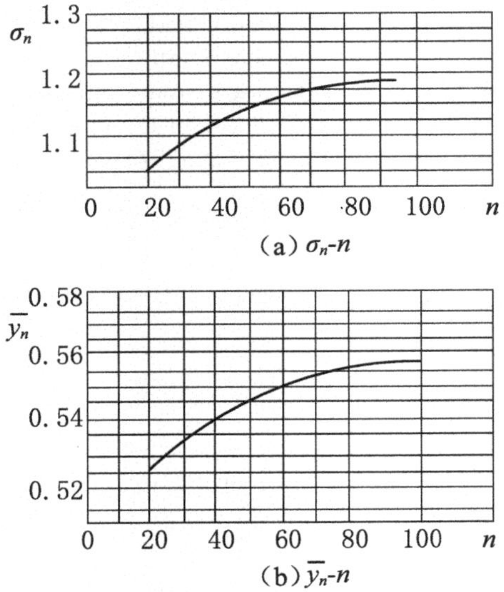

Fig. 3.6 σ_n, $\overline{y_n}$ values in the Gumbel distribution

$P = m/(n+1)$. m is the rank of x when the data are arranged in descending order. The relationships between n and between $\overline{y_n}$, σ_n may be obtained from a curve or in tabulated form, as shown in Fig. 3.6.

The cumulative frequency for the specified variable x_P is

$$P = 1 - e^{-e^{-y_P}} \tag{3.31}$$

And

$$y_P = \alpha(x_P - u) \tag{3.32}$$

Substituting Eq. (3.32) into Eq. (3.31), we get

$$x_P = u - \frac{1}{\alpha}\ln[-\ln(1-P)] \tag{3.33}$$

Let $u = \overline{x} - \frac{\overline{y_n}}{\alpha}$, $\alpha = \frac{\sigma_n}{S_n}$, and $S_n = \sqrt{\frac{1}{n}\sum x_i^2 - \left(\frac{1}{n}\sum x_i\right)^2} = \sqrt{\frac{1}{n}\sum (x_i - \overline{x})^2}$. Substituting these expressions into Eq. (3.33), we can get

$$x_P = \overline{x} - \frac{y_n S_n}{\sigma_n} - \frac{S_n}{\sigma_n}\ln[-\ln(1-P)] = \overline{x} + \frac{1}{\sigma_n}\{-\ln[-\ln(1-P)] - \overline{y_n}\}S_n \tag{3.34}$$

Make

$$\lambda_{P_n} = \frac{1}{\sigma_n}\{-\ln[-\ln(1-P)] - \overline{y_n}\} \tag{3.35}$$

Then

$$x_P = \overline{x} + \lambda_{P_n} S_n \tag{3.36}$$

Equation (3.36) is called the Type I extreme value distribution law, where λ_{P_n} is related to P and n only, and is tabulated in Schedule 3. $\overline{x}$, S_n can be calculated from measured data.

Similarly, the minimum term extreme value distribution law can be derived as

$$x_P = \overline{x} - \lambda_{P_n} S_n \tag{3.37}$$

The sample variables should be arranged in increasing order when calculating the empirical cumulative frequency of the minimum term.

The *Code of Hydrology for Harbour and Waterway* (JTS 145-2015E) states that the Gumbel distribution is used for the tidal frequency distribution curve. The specific application can be found in Chap. 6. In practice, the annual maximum flood flow also follows this distribution.

3.2.4.4 Weibull Distribution

The Weibull distribution was introduced by the Swedish physicist W. Weibull in 1939 and was originally used for the distribution of the breaking strength of materials, then in 1954, Gumbel applied it to the statistical analysis of the hydrology of the annual minimum flow, also known as the Type III minimum value distribution. The random variable distribution function is

$$P(X \geq x) = F(x) = \begin{cases} 1 - \mathrm{e}^{-(x-\gamma)^{\beta}/\alpha} & (x \geq \gamma) \\ 0 & (x < \gamma) \end{cases} \tag{3.38}$$

where α is the scale parameter; β is the shape parameter; γ is the position parameter.

If the random variable follows the Weibull distribution, the distribution curve can be determined by knowing the above three parameters.

The shape of the Weibull distribution function curve, when taking different parameters, is shown in Fig. 3.7.

When dealing with dry water flow problems, it is usually considered better to have a lower bound of zero for the distribution, i.e., when $\gamma = 0$, then the distribution function is simplified to two parameters, i.e.

$$P(X \geq x) = F(x) = 1 - \mathrm{e}^{-x^{\beta}/\alpha} \tag{3.39}$$

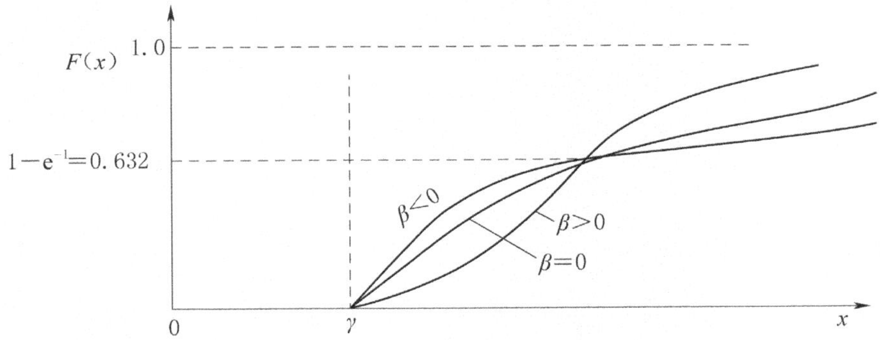

Fig. 3.7 Weibull distribution function curves for identical values of α and γ but different values of β

The parameters β, α, can be estimated by using the intercept method. However, determining the Weibull distribution parameters by this analytical approach is computationally cumbersome. Therefore, the parameters are usually obtained by graphical fitting on the standard Weibull probability density curve chart, using the empirical cumulative-frequency points of the variable, or alternatively by least-squares fitting.

3.2.5 *Return Period*

As the hydrological characteristics of hydrological phenomena are often random variables, it is necessary to apply stochastic analysis to these basic hydrological characteristics in engineering planning and design, and to obtain the cumulative frequency curves of these characteristics at different line types, so as to know the quantity of a certain characteristic at different cumulative frequencies. In engineering design, the concept of "return period" is often introduced for those hydrological features with annual cycles, such as design floods, design waves, etc., and the return period is specified in the engineering design standards according to the scale of the project, the characteristics of a hydrological feature, and the importance of the project. The recurrence period is the average time interval between the repeated occurrence of a value in a series of random variables, i.e., the average recurrence interval. The recurrence period T corresponds to the cumulative frequency P and can be expressed as

$$T = \frac{1}{P} \tag{3.40}$$

where T is the recurrence period in years; P is the cumulative frequency, in decimals or percentages.

If the cumulative frequency of the hydrological characteristics (e.g., flood peaks or floods) of a given year is $P = 0.\ 1\%$, substituting it into Eq. (3.40) gives $T = 1/0.001 = 1{,}000$ years, i.e., the design standard can be said to be a 1,000-year event.

If the hydrological characteristics (e.g., the dry water level) of a year to be studied have a cumulative frequency of $P > 50\%$, the expression for the relationship between the recurrence period and the cumulative frequency is

$$T = \frac{1}{1 - P} \tag{3.41}$$

If the design cumulative frequency of the lowest water level in a section of river is $P = 95\%$, substituting it into Eq. (3.41) gives $T = 1/(1–0.95) = 20$ years. That is, the design standard can be said to be a 20-year event.

It should be noted that the cumulative frequency referred to above represents the long- term average probability of occurrence over many years, whereas the return period denotes an average recurrence interval; that is, an event occurs once on average every specified number of years rather than at a fixed cycle. A 100-year event does not mean that the event necessarily occurs once every one hundred years. In practice, it may occur several times within a particular one-hundred-year period, or it may not occur at all. This expression merely indicates that, over a sufficiently long time scale, the event has an average return period of about one hundred years. In general, the term unprecedented in a hundred years refers to the maximum value of a hydrological feature (e.g., a flood peak) in the past 100 years. This concept differs from the hydrologic characteristic value associated with a 100-year return period event, namely the 100-year design value (or 100-year quantile). The observed 100-year maximum may be smaller than, equal to, or greater than the corresponding 100-year standard value.

3.3 Estimation of Statistical Parameters

3.3.1 Overall and Sample

Statistical mathematics refers to the sum of all the individuals in an object of study as the total, and the number of individuals contained in the total is called the capacity of the total.

In practice, the totality of a random phenomenon is often not known, and in order to simplify the problem so that the totality need not be studied, a portion of the total volume is often taken for analysis. A portion of the individuals in the total is called a sample. The number of individuals contained in the sample is called the volume of the sample. The same overall population can be divided into multiple samples. If a sample is drawn arbitrarily without any subjective component, it is called a random sample and the sampling process is called random sampling.

The totality of hydrological phenomena should include all the variables of river processes and can be considered as an infinite totality. The existing series of observations of hydrological elements is only a small part of this, a random sample.

As the sample is part of the total, the characteristics of the sample reflect, to some extent, those of the total. In cases where the overall population is difficult to obtain, the patterns of the sample are often used to infer the regularity of the overall population or to simplify its analysis. As to whether or not the patterns inferred from the samples truly reflect the overall patterns, it depends on how representative the samples are of the overall population. For example, if a section of the annual runoff sample series is to be representative, it should include a variety of information on abundant water, flat water, and dry water. Usually, a larger sample size is more representative than a smaller one. In practice, the selection of a representative series should be preceded by a field survey, a review of historical literature, and a comparison of hydrological data from neighboring areas.

3.3.2 Statistical Parameter Estimation

Corresponding to the parameters of the probability distribution of the overall random variable, there are also three cumulative frequency curves for a sample of random variables, namely the mean, the coefficient of variation, and the coefficient of skewness.

These parameters are often calculated using the method of moments, and specific calculations and descriptions of the introduction to the method of moments, the statistical parameters of the sample series, the unbiased estimates of the sample parameters, and the sampling errors are detailed in ESM 3.5.

3.4 Hydrologic Frequency Calculation Method of Moment Fitting Line

3.4.1 Overview

In the planning and design of hydraulic, hydropower, and shipping projects, it's often essential to determine certain hydrological characteristics (e.g., peak flood flow, minimum water level) and their probabilities during the project's service life (cumulative frequency). Alternatively, it's necessary to provide hydrological characteristics with a specific reliability standard (cumulative frequency) to serve as a basis for determining the project's scale. These characteristic values are generally obtained through frequency analysis, the core of which is the cumulative frequency curve.

Assuming the hydrological characteristics are based on a sufficiently long and comprehensive dataset, it is possible to calculate the cumulative frequency of the series using the method outlined in the example in ESM 3.2. By using the equation $P = m/n \times 100\%$, the cumulative frequency can be calculated (as shown in Table 1 in ESM 3.2), and the cumulative frequency curve can be plotted against the data. Alternatively, if the dataset is consistent with a probability distribution (e.g., Pearson Type III distribution), the cumulative frequency can be calculated by integrating Eq. (3.19) and plotting the cumulative frequency curve, as in the example in ESM 3.4. However, hydrological data series are often short, making it challenging to meet the assumptions above. If the design values for rare cumulative frequencies (e.g., design standard cumulative frequency $P = 0.1\%$, calibration standard cumulative frequency $P = 0.01\%$) are to be derived, the empirical cumulative frequency curves plotted at the limited data points must be extrapolated. This process introduces subjectivity and arbitrariness. To minimize arbitrariness, a probability distribution curve (such as those described in Sect. 3.2) often serves as an essential reference for determining the cumulative frequency curve and the extrapolated empirical cumulative frequency curve.

Currently, in engineering planning and design, methods for plotting cumulative frequency curves include the method of finding the appropriate line of moments, the three-point method, and the method of reading points to create moments. The most widely used approach is the method of finding the appropriate line of moments, commonly referred to as the appropriate line method.

3.4.2 Moment-Adapted Line Method for Plotting Cumulative Frequency Curves

3.4.2.1 Basic Points of the Moment-Adapted Line Method for Plotting Cumulative Frequency Curves

Empirical cumulative frequency points are calculated from the measured data and plotted on a probability grid. Statistical parameters such as $\overline{x}$, and C_v are then calculated using the method of moments, treating the actual data as samples, and the probable error is estimated. A suitable probability distribution curve, also known as a theoretical cumulative frequency curve (e.g., Pearson Type III curve), is then selected based on statistical parameters like $\overline{x}$, and C_v and C_s. The theoretical cumulative frequency curve is plotted against the empirical point distribution to assess the fit. If discrepancies exist, the fit can be adjusted by modifying C_s or the other two statistical parameters, $\overline{x}$ and C_v, within a reasonable margin of error, to achieve a better fit with the empirical cumulative frequency data. The goal is to identify a satisfactory cumulative frequency curve.

Ultimately, the statistical parameters and corresponding cumulative frequency curves used for project planning and design should not solely rely on hydrological statistics but also be informed by a comprehensive analysis of the causes of hydrological phenomena and the regional distribution of their characteristic values.

3.4.2.2 Equation for Calculating the Cumulative Frequency of Experience

Given the brief duration of the measured hydrological characteristics, the grouping and direct product methods cannot be used to obtain the empirical cumulative frequency. In China, the empirical cumulative frequency formula is commonly used as the mathematical expectation formula in hydrological analysis, as detailed in Sect. 3.2, where Gumbel utilized the following formula:

$$P = m/(n+1) \times 100\% \tag{3.42}$$

where P is the cumulative frequency of experience; n is the total number of items in the series (sample size); m is the ordinal number of the variables in the sample series sorted from largest to smallest.

3.4.2.3 Specific Steps for Plotting Cumulative Frequency Curves Using the Fit-Line Method

(1) Plot the empirical cumulative frequency curves. Arrange the measured hydrological data in order of variable size, from largest to smallest, and substitute the variable number m into Eq. (3.42). Use the variable value as the vertical coordinate and P as the horizontal coordinate to calculate and plot the empirical cumulative frequency curve.

(2) To find the sample statistical parameters using the method of moments, calculate the mean of the sample statistical parameters using Eqs. (3.9, 3.11 and 3.16) in ESM 3.5, obtaining $\overline{x}$, the coefficient of variation C_v, and the mean square errors $\sigma_{\overline{x}}$ and σ_{Cv}. If the sample series is long, the series skewness coefficient C_s can be calculated from Eq. (3.15) in ESM 3.5 as reference data. If the series is short, calculating C_s directly may result in a high sampling error, making it impractical. Instead, several C_s values can be assumed. For annual runoff series, $C_s = (2{-}3)C_v$, and for storm flood series, $C_s = (2.5{-}4)C_v$, which are determined using the fit-line method.

(3) Select a line type. Typically, a Pearson Type III curve is chosen. If this does not fit well, another line type should be selected.

(4) Fit a suitable line. Based on the calculated $\overline{x}$, C_v, and several assumed C_s values (e.g., 2.0 C_v, 2.5C_v, 3.0C_v), different scenarios are formed. The deviation coefficients of Pearson Type III cumulative frequency curves, Φ_P, are checked separately (refer to Schedule 1). From C_s and P, the corresponding Φ_P values

are derived and substituted into Eq. (3.22) to compute the corresponding x_p. Multiple theoretical cumulative frequency curves are then plotted on the same probability grid as the empirical cumulative frequency curves, using x_p as the vertical coordinate and P as the horizontal coordinate. Analytical arguments and comprehensive comparisons are made to assess the fit.

(5) By comparing the regional parameter patterns, a curve that best matches the empirical data is selected, and the corresponding statistical parameters $\overline{x}$, C_v, and C_s are determined. This frequency curve is then used to address the design-related questions. Please refer to ESM 3.6.

3.4.3 Handling Special Cases in Frequency Analysis

3.4.3.1 Effect of Datum Changes on Parameters

The cumulative frequency calculation for water levels is similar to that for flow, rainfall, etc. The difference lies in the cumulative frequency calculation of water levels, where the statistical parameters $\overline{x}$, σ, C_v, and C_s are tied to the water level datum. If the datum is obtained differently, the series mean value $\overline{x}$ and coefficient of variation C_v will vary, but the mean squared deviation σ and skewness coefficient C_s will remain unchanged. Since $\overline{x}$ reflects the location characteristics of the random variable, an increase or decrease in the baseline shifts the density curve's origin horizontally, moving the curve laterally while leaving its dispersion and symmetry unchanged. Thus, if $\overline{x}$ changes, σ and C_s remain the same, but $C_v = \sigma/\overline{x}$ will change.

The equation for the change relationship is

$$\overline{x}_{\pm a} = \overline{x} \pm a \tag{3.43}$$

$$(C_v)_{\pm a} = \frac{\overline{x}}{\overline{x} \pm a} C_v \tag{3.44}$$

where $\overline{x}_{\pm a}$ is the mean value after increasing or decreasing the constant a in the datum; $\overline{x}$ is the mean value before the change; $(C_v)_{\pm a}$ is the coefficient of variation after adding or subtracting a constant a from the base plane; C_v is the coefficient of variation before the change.

When reviewing and utilizing water level data, it's sometimes necessary to convert the vertical datum, for example, from a station datum to an absolute datum, from the original datum to the nationally unified Yellow Sea datum, to unify water levels from different gauging stations along the same river to a common datum. In such cases, a constant a can be uniformly added to or subtracted from the water-level data. If the original parameters have already been calculated, the corresponding parameters can be obtained using Eqs. (3.43) and (3.44).

In practice, frequency calculations are often performed by taking a reference level close to the lowest water level (or breakwater level). If the datum is too low, the mean value decreases, C_v becomes smaller, and the relative error increases.

3.4.3.2 Negative Bias Line Shape ($C_s < 0$) for Frequency Calculation

In general, empirical cumulative frequency curves derived from hydrological data samples have a positive bias factor ($C_s > 0$) when the Pearson Type III cumulative frequency curve is used as the appropriate line. However, in some cases, such as during frequency analysis of annual minimum eigenvalues (e.g., minimum flow and minimum water level), the empirical points can be negatively skewed ($C_s < 0$). In such cases, the empirical points require fitting with a negatively skewed cumulative frequency curve. The existing k_P and Φ_P value tables are positively skewed and cannot be used for negative skewness, so adjustments need to be made.

For negatively skewed frequency density, a case can be imagined where the frequency density is identical but in the opposite direction (positive skew), as shown in Fig. 3.8.

Two corresponding density curves are denoted by negative bias $f(x)$ and positive bias $f(x')$. As can be seen from the graphs, the two distributions have identical statistical parameters except for the sign of the skewness coefficient; all other parameters are the same (the curves are symmetric with respect to point A, where A is the midpoint of x and x'). Their density functions satisfy the following relationship:

$$P(x) = \int_{x_i}^{x_0} f(x)\mathrm{d}x = 1 - \int_{x_i'}^{\infty} f(x')\mathrm{d}x' = 1 - P(x') \quad (3.45)$$

where $P(x)$ is the original negative bias series density plot, under the solid line from x_i integral to x_0 of the cumulative frequency, right area of the shaded part; $P(x')$ is the dashed line graph from x_i' integral to the right end (infinity).

The cumulative frequency of the total area under the density curve is 1, the shaded area on the left is equal to $1 - P(x')$, but the directions of integration are opposite.

As can be seen from Fig. 3.8:

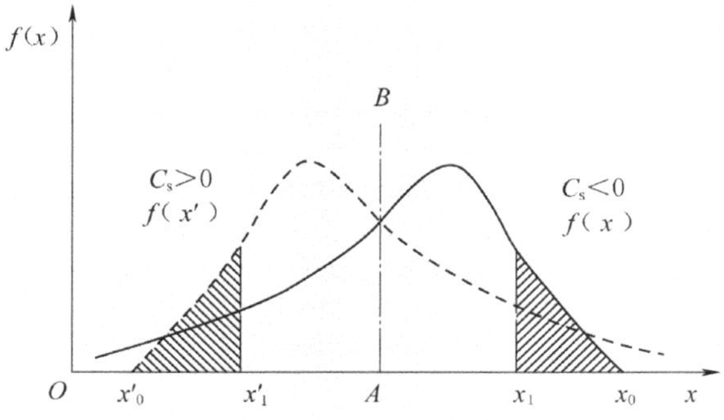

Fig. 3.8 Schematic diagram of a negatively skewed density curve

$$\overline{x} = \overline{x'}$$
$$x_i = x_i' + 2(\overline{x} - x_i') = 2\overline{x} - x_i' \tag{3.46}$$

The deviation coefficient for positive deviation is known:

$$\Phi' = \frac{x_i' - \overline{x}'}{\overline{x}' C_v} \tag{3.47}$$

So for a negative bias, substitute Eq. (3.46) into Eq. (3.47):

$$\Phi = \frac{x_i - \overline{x}}{\overline{x} C_v} = \frac{2\overline{x} - x_i' - \overline{x}}{\overline{x} C_v} = -\frac{x_i' - \overline{x}}{\overline{x} C_v} = -\Phi' \tag{3.48}$$

Therefore, Φ and Φ' are equal in absolute value but opposite in sign. Hence, for a given cumulative frequency, if the Φ' value corresponding to a negative skewed distribution $P(x_P)$ is required, one may first obtain the corresponding Φ' value from the table of frequency factors for positively skewed distributions using 1–$P(x_P)$, and then take its negative value, i.e., $\Phi = -\Phi'$, thereby the corresponding value of Φ in $P(x_P)$ can be obtained.

For example, $C_s = -0.3$ for a Φ value of $P = 1\%$, we first look up in Schedule 1, $\Phi_P' = -2.10$ corresponding to $C_s = 0.3$, $P' = 1 - P(x) = 1 - 1\% = 99\%$. Then change the negative value to a positive one, which gives $C_s = -0.3$ corresponding to $P = 1\%$, $\Phi_P = 2.10$. Thereafter, the general frequency calculation method can be applied to obtain the required cumulative frequency curve.

3.4.3.3 Very Large Value Treatment Issues

In hydrological frequency calculations for port and navigation engineering planning and design, it's often found that the highest annual flood level or flow within a continuous series contains an exceptionally high value (mega-value). When plotting the empirical cumulative frequency line, it's easy to detect a significant, abrupt discontinuity between this extreme value and other values. The recurrence period of this extreme point is much longer than the number of observation years available (i.e., the number of terms in the series).

Therefore, in frequency analysis, this extreme value should not be treated the same as other values in the series but should be appropriately adjusted or handled based on its return period, a concept known as the treatment of extreme values.

Exceptional values, shown in Fig. 3.9, can appear outside the measured series (Fig. 3.9a), within the measured series (Fig. 3.9b), or both inside and outside the measured series (Fig. 3.9c).

Information on mega values. Some stations can be observed, while others can only be obtained through surveys. The treatment of mega-values almost extends the original measured series from a few years to the equivalent of the mega-values test

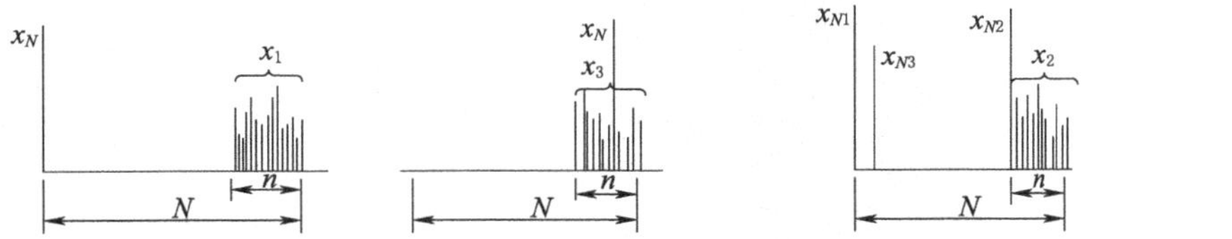

(a) Beyond observations (b) Inside observations (c) Both beyond and inside observations

Fig. 3.9 Schematic diagram of the series considering the value to be large

period N, and thereby increasing the representativeness of the data and the reliability of the calculated results.

For example, a station on a river in North China was planned in 1955 with $n = 20$ years and the calculated design standard $P = 0.1\%$, flood peak $Q_{0.1\%} = 7500\ \text{m}^3/\text{s}$; in 1956, a very large flood occurred and the measured flood flow was 13,700 m^3/s > $Q_{0.1\%}$. If the 1956 information is added and $n = 21$ is not treated, the resulting $Q_{0.1\%} = 25,900\ \text{m}^3/\text{s}$, which is three times the original calculation; if the year 1956 is treated as a mega-flood, with a certified return period of $N = 162$ years, then the calculation $= 19,700\ \text{m}^3/\text{s}$. As can be seen, the effect of the mega-flood on the calculated results is very significant.

The key to the treatment of very large values is to determine the return period N. The period before the year for which measured flood data are available and up to the furthest point in the history of the flood is usually referred to as the investigation period. The period prior to the survey period can be further verified by means of historical documentation, and the period for which historical documentation is available is usually referred to as the verification period. The period for which the evidence is combined with the survey is called the survey period, and is generally obtained through historical flood surveys and historical documentation.

The empirical cumulative frequency calculations and statistical parameters for the consideration of very large floods are as follows.

(1) Method of calculating the empirical cumulative frequency of very large floods. Let there be l mega-floods in the measured series and b mega-floods outside the measured series (l, b = 0, 1, 2...), then the empirical cumulative frequency of mega-floods is calculated as

$$P = \frac{M}{N+1} \times 100\%\ (M = 1, 2, \cdots, b + l) \tag{3.49}$$

where N is the period of investigation and examination (return period) of the first mega-flood; M is the number of the mega-flood in descending order of magnitude.

The general empirical cumulative frequency calculation still uses Eq. (3.42).

For example, a station has 44 years of measured flood flows from 1953 to 1996, and through investigation and verification, the historical flood information of the station is obtained: 3,890m^3/s (1829); 4,980m^3/s (1935); 2,960m^3/s (1960); 4,980m^3/s (1935); 2,960m^3/s (1960).

Based on interviews with local residents aged 90 and above and verification against historical documents (e.g., the local county gazetteer), we get that the floods of 1935 and 1829 were two extraordinary floods that have occurred since 1726, and that there were 271 years between 1726 and 1996 ($N = 1996–1726 + 1$), so Eq. (3.49) can be used to calculate the cumulative frequency of the three large floods, in order of 1935, 1829 and 1960 ordered by $M = 1, 2, 3$, so

$$P_1 = \frac{1}{N+1} \times 100\% = \frac{1}{271+1} \times 100\% = 0.37\%$$

$$P_2 = \frac{2}{N+1} \times 100\% = \frac{2}{271+1} \times 100\% = 0.74\%$$

$$P_3 = \frac{3}{N+1} \times 100\% = \frac{3}{271+1} \times 100\% = 1.10\%$$

(2) The calculation of statistical parameters after considering a very large flood. When there is no mega-flood, the variables are ordered from largest to smallest in consecutive order, so that the samples are called consecutive (continued) series. If the historical flood investigation proves that there is a very large flood, with some gaps in the chronological order between the very large flood and other flood values (missing measurements), the ranking numbers from largest to smallest are discontinuous, and such a sample is called a discontinuous series.

If the n-year measured flood information has flood characteristics representative of the general year, it can be assumed that the mean value of the missing years between the very large flood and the measured series is equal to the mean value of the n-year measured information (Let the observed record contain l extraordinary values, and let b extraordinary values lie outside the observed record, $l + b = a$), so that the mean value of the n-year series including the very large flood and the general flood is

$$\overline{x_N} = \frac{1}{N}\left[\sum_{j=1}^{a} x_j + \frac{N-a}{n-l}\sum_{i=l+1}^{n} x_i\right] \tag{3.50}$$

Similarly, assuming that the variance of the missing years is equal to the variance of the n years of measured information, the mean variance, including the extra-large values, is

$$\sigma_N = \sqrt{\frac{1}{N-1}\left[\sum_{j=1}^{a}(x_j - \overline{x_N})^2 + \frac{N-a}{n-l}\sum_{i=l+1}^{n}(x_i - \overline{x_N})^2\right]} \tag{3.51}$$

The coefficient of variation is

$$C_{vN} = \frac{\sigma_N}{\overline{x_N}} = \sqrt{\frac{1}{N-1}\left[\sum_{j=1}^{a}(k_j-1)^2 + \frac{N-a}{n-l}\sum_{i=l+1}^{n}(k_i-1)^2\right]} \tag{3.52}$$

where k_j is the modulus ratio coefficient for an extraordinary flood, $k_j = \frac{x_j}{x_N}, j = 1,$ 2, …, a; k_i is the modulus ratio coefficient for a measured flood, $k_i = \frac{x_i}{x_N}, i = 1, 2,$ …, n-1; $\overline{x_N}$ is mean values after considering mega-floods.

When solving for $\overline{x_N}$ and C_{vN} then the line fit method can be used to derive the cumulative frequency curve for the consideration of very large floods.

(3) Sensitivity analysis of the examination period N. That is the analysis of the sensitivity of the examination period N to the statistical parameters $\overline{x_N}$, C_{vN}. From the mean squared errors (relative error) of the continuous series of parameters $\overline{x}$ and C_v (Table 1 in ESM 3.5), it can be seen that when $C_s = 2\ C_v$, $C_v <$ 1.0, if the sample size $n > 25$, then relative error of $\overline{x}$ and C_v is less than 20%; if $n \geq 100$, the relative error is less than 10%. At present, the number of stations and sites with a sample size of 100 is very small, so in order to reduce the relative error of the parameter estimation, people propose to verify the historical flood or the actual measurement of the mega flood. The method for the reproducibility of water, N, applying Eqs. (3.50 and 3.52) to calculate the statistical parameters of the discontinuous series $\overline{x_N}$ and C_{vN}. It is clear that there is an error in the determination of N, particularly in the determination of N from the recollections of nearly 100 years old, and the extent to which this error affects the parameters can be illustrated by arithmetic examples. Please refer to ESM 3.7.

3.4.3.4 Dealing with Data Shortage

In hydrological statistical calculations, data shortage is often encountered, or the data sequence of the same station is too short, or there is a lack of data in the middle, etc., and the data of adjacent stations is often used for calculation through correlation analysis, or different elements of the same station are analyzed for calculation, such as calculating flow from water level. For knowledge of correlation analysis, refer ESM 3.8.

Chapter 4
River Engineering Design Water Levels and Flow Derivation

East of the water gate lies the Sui Canal.
水门东去接隋渠,
The town's layout, like fish scales, is unmatched in order.
井邑鱼鳞比不如。
Ships from Chu and Wu sail thousands of miles.
楚柂吴樯万里舡,
North and south of the bridge, the scenery is splendid.
桥南桥北好风烟。
Inscription on *Along the River During the Qingming Festival*, Zhang Gongyao, Jin Dynasty
—题《清明上河图》,张公药,金代
▲
This poem depicts the water gate, Sui Canal, orderly town's layout, and bustling scenes around the bridge, reflecting the importance of river engineering design—such as water level control and flow derivation—in urban development and navigation during ancient times.

Abstract This chapter introduces the characteristics, elements, and quantification of floods, as well as design flood criteria and calculation methods for flood control and reservoir regulation projects. It also focuses on navigable design water levels, which serve as key parameters for waterway structures, ship locks, and cross-river projects, and directly affect waterway construction, dredging, and water transport economics.

For a visual overview of the chapter's key concepts, refer to the knowledge map below.

Supplementary Information The online version contains Electronic Supplementary Material (ESM) available at https://doi.org/10.1007/978-981-95-9089-6_4.

X. Feng (ed.), *Engineering Hydrology*, https://doi.org/10.1007/978-981-95-9089-6_4

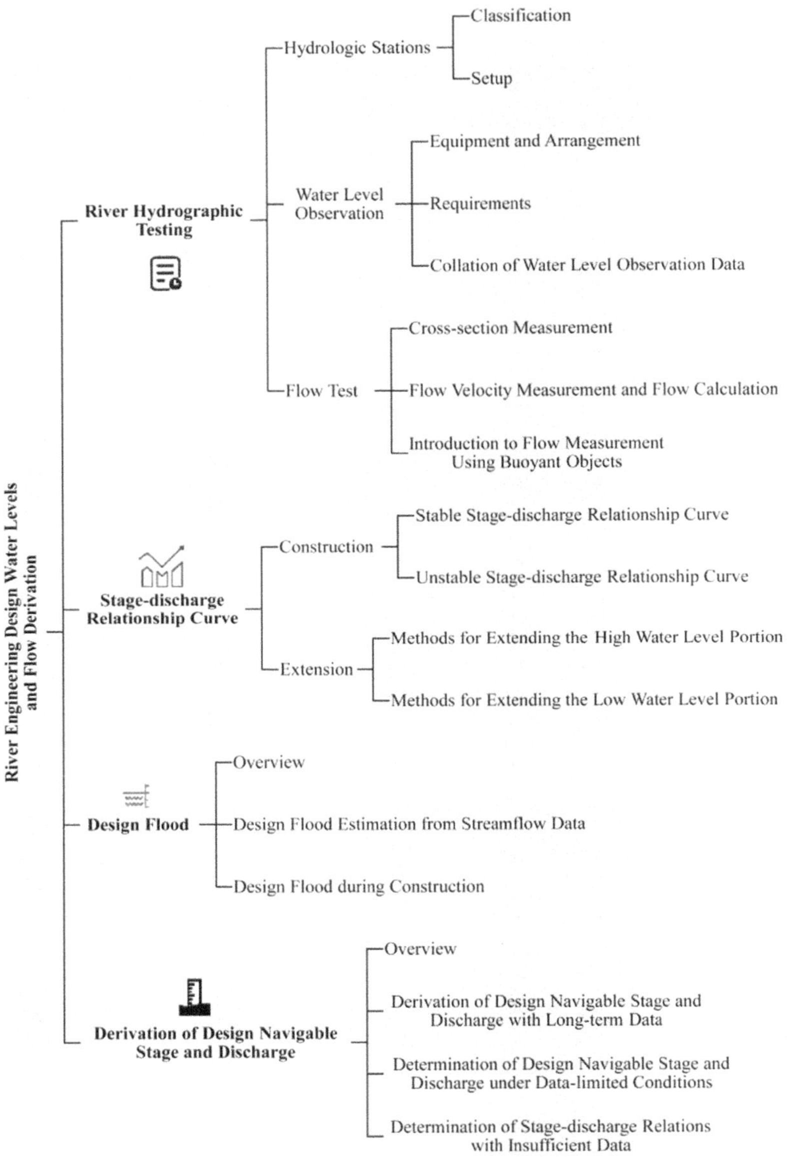
River Engineering Design Water Levels and Flow Derivation
River Hydrographic Testing
Hydrologic Stations
Classification
Setup
Water Level Observation
Equipment and Arrangement
Requirements
Collation of Water Level Observation Data
Flow Test
Cross-section Measurement
Flow Velocity Measurement and Flow Calculation
Introduction to Flow Measurement Using Buoyant Objects
Stage-discharge Relationship Curve
Construction
Stable Stage-discharge Relationship Curve
Unstable Stage-discharge Relationship Curve
Extension
Methods for Extending the High Water Level Portion
Methods for Extending the Low Water Level Portion
Design Flood
Overview
Design Flood Estimation from Streamflow Data
Design Flood during Construction
Derivation of Design Navigable Stage and Discharge
Overview
Derivation of Design Navigable Stage and Discharge with Long-term Data
Determination of Design Navigable Stage and Discharge under Data-limited Conditions
Determination of Stage-discharge Relations with Insufficient Data

4.1 Introduction

Floods have been a severe challenge throughout human history. Over the long-standing years, different countries and ethnic groups have demonstrated unique wisdom and tenacity in the fight against floods, giving rise to many touching stories and profound cultural connotations.

In ancient Chinese times, a great flood ravaged the land. The floodwaters were overwhelming, submerging large areas of land and bringing destruction to all living things. According to legend, after the flood receded, only Nüwa and Fuxi were left in the world. In order to continue the human race, the two intermarried, allowing humanity to reproduce again. Nüwa and Fuxi have become the cultural ancestors of the Chinese nation and are forever remembered and revered by later generations.

Coincidentally, in the West, the story of Noah's Ark is recorded in the Bible. God saw the depravity of humanity and decided to destroy the world with a flood. Only Noah was a righteous man, and God instructed him to build an ark. Noah, his family, and various animals boarded the ark and drifted in the flood for a long time. After the flood receded, humanity and all things were able to reproduce again. These stories not only carry the memories of ancient humans of flood disasters but also reflect the tenacious survival instincts of humans and their unremitting efforts to continue civilization.

As history has progressed, the stories of humans' battles with floods have continued. Yu, the founder of Xia, China's first hereditary dynasty, is well-known for his immortal feats in flood control. It is said that he conducted in-depth research on the floods of the Yellow River, abandoned the simple blocking method, and innovatively adopted the strategy of dredging. He led the people to dig ditches and canals to guide the floodwaters into the sea. After thirteen years of hard work, he passed by his home three times without entering. Finally, he successfully controlled the flood. His spirit has become a symbol of the perseverance of the Chinese nation, and he has also become a flood control hero in the hearts of the people, highly respected by later generations. Today, his name, Yu, or the more well-known Da Yu, has become an icon of China's water conservancy projects.

In ancient Egypt, pharaohs also faced the problem of the flooding of the Nile River. Whether it was due to the great personal charm of the pharaohs at that time or their impressive organizational abilities, thousands of ancient Egyptians came together to build dams and dig canals, greatly reducing the probability of floods and droughts. As leaders, pharaohs were regarded as incarnations of gods and played a key role in flood control and agricultural development, being deeply respected by the people.

A similar story of unity also took place in modern times. In 1998, the Yangtze River Basin in China experienced an unprecedented severe flood. At that time, the construction of the Three Gorges Dam had just started four years ago. Countless embankments on both sides of the Yangtze River were breached. In the face of the surging floodwaters, the soldiers of the Chinese People's Liberation Army charged forward and jumped into the torrential floods, building a solid embankment with their flesh and blood. They carried sandbags and blocked the breaches. Their heroic deeds have forged the great flood-fighting spirit and made them heroes in the hearts of the people in modern times. In 2006, the Three Gorges Dam was finally completed. The construction of the Three Gorges Dam has greatly enhanced the ability to control the floods of the Yangtze River. It is a monument in China's water conservancy history and is world-famous. Recently, many people have been concerned about the potential negative impacts of building embankments and dams on inland rivers, so there have been many different voices internationally regarding the completion of the Three Gorges Dam. However, many people do not know that the Three Gorges Dam was precisely built to resist the frequent floods of the Yangtze River, and its completion has indeed ensured the safety of the middle and lower reaches of the Yangtze River for decades up to now.

Returning to the long river of history, after the flood receded, the land was covered with silt and became fertile. The pharaohs organized the people to cultivate the farmland, making use of the nutrients brought by the regular flooding of the Nile River to develop irrigation agriculture and giving birth to a brilliant agricultural civilization. Today, the pyramids lying quietly on the banks of the Nile River are also witnesses to this great history. And the struggle between humans and water has shifted from being in opposition to nature to a more skillful and circuitous one. How to "turn water disasters into water benefits" has become a new goal for human progress. In the Warring States period of China, a father-son pair was outstanding representatives of achieving this goal. From 256–248 BCE, Li Bing and his son harnessed the Min River in Shu (in today's Chengdu and surrounding areas in Sichuan, China) and built the Dujiangyan Irrigation System. Adhering to the scientific concept of "dredge the shoal deeply and build the weir low", they finally created the "Land of Abundance". The Dujiangyan Irrigation System still benefits the area to this day, and it has been praised by the people from generation to generation.

In Europe, the Dutch, through generations of efforts and wisdom, have also given a perfect answer to this issue. The Netherlands, with its low-lying terrain, has long been invaded by seawater. They built dikes, sluices, reclaimed land from the sea to expand their living space, and established a complete water conservancy management system. Their flood control heroes and engineers have also become the pride of the Dutch people.

On the other hand, China has also achieved remarkable results in the utilization of waterways. First, Emperor Yang of Sui, who laid the foundation for the prosperity of the Sui and Tang dynasties in ancient times, constructed the Grand Canal, connecting five major water systems and becoming the main artery of ancient north–south traffic. It promoted the prosperity of cities along the route and drove the economic and cultural exchanges and development, and it is still playing an important role today. Now, the world's largest Three Gorges Ship Lift allows ships to pass through the dam efficiently, improving the transportation capacity of the Yangtze River waterway. The Pearl River Water System in China has built a smart waterway, using big data and Internet of Things technologies to achieve intelligent ship dispatching and improve navigation efficiency; while the Huaihe River ecological waterway has built green slopes to reduce soil erosion and protect the water ecological environment, providing an example for the sustainable development of shipping.

Throughout history, floods have not only brought disasters but have also promoted the progress of human civilization. Whether Nüwa, Fuxi, Yu the Great, Li Bing and his son in China, or Noah, the pharaohs of ancient Egypt, and the flood control heroes of the Netherlands are considered, all have become legendary for the wisdom and courage they showed in flood control and are admired by later generations. These stories and figures inspire us to keep exploring and innovating to achieve harmonious coexistence between humans and nature in our interaction with nature, and to promote the continuous development of human civilization.

4.2 River Hydrographic Testing

4.2.1 Hydrologic Stations

4.2.1.1 Classification of Hydrologic Stations

Hydrologic data is fundamental for hydrological analysis, computation, and forecasting. Most of this data is collected through hydrological stations and compiled annually into databases stored on computers of hydrological survey departments, ready to be accessed at any time or published in the form of *Hydrological Yearbooks*.

Hydrologic stations are classified into three categories based on their purpose and nature:

1. Basic Stations

Basic stations are permanent hydrologic stations uniformly deployed and hierarchically managed by the national hydrological authorities across major rivers nationwide. In accordance with the *Specification for Geodesic Survey in Hydrology* (SL 58-2014), these stations conduct long-term, systematic observations of water levels, discharge, sediment, precipitation, and evaporation. They compile and store data to meet the comprehensive needs of various relevant departments.

Basic stations are further divided into sub-categories based on the type of observation they conduct, such as discharge stations, water level stations, rainfall stations, and sediment stations.

2. Experimental Stations

Experimental stations are stations set up for specific experimental research tasks, such as runoff experimental stations, riverbed evolution experimental stations, estuarine experimental stations, and reservoir evaporation experimental stations.

3. Special-Purpose Stations

Special-purpose stations are established for specific purposes or particular projects. Examples include temporary hydrological stations set up for port and navigation surveys, temporary hydrological stations during the construction phases of reservoirs and hydroelectric stations, and specialized rainfall auto-reporting stations for reservoir flood forecasting.

4.2.1.2 Setup of Hydrologic Stations

The setup of hydrologic stations encompasses the selection of test river sections and the deployment of basic hydrographic stations.

1. Selection of Test River Sections

Before establishing a hydrological monitoring station, it is crucial to select an appropriate test river section, as this choice significantly affects the testing work. The principles for selecting test river sections are explicitly stated in the *Hydrological Testing Provisional Standards*: the river section for the station should meet two conditions: ① It must fulfill the purpose and requirements of the station; ② It should facilitate the simplification of data processing while ensuring that the results have the necessary precision. Specifically, the chosen river section should ideally have a simple relationship curve between water level and flow rate, making it easier to estimate flow rate from observed water levels and thus simplifying the testing process. According to the principles of hydraulics, the main factors affecting flow rate in a river section include cross-sectional area, roughness, hydraulic radius, and water surface slope. Therefore, in plain river sections, the chosen test river should be as straight and uniform as possible, with a length of 3–5 times the width of the river to ensure consistent slope; banks and riverbeds should be stable and not prone to aquatic vegetation growth; and should not be affected by backwater causes such as the influx of tributaries or lakes downstream. In mountainous river sections, avoid areas affected by rock beams, rapids, constrictions, and chaotic rock blockages, and select sections where the river channels are straight, uniform, and have a steady flow regime.

2. Deployment of Basic Hydrographic Stations

Basic hydrological stations primarily monitor water levels, flow rates, sediment, precipitation, and evaporation. For this purpose, the stations are typically equipped with benchmarks, staff gauges, flow measurement cross-sections, and baselines. Various survey markers are established, and necessary instruments and tools are deployed. Additionally, the river section's topography is measured, and station logs are maintained.

(1) Establishing baselines and primary flow measurement cross-sections. A primary flow measurement cross-section is typically established at the midpoint of the river segment selected for the station. This cross-section is aligned perpendicularly to the average flow direction, as in Fig. 4.1. Subsequently, a baseline perpendicular to this primary flow measurement cross-section is positioned along the riverbank. The length of the baseline should be sufficient to ensure that the angle

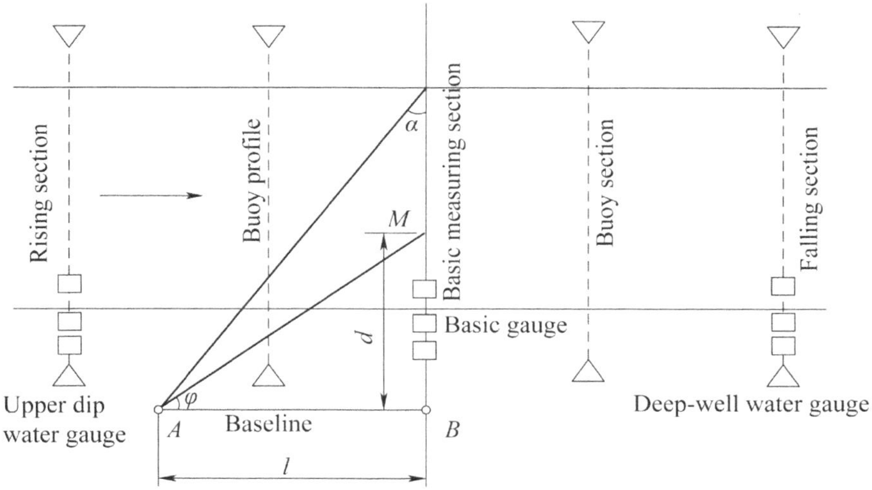

Fig. 4.1 Schematic diagram of the station layout

between the farthest instrument on the primary measurement cross-section and the cross-section itself exceeds 30°.

(2) Establishing buoy flow measurement sections and gradient sections. Buoy sections include upper, middle, and lower buoy sections. The middle section coincides with the primary flow measurement section, while the spacing between the upper and lower buoy sections is generally not less than 80 s multiplied by the maximum average flow velocity of the section.

The purpose of gradient observations is to facilitate the calculation of the river's roughness using flow data. Gradient sections are positioned both upstream and downstream of the primary flow measurement cross-section. The segment of the river between these two sections should be straight. The shapes of the cross-sections should remain essentially identical, and there should be no significant shifts in riverbed slope or water surface slope. The spacing between the two sections should be such that the error in the measure is less than 15%. Specific calculation methods can be referenced from standards or guidelines.

(3) Establishing leveling benchmarks and water gauges. Leveling benchmarks are the primary references for determining the elevations of hydrological station structures and water gauges. They should be located near the station with a stable foundation and placed above the highest flood level to avoid being submerged.

For observing water level fluctuations in the river segment, a primary water gauge should be installed at the main flow measurement section. To observe the water surface elevation changes, water gauges also need to be installed at the upstream and downstream gradient sections.

(4) Instrumentation and tool setup. This primarily includes the use of survey boats or automatic observing cableways, velocity meters, automatic computing microcomputers, and control devices. Additionally, devices for measuring precipitation and evaporation are also installed.
(5) Other procedures. Topographic measurements (including subaqueous topography) of the river segment, as well as longitudinal and transversal cross-sectional surveys of the river channel, should be conducted; creating diagrams of water surface velocities and flow directions of the river segment is also necessary.

3. Deployment of temporary hydrological stations for port and navigation surveys

During the development of navigation projects, it is often necessary to establish temporary hydrologic stations to conduct short-term essential hydrological observations using data from basic hydrological stations located far from the project site. This data assists in relevant analysis.

The requirements for setting up such temporary hydrologic stations are: ① The station must be located at the project site; ② The layout and scope of the station should align with the observational goals required by the project. Observations typically focus primarily on water levels, or may include water surface slopes and flow directions. When the temporary station is far from a basic hydrological station, or there is a significant tributary between the basic station and the project site, it is necessary to measure water levels and flow rates to establish a relationship between water level and flow rate.

4.2.2 Water-Level Observation

4.2.2.1 Equipment and Arrangement

Water level observations primarily involve rivers, hydroelectric reservoirs, sluice gates, lakes, and ports. The main devices used for observing water levels are staff gauges and automatic water-level recorders.

1. Staff Gauges

The staff gauges are the simplest and most reliable devices for measuring water levels and are commonly available in vertical, inclined, and short types, with the vertical type being the fundamental variant. It is generally required that staff gauges be installed in such a manner that they can measure both the highest and lowest flood, with an additional 0.5 m of leeway above and below.

2. Automatic Water-Level Recorders

At present, there are a variety of automatic water level recorders. Classified by the measuring method, there are three types: mechanical, capacitive, and ultrasonic. By transmission distance, there are two types: local and remote recording. By transmission mode, there are two types: wireless and wired. The recording frequency has evolved from daily to quarterly. The recording methods have progressed from analog recording on paper to storage, display, or printing via computers.

4.2.2.2 Requirements

1. Water level observation requirements at hydrological stations

Water level observations at hydrological stations include measurements using basic staff gauges and gradient staff gauges, with the following specific requirements.

(1) Observation period requirements. In general, it is observed twice a day at 08:00 and 20:00, which is called the two-segment observation. During the dry season, the observation is conducted once a day at 08:00. In flood seasons, when water levels change significantly, observations are increased to four times a day (at 02:00, 08:00, 14:00, and 20:00) or even to eight times a day (at 02:00, 05:00 … until 23:00). During flood events, to accurately capture the rise and fall of the peak flood levels, it may be necessary to observe water levels every half hour, or even every few minutes.
(2) Requirements for measurement using a gradient staff gauge. Gradient water levels are essential for calculating surface slopes, analyzing riverbed roughness, and are crucial for analyzing the relationships between water levels and flux. Observations should be frequent enough to capture the gradient across all water levels during differential water level observations. It is mandatory to observe the basic staff water levels synchronously.
(3) Accuracy requirements. For gauges affected by wind and water surface fluctuation, the average value of water surface fluctuation should be read when there is no stilling device. The basic staff gauge is read to l cm, and the reference staff gauge is estimated to read to 0.5 cm.
(4) When measuring flow velocity, direction, and sediment, it is also necessary to observe the basic water level at the same time.

If self-recording water level gauges are used, water level observations will be greatly simplified. The observer's task is to monitor the normal operation of the computer, change recording paper on time, extract water levels for specified periods as required, conduct regular water level calibration, and check equipment.

2. Temporary water level observations in port and navigation surveys

Temporary water level observations are characterized by short observation times and do not require continuous process measurements. Staff gauges are commonly used as observation devices, with the main requirements being as follows:

(1) Temporary water level observations for shoal risk control and newly built port areas. For ship lock structures, in order to establish the water level-flux relationship downstream of the dam, staff gauges need to be set up below the dam. For predicting changes in shoal water levels, staff gauges need to be set up at three locations along the shoal: upper, middle, and lower. In bend areas, staff gauges

should be set up on both banks to observe the transverse slope. For port areas, it is generally sufficient to set up staff gauges near the docks.

(2) Synchronous water level observations along the river. This involves setting up a series of water level stakes along a relatively long stretch of the river and simultaneously observing readings to obtain the instantaneous water surface profile of that river section at the same moment. It serves as the basis for transferring the lowest designed navigable water level of the base station to the remediated river section and provides essential data for river navigation planning. To measure it accurately, staff gauges should be set up at every point where there is a change in the water surface, and observation timings should be during the relatively stable low-water period of the river, preferably near the lowest water level. Temporary water level observations are usually required 1–3 times before they can be removed.

4.2.2.3 Collation of Water Level Observation Data

The organization of water level observation data includes the calculation of daily, monthly, and annual average water levels, compilation of daily average water level tables, the plotting of daily average water level variation curves and daily average water level duration curves, etc. Please refer to ESM 4.1 for more detailed information.

(1) Arithmetic averaging method. This method is used when water level changes within a day are slow or at equal time intervals for observations (or excerpts). The calculation formula is

$$\overline{H} = \frac{1}{n}\sum_{i=1}^{n} H_i \tag{4.1}$$

where $\overline{H}$ is the daily average water level, *m*; H_i are the observation values of water level each time, *m*; *n* is the number of times the water level was observed in a day.

(2) Area bracketing method. If the water level changes significantly within a day and the observation times are uneven, then the average daily water level is calculated by dividing the area enclosed by the water level process line from 0 to 24 h by the total time of the day. Referring to Fig. 4.2, the formula for calculation is

$$\overline{H} = \frac{1}{2\times 24}\left[H_0 a + H_1(a+b) + H_2(b+c) + \cdots + H_{n-1}(m+n) + H_n n\right] \tag{4.2}$$

where $a, b, c, \ldots, m, n$ are time intervals between each observation, *h*; $H_0, H_1, \ldots, H_{n-1}, H_n$ are water levels at each observation time, *m*.

In calculations, if there are no measured values at 0 and 24 h, the values should be interpolated based on the adjacent water levels.

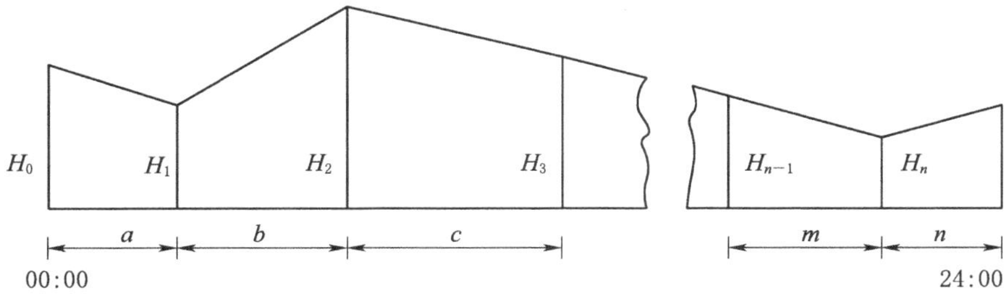

Fig. 4.2 Schematic diagram of the area bracketing method

Based on the daily average water level obtained, the monthly and annual mean water levels can be further calculated.

4.2.3 Flow Test

4.2.3.1 Cross-Section Measurement

The flow rate in a river is typically determined by measuring the cross-sectional area of the river and the water velocity through that section.

Cross-section measurement primarily involves measuring the water depth, starting point distance, and water level.

1. Water Depth Measurement

Water depth measurement involves taking measurements along vertical lines known as sounding lines. The methods used can vary depending on the water depth, flow velocity, required accuracy, and the flow measurement technique employed. Common methods include:

(1) Using depth measuring devices. This includes using a measuring stick, sounding weights, or sounding leads. The stick is accurate and suitable for depths under 5 m and flow speeds less than 3 m/s. Sounding weights, typically 5–10 kg, are used for greater depths, while the leads, weighing about 10–50 kg and exceptionally up to 700 kg, are suitable for deep, fast-flowing rivers. Measurements should be taken twice along each sounding line, and the average of these two measurements is used as the water depth.
(2) Cableway depth measurement. A lead weight is hung from a cable suspended across the river and moved by a cable car. The depth is determined by counting the ingress of the cable into the water using a sensor device attached to the cable.
(3) Ultrasonic depth measurement. This method utilizes an echo sounder, which is among the most advanced depth-measurement instruments. The principle is based on calculating the depth from the speed of sound in water and the time it takes for the sound to travel to the riverbed and back. Echo sounders are fast,

precise, and not limited by water depth or flow speed. However, in turbulent waters or areas with high sediment content, sand particles and air bubbles can reflect the sound waves and cause distorted recordings. In such conditions, the use of an echo sounder should be avoided.

It is crucial to observe the water level at the beginning and end of depth measurements. For large cross-sections, it is best to take measurements when water levels are stable. Rapid changes in water level should be recorded for each sounding line to calculate the river bottom elevation accurately from the depths and water levels measured at each sounding line.

2. Starting Point Distance Measurement

The horizontal distance from a sounding line to a specified point (usually the intersection of the baseline and the flow measurement section) is referred to as the starting point distance. Common methods for measuring the starting point distance include:

(1) Section cable reading method. When the river is narrow enough, and conditions permit the setup of a section cable, the distance markers on the cable can be directly used to read the starting point distances for each sounding line.
(2) Instrumental intersection method. This can be performed using either a theodolite with the forward intersection method or a sextant with the back intersection method.

When using a theodolite for intersection, refer to Fig. 4.1. Set up the instrument at one end of the baseline (Point A), and observe the angles formed between each sounding line and the baseline. The starting point distances can then be calculated using the following formula:

$$d = l \tan \varphi \tag{4.3}$$

where d is the starting point distance, m, i.e., BM length; l is the length of the baseline, m; φ is the angle between the baseline and the sounding lines.

3. Drawing Cross-Section Diagrams

Based on the water level during the sounding, the measured water depths and starting point distances can be used to draw a water cross-section diagram. If elevations and starting point distances above the water surface are also measured, a larger cross-section diagram can be drawn, as shown in Fig. 4.3.

4.2.3.2 Flow Velocity Measurement and Flow Calculation

1. River Channel Cross-Section Velocity Distribution and Flow Measurement Principles

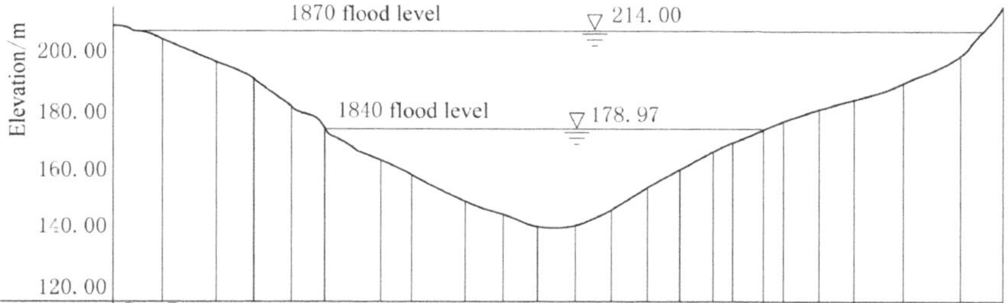

Starting point distance/m	−30.00	−15.00	0	10.00	20.00	30.00	35.50	45.00	55.00	70.00	80.00	90.00	100.00	110.00	120.00	130.00	140.00	145.00	155.00	160.00	170.00	180.00	195.00	210.00	220.00
Water depth/m							3.36	9.40	16.70	23.20	29.00	32.80	33.10	29.60	22.60	14.50	8.60	6.20	0.01						
Elevation/m	215.26	209.39	202.07	196.69	186.97	181.32	175.61	169.57	162.27	155.77	149.97	146.17	145.87	149.37	156.37	164.47	170.37	172.77	178.96	180.68	185.10	188.42	194.58	203.63	222.52

Fig. 4.3 Cross-section and large cross-section diagram of × × station

(1) Cross-section velocity distribution. Due to the influence of river topography, water depth, and riverbank morphology on the flow regime, velocity distribution across a natural river cross-section is uneven, changing along the depth and horizontal direction of the cross-section.

The distribution of velocity across the section is shown in Fig. 4.4. Typically, the velocity increases gradually from the riverbanks towards the river center, with the velocity generally near the water surface being at the point of maximum depth, as seen in Fig. 4.4a. The distribution along a vertical line, as shown in Fig. 4.4b, indicates that the velocity gradually decreases from the water surface to the riverbed, with the velocity near the riverbed being zero. The maximum velocity occurs at a depth of 0.2 times the water depth below the surface, and the average velocity along the vertical line is about 0.6 times the depth below the surface.

(2) Flow measurement principles. Since the distribution of flow velocity across a river cross-section is non-uniform, if *h* represents the water depth at any point

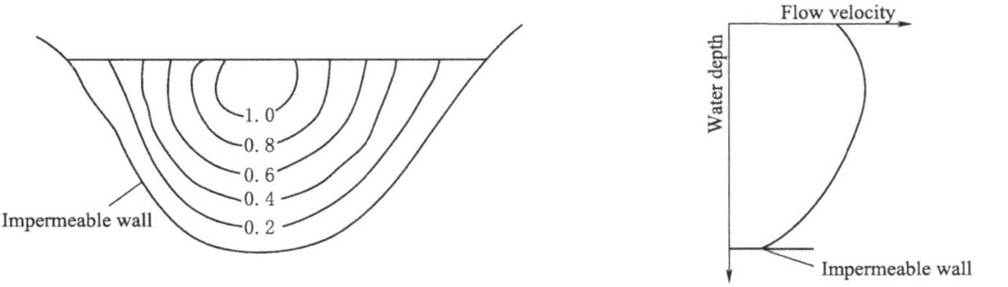

(a) Cross-sectional isovel map (b) Velocity distribution diagram along a vertical

Fig. 4.4 Distribution of the flow velocity in a river cross-section

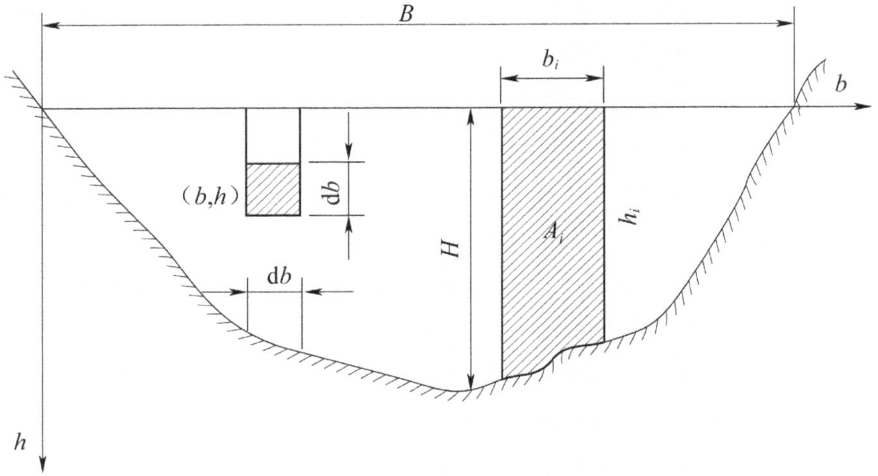

Fig. 4.5 Schematic diagram of the flow measurement principle

within the cross-section and b represents the horizontal distance from that point to the water's edge, as shown in Fig. 4.5, and the flow velocity at that point is v, then the flow rate through the entire cross-section can be calculated using the integral method:

$$Q = \int_0^A v\mathrm{d}A = \int_0^B \int_0^H f(h, b)\mathrm{d}h\mathrm{d}b \tag{4.4}$$

The relationship $v = f(h, b)$ in Eq. (4.4) is quite complex, and the integral formula is rarely used to calculate flow rate in practice. Instead, the flow rate is estimated using the measured cross-sectional area and measured flow velocities, replacing Eq. (4.4) with a summation Eq. (4.5).

$$Q = \sum_{i=1}^{n} v_i A_i \tag{4.5}$$

where A_i is the area of block i, m^2; v_i is the average flow velocity over i area section, m/s; n is the total number of blocks into which the full cross-sectional area is divided.

2. Methods for Measuring Flow Velocity with a Flowmeter

A flowmeter is an instrument used to measure the flow velocity at any point in a water flow. There are many types and models, with the main ones being the rotor type (including cup-type and propeller-type) and the electromagnetic-acoustic type. A flowmeter primarily consists of a rotor (either cups or propellers) that is activated by the water, a signal recorder or a microcomputer that directly calculates and displays the flow, and a tail fin that keeps the instrument aligned with the water flow.

The cups or propellers are propelled by the water flow; the greater the flow velocity, the faster they rotate. The flow velocity at any measuring point can be deduced based on the relationship between the rotations per second and the flow velocity. Based on this principle, each device is required to be calibrated in the laboratory before it leaves the factory and comes with a calibration formula and curve. The form of the formula is

$$v = kN + c \tag{4.6}$$

where v is the flow velocity at the measurement point, m/s; k is the instrument calibration constant, m; N is the number of rotations per second, also known as the rotation rate, is given by $N = n/T$, where n is the total number of rotations and T is the duration of the velocity measurement, s; c is the friction coefficient of the instrument, m/s.

During flow measurement, it is only necessary to record n and T, and then the flow velocity at the measurement point can be calculated using Eq. (4.6). Generally, the measurement time T should be more than 100 s.

Typically, cup-type velocimeters are suitable for rivers with little sand, with a measurement range of 0.2–3.5 m/s; propeller-type velocimeters can also be used in rivers with more sand, with a measurement range of 0.2–5.0 m/s.

Generally, the number of verticals and velocity-measuring points is arranged according to the width of the river and the depth of flow; three procedures—detailed, regular, and simplified—are thereby distinguished. Detailed measurement involves using more vertical lines and measurement points to precisely determine the cross-sectional flow velocity. Regular measurement is based on the detailed method but utilizes fewer vertical lines and measurement points after streamlining analysis, provided the errors meet the specifications. Simplified measurement, suitable for flood surges, backwater effects, or ice flow conditions, uses a minimal number of vertical lines and measurement points to quickly measure the flow velocity, ensuring errors remain within allowable limits.

The primary objective in selecting a method is to reduce measurement time while ensuring accuracy. Typically, the difference in water level during a single flow measurement should be less than 10%–20% of the average depth, and between 20% and 30% during significant rises or falls.

The number of vertical lines for various velocimetry methods can be found in testing standards. The distribution of vertical lines should be roughly uniform, but denser in the main channel than on the riverbanks. Appropriately, additional lines should be set up in areas with uneven terrain, flow velocity distribution breakpoints, bifurcations, or cross channels; nearby shorelines should accommodate various water levels without being too close to the water's edge. Moreover, the positions of vertical lines should be as fixed as possible to facilitate comparison between different flow measurement results.

Many countries adopt the principle of using multiple lines with fewer points. International standards recommend no fewer than 20 velocity measuring vertical lines, with any part of the flow not exceeding 10% of the total flow, achieving a standard deviation of less than 2% in flow measurements.

The number and position of measurement points on vertical lines are determined based on water depth, instrument suspension method, and the characteristics of the velocity distribution on the vertical line. Table 4.1 lists only the detailed measurement requirements for measurement point setup; other requirements can be referred to in relevant specifications.

3. Flow Calculation

   (1) Average velocity along the vertical. The calculation is based on the point velocities measured along the vertical line to determine the average velocity along the vertical. The calculation formula and applicable conditions are listed in Table 4.2. In the formulas, the subscript number of v represents the relative depth of the measurement point.

Table 4.1 Distribution of flow velocity measurement points for the detailed measurement method

Water depth or effective water depth/m		Number and location of measurement points on the vertical lines	
Suspension rod	Suspension by cable	Free flow period	Ice period
> 1.0	> 3.0	Five points (surface, 0.2, 0.6, and 0.8 water depth, river bottom)	Six points (surface or ice bottom or ice splash bottom, 0.2, 0.4, 0.6, and 0.8 effective water depth, river bottom)
0.6–1.0	2.0–3.0	Three points (0.2, 0.6, and 0.8 water depth) or two points (0.2, 0.8 water depth)	Three points (0.15, 0.5, and 0.85 effective water depth)
0.4–0.6	1.5–2.0	Two points (0.2 and 0.8 water depth)	Two points (0.2 and 0.8 effective water depth)
0.2–0.4	0.8–1.5	One point (0.6 water depth)	One point (less than 0.5 effective water depth)
0.16–0.2	0.6–0.8	One point (0.5 water depth)	
	< 0.6	Switch to using a hanging rod for suspension or other flow measurement methods	Switch to using a hanging rod for suspension
< 0.16		Switch to the small buoy method or other methods	

Note The effective water depth is obtained by subtracting the thickness of submerged ice and frost from the total water depth

Table 4.2 List of equations for calculating the mean velocity of a vertical line

Formula names differentiated by the number of measurement points	Calculation formulas	Applicable water depth/m
One-point method	$v_m = v_{0.6}$	< 1.5
Two-point method	$v_m = \frac{1}{2}(v_{0.2} + v_{0.8})$	1.5–2.0
Three-point method	$v_m = \frac{1}{3}(v_{0.2} + v_{0.6} + v_{0.8})$	2.0–3.0
Five-point method	$v_m = \frac{1}{10}(v_0 + 3v_{0.2} + 3v_{0.6} + 2v_{0.8} + v_{1.0})$	> 3.0

(2) Partial area calculation. As shown in Fig. 4.6, it can be calculated using segmented trapezoids (there may be several trapezoids between adjacent velocity measurement verticals) or triangles.

(3) Partial area average velocity calculation. This is obtained by adding average velocities from adjacent velocity measurement verticals and dividing by 2. However, the average velocity of a partial area near the bank can be calculated by multiplying the average velocity of a vertical line close to the bank by a bank velocity correction (a). The value of a is typically between 0.6 (stagnant water edge) and 0.9 (smooth, steep bank), chosen according to the conditions of the bank.

(4) Partial and cross-sectional flow calculation. The partial flow is equal to the product of the partial area (A_i) and the partial average velocity (v_i), specifically

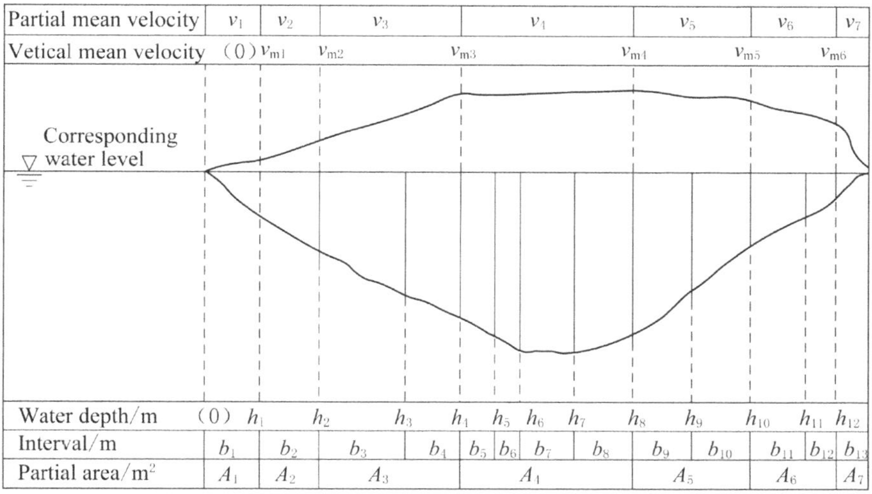

Fig. 4.6 Schematic diagram of part of the area and part of the average velocity at station ××

$$q_i = v_i A_i \tag{4.7}$$

The cross-sectional flow is the algebraic sum of the partial flows, calculated using Eq. (4.5).

$$Q = \sum_{i=1}^{n} v_i A_i = \sum_{i=1}^{n} q_i \tag{4.8}$$

(5) Calculation of other hydraulic elements. This mainly includes the cross-sectional area A, cross-sectional average velocity $\overline{v}$, water surface width B, average water depth $\overline{h}$, stream gradient I during the river section flow measurement, and average roughness n of the river section, with the formulas for each element as follows:

$$A = \sum_{i=1}^{n} A_i \quad (\text{m}^2) \tag{4.9}$$

$$\overline{v} = \frac{Q}{A} \quad (\text{m}^2/\text{s}) \tag{4.10}$$

B is equal to the difference between the starting point distance of the two banks (m).

$$\overline{h} = \frac{A}{B} \tag{4.11}$$

$$I = \frac{\Delta Z}{L_{\text{G}}} \tag{4.12}$$

where ΔZ is the difference in water levels between the upstream and downstream staff gauges, m; L_{G} is the distance between the upstream and downstream gradient staff gauges, m.

$$n = \frac{1}{\overline{v}} R^{2/3} I^{1/2} \tag{4.13}$$

where R is hydraulic radius, which can be taken as $R = \overline{h}$ for wide and shallow channels.

4.2.3.3 Introduction to Flow Measurement Using Buoyant Objects

1. Conditions of use and basic assumptions

During flood seasons when large floods occur, if the rapid water flow makes it difficult to anchor a measurement boat, or if river flow exceeds the operational range of current meters, the surface float flow measurement method can be applied.

The basic assumptions of this method are as follows: the velocity of a float on the water's surface substitutes for the surface flow velocity; the direction of the float's movement determines the flow direction at the surface; surface flow velocity is used in place of the mean velocity along a vertical line to calculate the cross-sectional flow (termed as virtual flow); simulated float coefficients are used instead of correction coefficients derived from virtual flow to calculate actual flow.

This method is simple and frequently employed in waterway and port engineering surveys and designs.

2. Key points of the buoy method of current measurement

For detailed steps and requirements of the float measurement method, refer to the relevant manuals and standards. Below, only the fundamental points are discussed.

(1) Measurement of surface flow velocity. Continuously deploy surface floats (such as bamboo tubes) at various positions upstream of the cross-section where the floats are measured. Record the time t (s) taken for each float to traverse the distance L_F between the upstream and downstream cross-sections, calculate the average velocity of the floats, and use a theodolite to measure the distance from the starting point of the cross-section that the float passes, substituting the average velocity of the float for the surface flow velocity at that vertical.
(2) Calculation of partial area between starting points. Based on the corresponding water level and depth, and the width of the water surface between the two starting points, calculate the partial area using trapezoidal or triangular methods.
(3) Calculating the average velocity for the partial area. Use the arithmetic mean of the surface flow velocities at two adjacent verticals as the average velocity for the partial area.
(4) Calculation of partial virtual flow and cross-sectional virtual flow. Although the equations from current meter methods are still utilized to compute partial and cross-sectional flows, since the surface flow velocity is used in place of the average velocity along a vertical, computed flows are higher than the actual flows, hence the term virtual is prefixed to the flow.
(5) Determination of buoy coefficients and calculation of cross-sectional flow. As the assumptions aforementioned result in cross-sectional virtual flow Q_f being greater than the actual flow Q, through numerous empirical comparisons between current meter methods and buoy methods, buoy coefficients $K_f = Q/Q_f$, typically between 0.8 and 0.9, are determined. Thus, selecting a K_f value allows the calculation of the actual cross-sectional flow using the equation $Q = K_f Q_f$.

4.3 Stage-Discharge Relationship Curves

4.3.1 Construction of Stage-Discharge Relationship Curves

The curve constructed from measured discharge data and corresponding stage measurements at a cross-section is known as the stage-discharge relationship curve. This curve has extensive applications in hydrological analysis and computation. Its primary uses include: estimating the discharge from observed water levels at the cross-section, thereby significantly reducing the need for labor-intensive discharge measurements and simplifying the process of monitoring water levels; converting hydrologically calculated design discharges (or stages) directly into design stages (or discharges); and compiling discharge data based on this relationship.

The stage-discharge relationships in natural river channels can generally be categorized into two types. One type is a consistent relationship, referred to as a stable stage-discharge relationship. The other type may associate more than one discharge value with a single stage, known as an unstable stage-discharge relationship. When establishing such a rating curve, its distinctive features must be carefully taken into account.

4.3.1.1 Stable Stage-Discharge Relationship Curve

Using water stage as the vertical axis and discharge as the horizontal axis, points corresponding to each pair of values are plotted on graph paper. Following the trend of the distribution of points, a smooth relationship curve is drawn through the center of the point cluster. To enhance accuracy, the method of least squares can be used for line fitting (see ESM 3.8 for details). Usually, on the same sheet where the stage-discharge (H-Q) curve is drawn, the stage-area (A) and stage-velocity (v) relationship curves are also plotted. These serve as auxiliary curves for analyzing the relationship between stage and discharge, as shown in Fig. 4.7.

A stable and good stage-discharge relationship generally occurs in mountain rivers with stable riverbeds, relatively steep channel slopes, and straight river segments.

4.3.1.2 Unstable Stage-Discharge Relationship Curve

In natural river channels, fluctuations in flood levels, backwater changes, changes in riverbed erosion and siltation, as well as the growth of aquatic plants or ice formation on the water surface, can cause changes in cross-sectional area, cross-sectional shape, hydraulic gradient, or roughness, leading to unstable stage-discharge relationship curves.

(1) Flood fluctuation impact. The characteristic of a flood propagating through a river channel is that it has a high gradient at the front of the wave, with a high

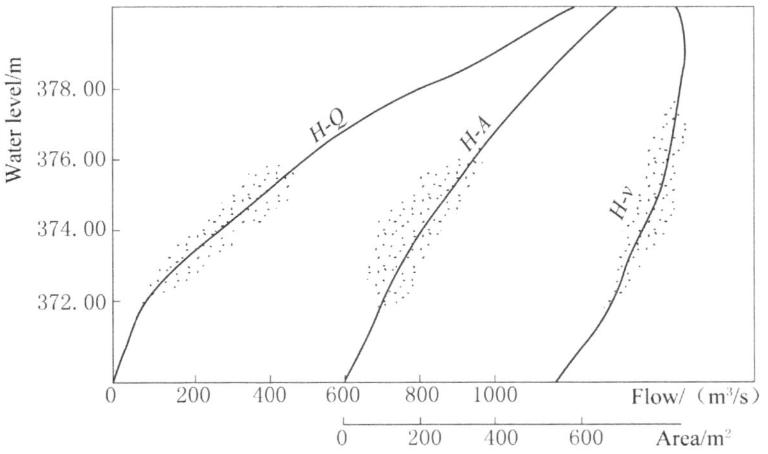

Fig. 4.7 Stable stage-discharge relationship curve

flow velocity, and a lower gradient at the back, with a corresponding decrease in flow velocity. Therefore, for the water level, the flow rate is higher during the rise of a flood, causing the stage-discharge relation points to shift toward the right side of the stable curve; during the fall of a flood, the flow rate is lower, causing the points to shift toward the left side of the stable curve. As a result, the stage-discharge relationship curve forms a loop, seen in Fig. 4.8.

(2) Impact of backwater changes. When the water level at the flow measurement section is influenced by factors such as the rising waters of tributaries, opening and closing of downstream gates, river freezing, blockages by driftwood or other floating debris, and tidal pushing, the water level at the flow measurement section will rise. This phenomenon is called the backwater effect. Although

Fig. 4.8 Stage-discharge relationship curve influenced by flood fluctuations

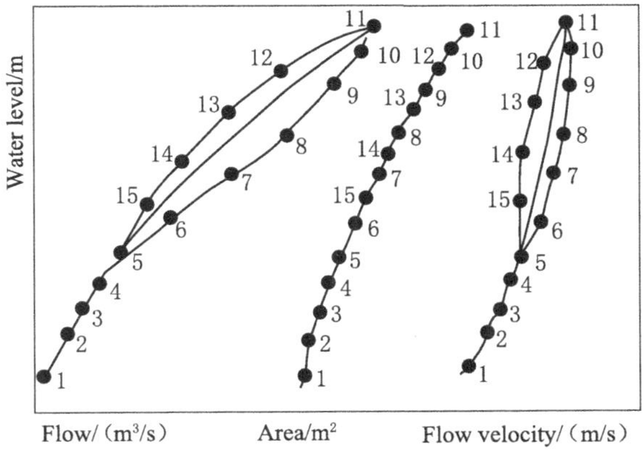

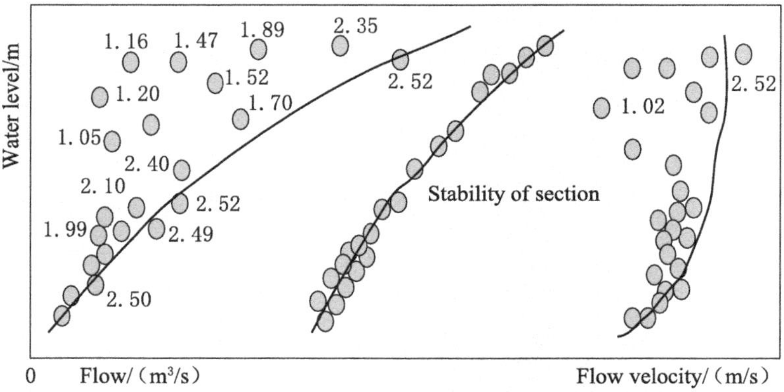

Fig. 4.9 Stage-discharge relationship curve affected by backwater influence

backwater raises the water level, the actual flow rate remains unchanged. Therefore, under the same flow rate, different degrees of backwater influence will result in different water levels, causing the relationship points between water level and flow rate, as well as water level and flow velocity, to scatter, as seen in Fig. 4.9.

(3) Influence of riverbed scouring and silting. Most notably, changes in the cross-sectional area and shape of the riverbed itself result in different flows under the same water level conditions, as shown in Fig. 4.10. When the riverbed undergoes general scouring, the cross-sectional area increases under the same water level, and the flow also increases, causing the *H*-*Q* curve to shift to the right. Conversely, when the riverbed experiences silting, the *H*-*Q* curve shifts to the left.

Localized scouring and silting of the riverbed generally occur in low-flow channels and only manifest their impacts during low water levels, causing the low-flow *H*-*Q*

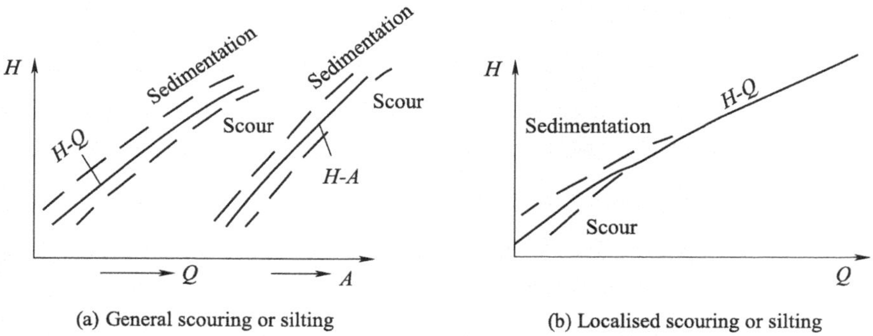

Fig. 4.10 Stage-discharge relationship curve affected by scouring and silting

curves to bifurcate. When the water level is higher and the discharge is greater, the impact is not significant, and the *H-Q* curve remains single.

In practice, the *H-Q* relationship may be simultaneously affected by various factors. Distinguishing primary and secondary influencing factors and establishing a definitive *H-Q* relationship requires research by hydrological professionals.

4.3.2 *Extension of the Stage-Discharge Relationship Curve*

River flow measurement is often limited by conditions or short observation periods, typically capturing only general magnitudes of flow for plotting stage-discharge relationships. However, in water resources planning and management, extreme flow values (either exceptionally high or low, or those specified by design standards) are often needed. Therefore, studying methods to extrapolate stage-discharge relationships under high and low water level conditions and subsequently estimating extreme flows is crucial.

4.3.2.1 Methods for Extending the High Water Level Portion of the Stage-Discharge Relationship Curve

There are several methods to extend the high water level portion of the *H-Q* relationship curve. Below are three commonly used methods, each suitable for different extents of extension:

(1) Stage-area-velocity extension method.
(2) Hydraulic formula extension method.
(3) Stevenson's method (Q–$A\sqrt{R}$ extension method).

The three methods described above are described in detail in ESM 4.2.

4.3.2.2 Methods for Extending the Low Water Level Portion of the Stage-Discharge Relationship Curve

Extension of the low water level of the *H-Q* curve requires extra caution compared to extending the high water level, as the smaller flow rate can lead to significant relative errors.

The extension method involves identifying the cut-off water level as the control point, and extending it downwards either from the measured section in a reverse manner, following the observed trend, or with the assistance of flow data from the upstream and downstream stations.

Low water extensions of the *H-Q* curve require more care as they are prone to large relative errors due to low flow rates compared to high water extensions.

The method of extension is to find the cut-off water level, as a control point, from the actual measured part of the downward extension of the inverse, or according to the trend of the extension of the inverse, or borrow the flow of upstream and downstream stations to assist. There are basically two methods for determining the cut-off water level.

(1) Method of determining the cut-off water level based on longitudinal and transverse cross-section data.
(2) Analysis method.

The two methods described above are described in detail in ESM 4.3.

4.3.3 Data Compilation on Flow Rates

Due to the time-consuming and challenging nature of flow measurement work, a large amount of water level data is obtained through extensive observations during a 1-year-long river flow change process, including flood fluctuations. Only by compiling and plotting the stage-discharge relationship curve can the river flow change process and flood flow process for one year be determined from the water level data. It is evident that the stage-discharge relationship is the foundation for organizing flow data.

The main content of flow data organization includes the calculation of daily, monthly, and annual average flow rates, and the compilation of daily average flow rate tables, observed flow rate results tables, and flood season flood elements summaries.

See ESM 4.4 for detailed collation results.

4.4 Design Flood

4.4.1 Overview

When natural rivers cannot meet navigation requirements, engineering interventions become essential. These include dredging sandbars, river training, canalization, and flood management, as well as constructing reservoirs and gates. These projects aim to regulate river discharge and augment water depth, thereby ensuring navigability.

The flow discharge or water level serving as the basis for the planning, design, and construction of harbor and navigation projects is referred to as the design flow or design water level. This design flow or design water level is generally classified into two categories. The first category is flood discharge and water level associated with engineering safety and economic considerations. For instance, determining the discharge capacity for a barrage requires a trade-off; excessive capacity is uneconomical, while insufficient capacity compromises safety. Similarly, setting the crest

Table 4.3 Design standards for quay front elevations

Classification of wharf elevations	Design standard (cumulative frequency of annual highest water level)/%		
	River network area	Plains	Mountain rivers
Category I	1	2	2–5
Category II	2	5	5–10
Category III	5	10	10–20

Note Piers are classified according to the amount of damage they would cause if flooded; those with significant damage are classified as Category I and those with some damage are classified as Category II. Lesser damage is classified as Category III

elevation of a jetty too low poses flooding risks, whereas setting it too high causes operational difficulties. This type of design flow and water level is often defined by cumulative frequency standards, as shown in Table 4.3. The second category is design discharge and water level for standard navigation, which ensures adequate river depth while preventing obstruction of the design flow and water level by cross-river structures or navigation buildings. Additionally, bridge clearance design during floods must not impede navigation.

To address the first category of problems, the design flood is determined using frequency analysis. For the second category, while frequency analysis is used to determine the navigable high water level, the design navigable minimum water level requires deduction through the use of duration curves and the guaranteed rate frequency method. A fundamental aspect in both cases is the derivation of the design flood, which involves several fundamental concepts and components as follows.

4.4.1.1 Characteristic Values of Flood Hydrograph

Chapters 2 and 4 addressed the causes of runoff and its flow tests. The final collected data from the flood season can provide a flood hydrograph, as depicted in Fig. 4.11.

The flood hydrograph is generally characterized by three elements: the peak discharge Q_m, the flood volume W, which is represented by the area enclosed by

Fig. 4.11 Flood hydrograph

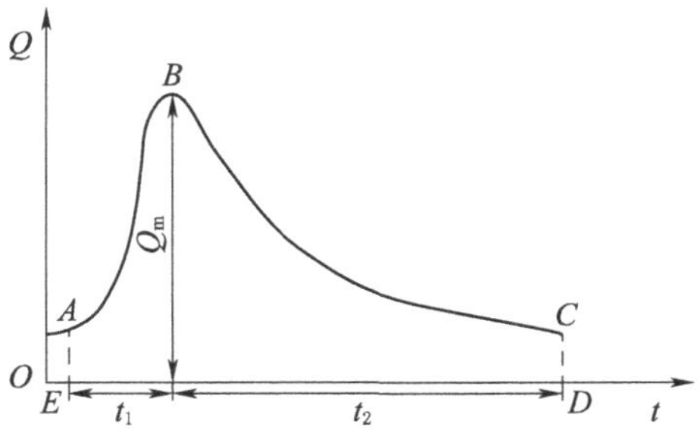

points *ABCDE* in the diagram, where segment *AC* represents the baseflow separation line; and the flood duration, which is the sum of the rising duration t_1 and the recession duration t_2. In practice, the actual recession limb is often prolonged. For the sake of simplicity, point *C* is selected such that its discharge equals the starting discharge. This simplification assumes that the recession volume occurring after point *C* is equivalent to the recession volume from the antecedent flood prior to point *A*.

4.4.1.2 Design Flood and Design Criteria

When designing hydraulic structures, selecting an appropriate design flood magnitude is crucial to guide the project's scale, planning, and structural calculations. Overestimating the flood magnitude leads to uneconomical over-design, while underestimation compromises safety and risks significant damage. Thus, it is essential to select a flood magnitude that ensures both safety and economic feasibility for hydraulic structure design.

The prevailing methodology involves selecting a specific cumulative frequency (e.g., 1, and 0.1%) of floods as a design basis or standard in compliance with regulatory requirements. Flooding is regarded as a random occurrence, and actual flood data distributions are factored in to estimate potential future floods probabilistically. The cumulative frequency of flooding serves as a design criterion, with different cumulative frequencies (or return periods) chosen based on the project's size and importance. For instance, higher design standards ($P = 0.2$ or 0.1%) are used for large reservoirs and significant protection targets, whereas lower standards ($P = 1$ or 2%) apply to small reservoirs that don't pose significant downstream flood control concerns. Once the design standard is established, the flood estimated based on this standard is termed the design flood.

To ensure the safety of reservoir dams, calibration criteria are outlined in codes. Generally, these criteria are expressed in terms of cumulative frequency or return period. However, for significant earth and rock dams, in addition to cumulative frequency being a design criterion, the probable maximum flood (PMF) is also designated as a check criterion to guarantee the dam's safety. The PMF is defined as the flood generated by the maximum precipitation a basin could produce under current climatic conditions.

The planning and design of hydraulic projects follow specific codes that categorize permanent hydraulic structures based on the type and significance of the project and the area. These codes establish distinct design standards and calibration criteria for each level (Tables 4.4 and 4.5). The design flood level of a hydraulic project is determined through adjustment calculations based on the design standard flood. Similarly, calibration flood levels are determined using calibration standard floods through adjustment calculations.

In the event of a calibration flood, the safe superelevation between the reservoir's highest water level and the dam's crest might be allowed to decrease, potentially disrupting the hydropower station's normal operations. However, the reservoir's main

Table 4.4 Flood standards for permanent hydraulic structures for water conservancy and hydropower projects in mountainous and hilly areas (return period: years)

Building class		1	2	3	4	5
Design		500–1000	100–500	50–100	30–50	20–30
Calibration	Earth and stone dam	PMF or 5000–10,000	2000–5000	1000–2000	300–1000	200–300
	Concrete dams, slurry masonry dams	2000–5000	1000–2000	500–1000	200–500	100–200

Table 4.5 Flood standards for permanent hydraulic structures for water conservancy and hydropower projects in plains and coastal areas (return period: years)

Item		Building class				
		1	2	3	4	5
Reservoir works	Design	100–300	50–100	20–50	10–20	10
	Calibration	1000–2000	300–1000	100–300	50–100	20–50
Barrage	Design	50–100	30–50	20–30	10–20	10
	Calibration	200–300	100–200	50–100	30–50	20–30

structures (e.g., dam, spillway) must remain intact. If the designed hydraulic structure serves to protect downstream areas from flooding, specific flood protection standards also apply to lower nuisance protection objects. These standards vary depending on the type (urban, agricultural, etc.) and importance of the protected object. However, they remain lower than the design standard for dams. The corresponding high-water level for flood protection is obtained through flood regulation calculations detailed in *Standard for Rank Classification and Flood Protection Criteria of Water and Hydropower Projects* (SL 252-2017).

4.4.1.3 Contents and Methods of Design Flood Estimation

(1) The content of design flood calculations primarily includes the design peak discharge, total design flood for various periods (e.g., 3 d, 5 d, 7 d, or 11 d), and the design flood hydrograph. The specific focus of the calculation varies based on engineering features. For projects with no storage capacity, such as embankments, bridges, culverts, and navigation-based canalization projects, as well as small reservoirs with minimal regulating capacity and run-of-river power stations, calculating the design peak discharge is crucial, as it determines the project's feasibility. In contrast, projects like flood storage areas and reservoirs possess regulating capacity, and the reservoir's outflow process is influenced by the entire inflow hydrograph. Therefore, in addition to considering specific

inflow flood characteristics (such as peak or volume for a specific time), calculating the design peak discharge or the volume for a specific period and deriving the entire design flood hydrograph are also necessary.

(2) The basic methods for deriving the design flood can be categorized into three approaches based on data availability and design requirements: ① Deriving design floods from streamflow data. When sufficient streamflow data is available, the design peak discharge and flood volume for each time period are derived for a specific cumulative frequency. A typical observed hydrograph is then selected, and the design flood peak and volume are adjusted to create a comprehensive design flood hydrograph. ② Deriving design floods from rainfall data. When adequate rainstorm data is available, the design storm can be designed using frequency analysis. Runoff yield and flow concentration calculations are then applied to derive the design flood hydrograph. If rainstorm data is obtained through weather analysis and statistical metrics like wind speed and dew point, these can be adjusted to estimate the maximum probable rainstorm. Runoff yield and flow concentration calculations then lead to the derivation of the maximum possible flood hydrograph. ③ Alternative methods. When data is sparse or lacking, empirical methods such as the hydrologic analogy method, isoline map method, or empirical formulas can be used to derive the design flood.

This book exclusively covers the method of deducing design floods from flow rates. For methods of deriving design floods from storm data, refer to Appendices B and C of *Regulation for Calculating Design Flood of Water Resources and Hydropower Projects* (SL 44-2006).

4.4.2 Design Flood Estimation from Streamflow Data

The method of deriving design flood from streamflow data primarily relies on the frequency analysis calculation approach introduced in Chap. 3. The following presents the procedure of the method along with relevant details.

4.4.2.1 Review of Flood Data

Before applying the data, the reliability, consistency, and representativeness of the source material are reviewed.

(1) Reliability review. This involves a focus on observation records prior to 1949 and station relocations, as well as the impact of different flood years in the design series. Key items to review include the location of the gauging station, gauge datum, and changes in the reference datum. Particular emphasis is placed on the reliability of the stage-discharge relation and high-stage extrapolation

methods, which can be verified by comparing historical stage-discharge relations (especially focusing on high water levels). It also includes examining upstream and downstream data, the consistency between the main stream and tributaries, and rainfall-runoff analysis. For investigated historical floods, their peak magnitudes and recurrence periods must be thoroughly analyzed and verified. A recent method for identifying paleofloods involves using radioisotopes, such as ^{14}C, to date ancient flood deposits, thereby determining the flooding period.

(2) Consistency review. The main aim is to examine whether flood-generating conditions are consistent over the record period in the basin. If significant changes occurred in certain years (e.g., due to reservoir construction, dike breaches, or flood diversions), the data should be adjusted to ensure consistency.
(3) Representativeness review. This review focuses on the representativeness of the sample series relative to the statistical population. This is often assessed using reference stations with concurrent records. The representativeness of the sample series at a specific site is evaluated based on its long-term records. Additionally, it can be used to determine whether the sample series includes high, medium, and low magnitude floods, as well as extreme wet and dry years. If the sample series lacks representativeness, series extension may be necessary.

4.4.2.2 Sample Selection for Flood Data

The objective of sample selection is to extract key variables (e.g., flood peaks, flood volumes for various durations) from flood data to develop a sample series for frequency analysis. Floods may occur several times within a year, featuring multiple peaks and varying durations. To satisfy the statistical requirements of frequency calculations, it is crucial that the samples are selected independently and randomly. Therefore, the annual maximum method is commonly employed for sample selection. This approach involves selecting the maximum peak discharge (or maximum flood volumes for durations of 1 d, 3 d, 5 d, etc.) for each year. If data spanning n years is available, a sample series with a sample size of n can be generated.

Basic data can be sourced from the hydrological yearbook and the hydrological information database.

4.4.2.3 Extension of Flood Data Series

When data is missing, insufficient, or poorly represented, interpolation and extrapolation can be performed using the correlation analysis method outlined in Chap. 3.

(1) Interpolation and extrapolation are performed using discharge data from upstream or downstream stations, main stream and tributary stations, or neighboring basin gauging stations. If measured data is unavailable at the river's design section, but is available from nearby upstream or downstream stations, it can be transposed to the design section under the following conditions: The

difference in catchment area between the two locations is less than 20%, rainfall distribution is uniform across the basin, and no specific storage effects occur within the interval.

$$Q_1 = (F_1/F_2)^k Q_2 \tag{4.14}$$

where Q_1, Q_2 are the values of the flood discharge at the design section and the reference station; F_1, F_2 are the catchment areas of the design section and the reference station; k is an exponent.

k can be determined from the analysis of measured data from the upstream and downstream reaches of the basin or neighboring basins. For large and medium rivers, $k = 0.5$ to 0.7; for smaller rivers, $k > 0.7$. For transposing flood data between nearby sections, take $k = 1$.

(2) Interpolation and extrapolation are performed using site-specific peak-volume relationships. Select several major floods, and plot the correlation between the peak discharge and flood volume as data points. Determine the peak-volume empirical curve or regression equation either graphically or through regression analysis. This equation allows for the interpolation or extrapolation of the flood volume based on the peak discharge. Typically, for investigated historical floods, only the peak discharge can be estimated from the flood marks, while the corresponding flood volume is derived from the established relationship.

4.4.2.4 Treatment of Extraordinary Floods and Parameter Estimation

The handling of extraordinary floods involves flood investigation and verification, as well as determining the return period. The empirical frequency of these floods is calculated according to Eq. (3.49). The parameters are then calculated using the method of moments [Eqs. (3.50) and (3.52)] while incorporating the impact of extraordinary floods $\overline{x_N}$, C_{vN}. The statistical parameters and frequency curves are determined using curve fitting, considering regional patterns. Based on these frequency curves, the corresponding design peak discharges and design flood volumes can be determined for different design standards.

4.4.2.5 Derivation of Design Flood Hydrograph

Once the design peak discharges and design flood volumes for various durations are known, the design flood hydrograph can be derived by scaling the typical hydrograph using the proportional scaling method or the same frequency amplification method.

(1) Proportional scaling method. This involves scaling each ordinate of a typical hydrograph by the same ratio to obtain the design flood hydrograph. For water conservancy projects with minimal regulating capacity, the design peak discharge is determined by scaling the typical peak discharge proportionally,

referred to as peak scaling. For projects with significant regulating capacity, the design flood volume is determined by scaling the typical flood volume proportionally, referred to as volume scaling. Although this method is computationally simple, it ensures that the peak reaches the design standard in the former case, whereas the latter does not guarantee this.

(2) Same frequency amplification method. Here, different scaling factors are used to scale a typical hydrograph for flood peaks and flood volumes at various durations. This ensures that the scaled design flood hydrograph matches the flood peaks and flood volumes of various durations at the same design frequency. For instance, in a 15-day flood hydrograph, the scaling ratios for peak discharges and flood volumes across different durations are as follows.

$$k_Q = Q_{\mathrm{mp}}/Q_{\mathrm{md}} \tag{4.15}$$

$$k_{\mathrm{w1}} = W_{\mathrm{1p}}/W_{\mathrm{1d}} \tag{4.16}$$

$$k_{\mathrm{w3-1}} = \left(W_{\mathrm{3p}} - W_{\mathrm{1p}}\right)/(W_{\mathrm{3d}} - W_{\mathrm{1d}}) \tag{4.17}$$

$$k_{\mathrm{w7-3}} = \left(W_{\mathrm{7p}} - W_{\mathrm{3p}}\right)/(W_{\mathrm{7d}} - W_{\mathrm{3d}}) \tag{4.18}$$

$$k_{\mathrm{w15-7}} = \left(W_{\mathrm{15p}} - W_{\mathrm{7p}}\right)/(W_{\mathrm{15d}} - W_{\mathrm{7d}}) \tag{4.19}$$

where Q_{mp}, Q_{md} are design flood peak discharge, and typical flood peak discharge; W_{1p}, W_{3p}, W_{7p}, W_{15p} are design flood volumes for 1 d, 3 d, 7 d and 15 d; W_{1d}, W_{3d}, W_{7d}, W_{15d} are typical flood volumes for 1 d, 3 d, 7 d, and 15 d; k_Q is peak flow scaling coefficient; k_{w1}, $k_{\mathrm{w3\text{-}1}}$, $k_{\mathrm{w7\text{-}3}}$, $k_{\mathrm{w15\text{-}7}}$ are flood volume scaling coefficients for 1-day, 3-day to 1-day, 7-day to 3-day, and 15-day to 7-day periods.

Discontinuity at the junction arises due to differing magnification ratios in process lines across different times, as shown in Fig. 4.12. This discontinuity needs to be smoothed to produce a continuous curve. When smoothing, it is important to maintain the design flood peak and design flood volume constant for each time period.

The selection of a typical hydrograph should be guided by the following principles:

(1) Select a hydrograph that features a high peak magnitude, is representative of major floods, and reflects the season of occurrence, the flood pattern, the timing of the main peak, and the peak-volume relationship.
(2) For safety considerations, select a hydrograph with a concentrated profile and a main peak that occurs late in the flood event.

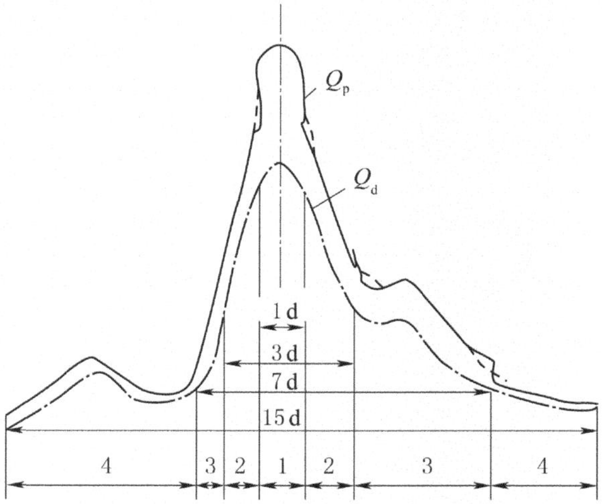

Fig. 4.12 Diagram of the magnification ratio and sawtooth process for each stage. 1—k_{w1}; 2—k_{w3-1}; 3—k_{w7-3}; 4—k_{w15-7}; Q_p—Design flood hydrograph; Q_d—Typical flood hydrograph

4.4.3 Design Flood During Construction

A standard approach for constructing hydraulic structures involves building cofferdams in stages to keep the pit within the cofferdam dry, enabling the construction of the hydraulic structure in the dry. To prevent water from entering the pit during construction, it's crucial to study how river water will flow through the narrowed channel (construction diversion). A proper solution to this challenge can reduce project costs and expedite the construction schedule; otherwise, it may delay the schedule or even cause significant damage by flooding the pit.

Construction diversion structures, which include cofferdams, open channels, tunnels, and bottom outlets, are determined by the anticipated flood magnitude. Determining a design flood to serve as the basis for designing a diversion structure is referred to as the construction design flood. The construction design discharge represents the maximum discharge anticipated during the construction period. While construction should ideally be scheduled during dry seasons, large projects often have construction periods exceeding a year or more, necessitating flood control measures.

Cumulative frequency is also used as a standard to determine the design flood for the construction period. This design standard should be significantly lower than that for the permanent structures and is established through project-specific studies conducted by the design entity.

The criteria for the construction design flood are closely tied to the construction schedule. When the construction period is prolonged, the flood resistance capacity varies across construction phases. For instance, as the dam height increases and diversion conditions change, the design flood for each phase needs to be determined accordingly.

4.4.3.1 Staging of Construction Design Floods

(1) Year-round design flood. For projects with a construction period exceeding one year, the cofferdam project must withstand flooding during the flood seasons. Therefore, the cofferdam's design flood calculations must be based on the annual maximum discharge.
(2) Seasonal design flood. The determination of seasonal periods should consider both construction requirements and flood seasonality. Seasons are generally categorized as dry, transition, and flood seasons. The construction planners propose several possible construction periods, and the frequency of discharge during each period of the calendar year is calculated separately to derive the design flood for each construction period. A longer construction period with lower discharge is recommended as the preferred option. To prevent anomalies, the statistical flood period can be shifted forward or backward by one month when defining the stages.

4.4.3.2 Calculation Methods for Construction Design Floods

The calculation of the construction design flood follows similar principles to those of the permanent design flood. Initially, data should be collected, interpolated, extended, and reviewed for reliability, representativeness, and consistency.

Subsequently, once the construction staging is determined, regular independent sampling can be conducted using the annual maximum method, which involves creating a sample series with the maximum values during each stage, year by year. For safety, extending the sample window by 5 to 10 days before and after each stage is also permitted. The statistical parameters of the staged floods are calculated, and curve-fitting principles are applied as they are for the annual maximum flood of permanent structures. Given the typically lower standards of construction design for floods, it's generally unnecessary to use a theoretical cumulative frequency curve; instead, an empirical cumulative frequency curve suffices.

For large projects, the initial stage of cofferdam flood control should be based on peak discharge, and an empirical frequency curve of peak discharge should be plotted. Similarly, small and medium-sized projects require a frequency curve of peak discharge to determine the design peak discharge. When closing the cofferdam or bottom outlet at the final stage of a large project, design flood discharge should focus on flood volume for a specific duration, and the empirical cumulative frequency curve for that duration should be plotted.

4.4.3.3 Reasonableness Analysis of the Results

(1) Analysis of the regularity of changes in the statistical parameters of floods across stages. By plotting the mean and coefficient of variation of the statistical parameters of floods in each stage, or the design value of each frequency, against the middle date of the corresponding stage, the distribution plots of each stage can be created. These are then used to analyze the seasonal variation pattern, which typically follows a smooth curve. Deviations from this pattern should be adjusted.
(2) Analysis of the reasonableness of the variation pattern of frequency curves of peak discharge and flood volume for each flood stage. These curves are plotted on the same probability paper as the frequency curve of the annual maximum flood to check the rationality. Generally, the frequency curves should not intersect, and the frequency curve of the annual maximum flood should be located at the uppermost position. If an intersection occurs, it is considered unreasonable and should be adjusted using the frequency curve of the annual maximum flood as a reference.

4.5 Derivation of Design Navigable Stage and Discharge

4.5.1 Overview

The design water level generally includes two components: the reservoir design water level, which corresponds to the design standard based on the cumulative frequency of the reservoir flood regulation curve, and the design navigable water level, which aligns with the navigation design standards. This section focuses on the derivation of the latter—the design navigable water level—under varying lengths of available data. The corresponding design discharge for this water level can be determined using the stage-discharge relation.

When the data is sufficient, frequency analysis can be used to derive the design maximum navigable water level and the design minimum navigable water level using the duration curve and frequency analysis method; when the data is insufficient, the correlation analysis method is used to derive the design navigable water level; when the data is not available, the data from neighboring stations can only be transferred.

The design navigable water level or discharge, regardless of the conditions and methods, corresponds to the navigability standard, which is also used as a probabilistic indicator. The design maximum navigable water level is based on the cumulative frequency or return period as a design indicator (standard); the design minimum navigable water level is based on the guaranteed rate or return period as a design indicator (criterion).

In accordance with the requirements of *Navigation Standard of Inland Waterway* (GB 50139-2014), the design maximum navigable water levels for river sections that

Table 4.6 Design maximum water level criteria for natural rivers

Waterway class	I–III	IV, V	VI, VII
Flood return period/years	20	10	5

Note For mountainous rivers where the water level exceeds the maximum navigable water level for a short time, Class II waterways requires a 10-year return level, Class IV and V waterways require a 5- to 3-year return level, Class VI and VII waterways require a 3- to 2-year return level

are not affected by tides and those that are not significantly affected by tides shall be as specified in Table 4.6; for river sections that are significantly affected by tides, the design maximum navigable water level shall be determined by using a tide level with an annual maximum tide frequency of 5%, calculated according to the Type I extreme value distribution.

In accordance with the requirements of GB 50139-2014, the design minimum navigable water levels for river sections that are either unaffected or only minimally affected by tides can be determined by the integrated duration curve method, which can be expressed in terms of the multi-year duration guaranteed rate; it can also be determined by the guaranteed rate frequency method, which is expressed in terms of the annual guaranteed rate and the return period, as detailed in Table 4.7. For river sections with significant tidal influence, the design minimum navigable water level shall be the tidal level with a cumulative frequency of 90% at low tide.

For rivers with construction, the design maximum water level should use the maximum value based on Table 4.8 (GB 50139-2014). When the hub is running normally, the storage water level has been raised and the increase in water level should be counted. When the silt accumulates, the water level increases correspondingly, which should also be taken into consideration.

On an annual basis, the guaranteed rate of navigation is expressed as a percentage of normal navigation calendar time, i.e.:

$$\text{Guaranteed rate of navigation} = \frac{\text{Days of normal navigation}}{\text{Annual total days}} \times 100\% \tag{4.20}$$

The number of normal navigable days in Eq. (4.20) refers to the durxation of the normal navigable calendar period in days, i.e., the calendar time during which the navigable water level H is below the maximum water level H_{max} and above

Table 4.7 Design minimum navigable water level standards for natural rivers

Waterway class	Multi-year duration guaranteed rate/%	Guaranteed rate frequency method	
		Guaranteed rate/%	Return period/years
I–III	≥ 98	98–99	5–10
IV, V	95–98	95–98	4–5
Vi, VII	90–95	90–95	2–4

Table 4.8 Maximum water level criteria for navigation constructions

Navigation construction class	I, II	III, IV	V–VII
Flood return period/years	20–100	10–20	5–10

Note [a] For mountainous rivers where the water level exceeds the maximum navigable water level for a short duration, Class IV and V waterways require a 5- to 3-year return period; Class VI and VII waterways require a 3- to 2-year return period
[b] For plain rivers with heavy navigation traffic, Class V–VII waterways (or structures) require a 20- to 10-year return period
[c] For overflow structures in mountainous areas, the upstream design water level may be determined according to local conditions, but shall not be lower than the standards prior to the navigation project

the minimum water limit $H_{\min}$, expressed as $t^* = t(H_{\max} \geq H \geq H_{\min})$. The remaining time $(365 - t^*)$ represents non-navigable conditions. As shown in Fig. 4.13, $365 - t^* = t_2 + t_4 + t_6 + t_8$.

In fact, the guaranteed rate is the percentage of the total number of days that allow for normal navigation during the statistical year, and it is an annual average concept. The guaranteed rate of navigation is supposed to take into account both high and low water conditions, but in the specific analysis and calculation, due to the different design objectives and requirements, the design maximum navigable water level and the design minimum navigable water level need to be calculated separately. The former determines the upper water level for the normal operation of navigable structures and the reference water level for the vertical clearance of cross-river structures. Examples include the determination of the vertical clearance of navigable bridges, the top elevation of lock gates, lock walls, and guide walls. The latter determines the controlling water depth of the channel and navigation structures and the reference water level for underwater crossings. Examples include the sill elevation of the lock head, the bottom elevation of the approach channel, and the elevation of underwater crossings in river regulation projects. It can be seen that

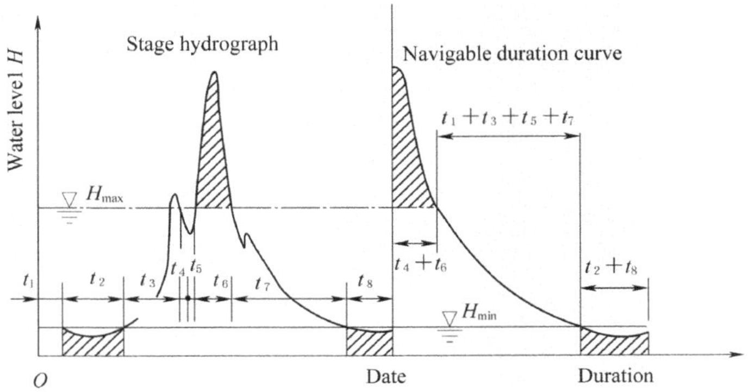

Fig. 4.13 Stage hydrograph and navigable duration curve

the design navigable water level is an important basis for the planning, design, and construction of shipping projects.

It should be noted that the design minimum navigable water level for a given guaranteed rate only indicates the degree to which navigation is guaranteed when the water level is greater than or equal to this level, and suspension due to high water is not taken into account; similarly, for the design maximum navigable water level at a given frequency, only the probability of meeting navigation requirements when the water level is less than or equal to this level is considered, and suspension due to low water is not taken into account.

As the design maximum navigable water level calculation method is basically similar to the design flood calculation method, this section will focus on the design minimum navigable water level derivation method.

4.5.2 Derivation of Design Navigable Stage and Discharge with Long-Term Data

When the natural river channel is relatively stable, the water level data can generally be used directly to derive the design navigable water level.

If the design section has measured discharge data, and the river channel changes significantly, or if a canalization project is planned, the design navigable discharge should first be derived from the discharge data, and then be converted to the design navigable water level through the stage-discharge relation curve. This is because after river channelization, flow conditions change, and water level data collected before construction will not apply after the project. However, since the natural geographic conditions of the basin remain largely unchanged, the discharge data vary little before and after the project. Therefore, using discharge data can maintain consistency. However, there are two kinds of errors in the water level data deduced from the discharge data: the error of the discharge measurement and the error of the stage-discharge relation curve.

4.5.2.1 Derivation of the Design's Highest Navigable Stage

The cumulative frequency curve method is usually used, and it follows the same procedure as that used to derive the design flood, but the following additional issues need to be considered.

(1) Navigability requirements for hydraulic complexes with storage capacity. The design maximum navigable water level shall not be higher than the design flood level of the complex and shall not be lower than the normal pool level or the flood control limit level.
(2) Frequency calculation requirements for mountainous rivers with low design standards. The daily mean discharge is used as the statistical variable for

frequency calculations, rather than the peak discharge. This is because, in this case, the duration of the flood peak is very short and has little impact on navigation projects that do not require high cargo volumes; otherwise, using peak discharge would result in uneconomic investment.

(3) Simplification of calculation procedures. When using water level data to derive the design of the maximum navigable water level, in order to facilitate the calculation, reduce the error, and avoid excessively small C_v values, the original water levels in the series of values are often reduced by a constant H_0 (taking the integer value below the lowest water level in the series), and then frequency calculations are performed. The final result needs to be added to H_0. Please refer to ESM 4.5 for an example.

4.5.2.2 Derivation of the Design Lowest Navigable Stage

At present, the design minimum navigable water level is mostly derived using the duration curve method and the guaranteed rate frequency method.

1. Duration curve method

Figure 4.14 shows a duration curve for a station for a given year, a characteristic curve that reflects the distribution of river streamflow. When plotting this curve, the year, season, or month can be taken as the time period, and the water level (discharge) can be taken as an instantaneous, daily mean, 10-day, or monthly mean value. The annual time scale is most commonly used, with water level (or discharge) averaged on a daily basis to obtain the daily water level (or discharge) duration curve. The horizontal coordinates of the curve are expressed in percentages, and the curves are called relative duration curves or guaranteed rate curves. The methods used to draw such curves are the integrated method, the average method, the representative year method, and the simplified method.

(1) Integrated method. ① Based on the range of daily average maximum and minimum water level changes during the statistical year, the daily average water level is divided into a number of intervals (Column I of Table 4.9). In the waterway sector, 5–20 cm is taken as one interval. In the water resources sector, the water level interval is determined according to the scale of the hydrograph; for example, with a vertical scale of 1:100, every 10 cm is taken as a level; with a scale of 1:200, every 20 cm is taken as a level, etc. ② In preparing the water level statistics table (Table 4.9), with a 10 cm interval over a 20-year period, the water level variables are filled between 12.00–29.99 m, as shown in Column I; the number of days for each water level grade is counted year by year based on the daily average values, and recorded in Column II of Table 4.9; the cumulative number of days is calculated from top to bottom, and the cumulative number of days at the lowest level should be equal to the total number of days included in the statistics (7305d), as shown in the last row of Column III; the cumulative number of days occurring at each level is divided by the total number of days in the statistical period to obtain the water level guaranteed rate

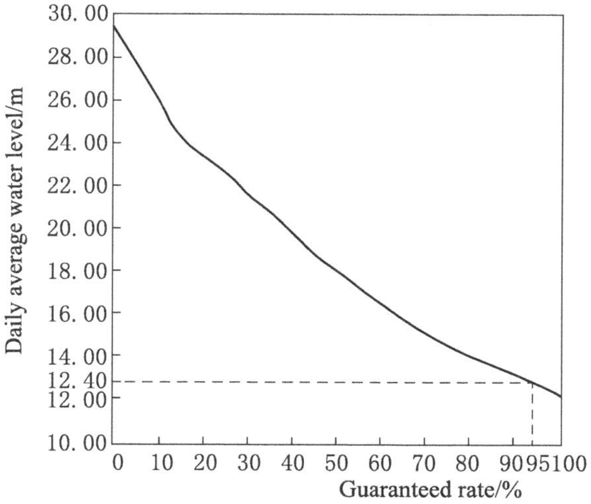

Fig. 4.14 Duration curve (guaranteed rate curve)

at each level, as shown in Column IV. ③ With the daily average water level as the vertical coordinate and the guaranteed rate as the horizontal coordinate, based on the values in Columns I and Column IV of Table 4.9, the points are plotted on the graph to obtain a composite duration curve (also known as the guaranteed rate curve), as shown in Fig. 4.14.

When plotting the points, all points represent the lower limit of the water level for each interval, except for the first point (the highest value) and the last point (the lowest value), and each point is connected to form a curve.

With the water level duration curve, the design water level can be investigated according to the design criteria (guaranteed rate). From Fig. 4.14, we can find the design water level of 12.40 m for the corresponding guaranteed rate of 95%.

The integrated duration curve method reflects the multi-year average of the hydrological variables and uses the daily navigational guaranteed rate to reflect the benefits of navigation, making it conceptually intuitive. It is suitable for conditions where river

Table 4.9 Water level duration statistics

I	II	III	IV
Water level/m	Time of occurrence/d	Cumulative number of days	Guaranteed rate/%
29.90–29.99	5	5	0.07
29.80–29.89	9	14	0.19
. . .	. . .	. . .	. . .
12.10–12.19	22	7294	99.70
12.00–12.09	11	7305	100.00

recharge sources are consistent, variables do not vary significantly from year to year, and the data series are long. The use of this method is limited when runoff is highly variable from year to year, the data series is short, and high design standards are required.

As rivers are affected by the size of the basin, climate, and other factors, the duration curves differ in their flatness and steepness. For flat curves, a 1%–2% difference in terms of guaranteed rate has little effect on water levels, while for steep curves, the water level can vary by tens of centimetres and must be carefully considered.

The flow duration curve is plotted in a similar way to the water level and will not be repeated.

(2) The average method and the representative year method. ① Average method. Based on the measured daily average water level (discharge) data at the design cross-section over many years, statistics are compiled annually according to the format of Table 4.9. The water level (discharge) duration curve for each year is plotted, resulting in n duration curves for n years. For any given duration p, there are n water level (discharge) values Z_i (p), $i = 1, 2, \ldots, n$. By taking their arithmetic mean $\overline{Z}(P) = \frac{1}{n}\sum Z_i(P)$ and plotting it against the corresponding duration p, the average duration curve of the water level (discharge) at the section is obtained. ② Representative year method. From the measured daily average water level (discharge) data over many years, three representative years—wet, normal, and dry—are selected. Statistics are compiled according to the format of Table 4.9, and the three water levels (wet, normal, and dry) at the same duration are averaged. The average values are plotted against their corresponding duration to obtain the representative year duration curve. This method is essentially a simplified version of the average duration curve method.

(3) Simplified method. Due to the heavy workload of the integrated method, only rough calculations are required during the planning stage; therefore, the following simplified method can be used to draw a composite duration curve. From the published daily average water level tables, select the water levels corresponding to annual durations of 15 d, 30 d, 90 d, 180 d, 270 d, and 330 d, as well as the annual maximum and minimum water levels. After filling these into Table 4.10, the arithmetic average is used to determine the mean water level for each duration.

Plot the average water levels from Table 4.10 (where the maximum and minimum water levels are replaced by the maximum and minimum daily averages of the actual statistical period) against the corresponding durations (in days) or percentages to obtain the duration curve.

For rivers with little change in runoff during the dry season, the duration curve method is a simple and applicable method.

2. Guaranteed Rate Frequency Method

This method introduces the concept of frequency into the guaranteed rate. For example, n guaranteed rate curves can be plotted for n years, and for any given guaranteed rate, each year has a corresponding water level $Z_i(p)$ $(i = 1, 2, \ldots, n)$.

Table 4.10 Table of water level characteristics at a station (simplified method calculation table)

Year	Average annual water level/m	Annual maximum water level/m	Various duration water levels/m						Annual minimum water level/m
			15 d	30 d	90 d	180 d	270 d	330 d	
1917	51.91	59.41	53.42	48.31	35.70	29.10	1&40	12.41	12.39
1918	52.73	70.39	62.30	50.32	38.80	28.43	17.59	12.48	12.40
.	.	.	.	.	.	.	.	.	.
.	.	.	.	.	.	.	.	.	.
.	.	.	.	.	.	.	.	.	.
1964	50.84	60.49	155.51	51.48	37.90	30.50	16.84	11.40	11.90
Average	51.73	70.10	58.35	49.40	38.10	30.50	17.10	12.10	12.10

The guaranteed rate frequency method treats these water levels as random variables, performs frequency analysis, and then determines the water level (discharge) at the specified frequency on the cumulative frequency curve as the design water level (discharge). The steps are: ① determining the frequency and guaranteed rate according to the specification and the navigational requirements; ② extracting the water level (discharge) value according to the selected guaranteed rate from the annual water level (discharge) duration curve; ③ using the selected water level (discharge) of each year as a sample for frequency analysis calculation. Finally, the corresponding design water level (discharge) is found on the curve according to the specified cumulative frequency. Please refer to ESM 4.6 for an example, in which the design minimum water level is derived for a project with a 10-year return period and a 95% guaranteed rate.

The guaranteed rate frequency method, which is also a frequency analysis calculation method, only differs in the way the samples are selected when the series is composed. This method takes into account both the duration of navigation interruption and load shedding and the frequency with which that duration occurs. Both intra- and inter-annual navigational reliability indicators are considered. It is also more suitable for the data than the calendar curve method. This method is more appropriate when the design standards are high.

4.5.3 Determination of Design Navigable Stage and Discharge Under Data-Limited Conditions

4.5.3.1 Determination with Limited Data

When there is insufficient data, correlation analysis is often used to interpolate and extend hydrological data to improve the representativeness of the sample series and reduce sampling error. By extending the series into a long-term one, it is still possible to derive the design navigable water level using the methods applicable under sufficient information conditions mentioned above; therefore, the procedures will not be repeated.

4.5.3.2 Determination Without Measured Data

This section focuses on methods for transferring design water levels from a reference station, including the slope projection method, the stage-correlation method, and the instantaneous stage method.

(1) Slope projection method. When the design section is close to the reference station, the river reach is straight, and the shape of the section does not change much (i.e., the variation in water surface slope is not significant), the design flood

level of the reference station can be used to obtain the design flood level of the design section based on the measured high-water surface slope. The formula is

$$H_{\mathrm{D}} = H_{\mathrm{P}} \pm LI \tag{4.21}$$

where H_{D} and H_{P} are the design flood levels of the design cross-section and the reference station, respectively, m; L is the distance from the design cross-section to the reference station's basic water gauge, m; I is the measured high-water surface slope at the reference station.

(2) Stage-correlation method. When the design section is far from the reference station, a temporary water level station can be established at the design section to observe the water level for at least one hydrological year (from the beginning of the flood period to the end of the dry period, including the fluctuations between flood and dry water levels). After establishing the water level correlation relationship (curve) of the two stations, the design water level of the design section can be determined from the correlation curve based on the design water level of the reference station. For the extrapolation of the design navigable high water level, care should be taken.
(3) Instantaneous stage method. If the water level data at the design section are insufficient, making it impossible to draw the stage-correlation curve between the design section and the reference station, then the instantaneous stage method can be used. For instance, to determine the design low water level, one can choose the dry period when the water level is stable (fluctuation does not exceed 5 cm) to collect water level data, and calculate the instantaneous average difference (fall) as listed in the last row of Table 4.11. The design water level of the design section is obtained by subtracting this difference value from the design water level of the reference station. The calculation method is included in the table description, so it is not repeated here.

Table 4.11 Table of instantaneous water levels at the reference and design stations

Date	Water level at reference station/m	Design station water level/m	Drop/m	Calculation method
2 January	6.30	5.50	0.80	The design water level at the reference station is 6.00 m, therefore, the design section water level is calculated as 6.00 − 0.81 = 5.19 m, which is adopted as 5.00 m. Note that in this case, the reference station is located upstream of the design section
3 January	6.28	5.47	0.81	
4 January	6.24	5.42	0.82	
5 January	6.20	5.39	0.81	
Total	25.02	21.78	3.24	
Average			0.81	

4.5.4 Determination of Stage-Discharge Relations with Insufficient Data

In the design and application of harbor and navigation projects, it is often necessary to determine the stage-discharge relation curve for a series of cross-sections. If there is a hydrological station near the engineering design section, the stage-discharge relation curve of the hydrological station can be directly transferred. However, most of the sections involved lack these data. In order to solve this problem, the following two methods are commonly used. First, conduct short-term mobile measurements, according to the method of plotting stage-discharge relations described in Chap. 3. However, since the observation duration is short and the measured stage fluctuation is small, it is difficult to establish the stage-discharge relation based solely on the results of actual measurements. The second is the indirect method. It uses the actual measurement data and the stage-correlation or hydraulic methods to determine the stage-discharge relation curve.

4.5.4.1 Stage-Correlation Method

A correlation is established between the short-term measured water level data at the design station and the water levels at the adjacent hydrological stations. Through this relationship, the stage-discharge relation curve at the reference stations is converted into the stage-discharge relation curve at the design section.

(1) Specific steps. ① Set up a water gauge at the design section to observe water levels. Focus on observing both peak and trough (low) water levels. ② Select the reference station. The principles are to choose a reach without a large tributary confluence, where the intervening catchment area is small, the river channel storage (or routing effect) is not large, and it is not subject to backwater effect, while being close to the design section. ③ Plot and extrapolate the stage-discharge relationship of the reference station. ④ Using the water level data of the design section and the reference station for the same period, and taking into account the impact of flood propagation, find the corresponding water levels (i.e., the corresponding peak and trough levels) to plot the stage-correlation curve. The high water part can be extrapolated through historical flood investigation. ⑤ Derive the stage-discharge relation curve at the design section by converting the stage-discharge curve of the reference station through the stage-correlation curve.

(2) Applicable conditions. ① Hydrological data are available at the proposed project site, and there are hydrological stations nearby for reference. ② The distance between the project site and the reference station is short, and the intervening inflow is minimal. ③ The variations in hydrological elements at the project site and the reference station are consistent and are not affected by backwater. ④ The hydraulic factors of the river channel are similar or identical, especially the bed slope, which should be close or equal.

4.5.4.2 Hydraulic Method

This method uses the hydraulic Manning formula to calculate discharge, as shown in ESM 4.2 [Eqs. (1)–(3)]. The key is to determine the roughness coefficient n and the energy slope I. The value of n can be back-calculated from the data obtained from measurements, and the roughness-stage curve can be plotted by referring to the roughness table according to the bed composition. As the slope varies from section to section and with water level, it is not possible to simply transfer the data from other sections; instead, one should use the results of the actual measurements to plot the relationship between water level and slope with reference to the flood mark survey. Using the section data to calculate the section area A and hydraulic radius R, and substituting them into Manning's formula to determine the corresponding discharge value for various water levels, one can plot the stage-discharge curve.

If there is a variable backwater or flood influence, the curve should be determined based on the trend, with reference to the shape of the relation curves of the adjacent upstream and downstream hydrological stations, and the appropriate parameters selected.

It should be noted that the hydraulic method is less accurate than the stage-correlation method. However, when the design section does not meet the conditions for the stage-correlation method (e.g., significant influences of intervening inflow, backwater effect, etc.), this method remains applicable.

Chapter 5
The Waves

The force of the waters was split apart.
分开水势,
Waves rose like collapsing mountains.
浪如山倒,
Torrents surged uncontrollably.
波涛横生,
And the flat ground was flooded with water several feet deep.
平地水长数尺。
The Investiture of the Gods: Nezha Conquers the Dragon King,
Xu Zhonglin, Ming Dynasty
—《封神演义·哪吒闹海》, 许仲琳, 明代
▲
This passage describes the scene of Nezha conquering the Dragon King, where the water is split, waves rise like collapsing mountains, and torrents surge uncontrollably, showcasing the immense energy and destructive power of waves.

Abstract This chapter introduces wave characteristics, elements, and classification, as well as wind conditions and statistical properties. It presents wave observation methods and random wave theory, with emphasis on wave spectra and engineering spectral models. It also describes design wave estimation methods for coastal engineering under different data conditions. Finally, it discusses wave propagation, deformation, and related theories in nearshore areas.

For a visual overview of the chapter's key concepts, refer to the knowledge map below.

Supplementary Information The online version contains Electronic Supplementary Material (ESM) available at https://doi.org/10.1007/978-981-95-9089-6_5.

X. Feng (ed.), *Engineering Hydrology*, https://doi.org/10.1007/978-981-95-9089-6_5

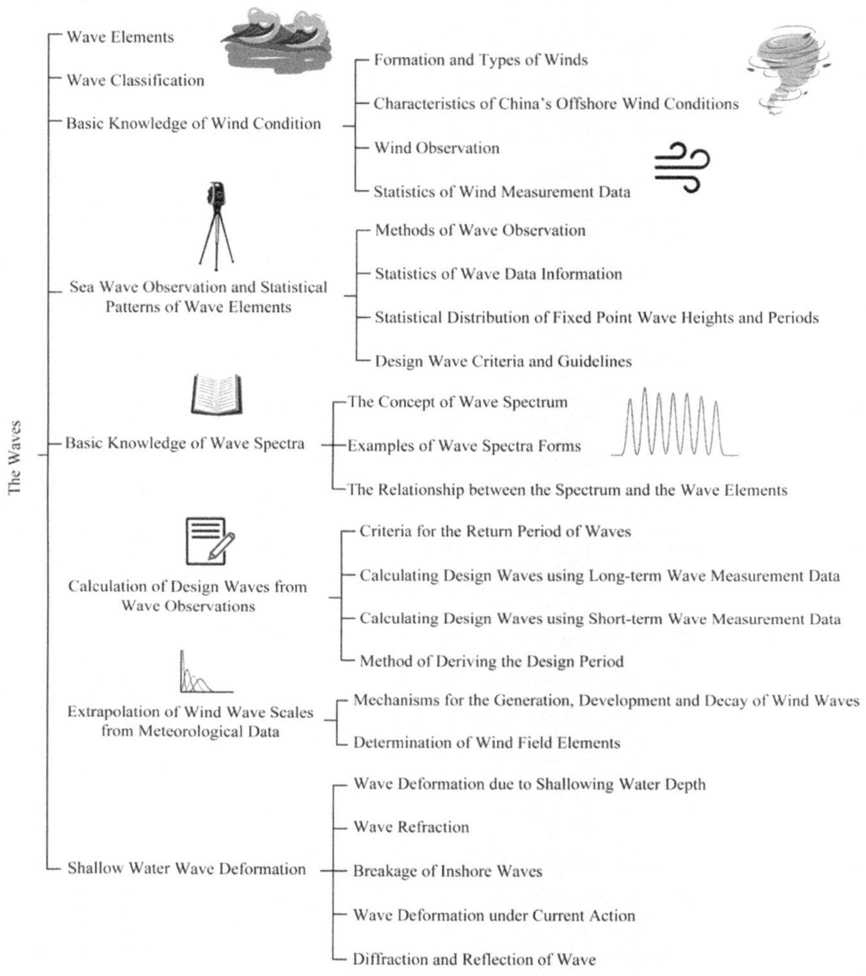

5.1 Introduction

Waves are a crucial environmental factor influencing maritime structures, such as breakwaters, seawalls, and offshore platforms. These structures must withstand the impact of various forces, including winds, waves, tidal currents, and earthquakes. To ensure their functionality, a thorough investigation of environmental conditions and their impact on design is essential. However, due to the complexity of waves, understanding their characteristics and behaviors is challenging.

Waves are created when wind blows, growing into larger waves during storms and dissipating once the wind stops. Their irregularity is a key feature, with varying sizes and directions observed offshore. When they reach the shore, they break, giving the impression of regular repetition. Basic theories of water waves were established in the late nineteenth century, but their engineering application took decades.

During World War II, coastal engineering emerged by combining mathematical theories and engineering practices. Sverdrup and Munk introduced wave-prediction methods, which evolved into the more complex S-M-B method. The significant wave concept was developed, representing the average height of the highest one-third of waves in a wave group.

In 1952, Pierson and other oceanographers recognized the irregularity of sea waves as a fundamental property, leading to the P-N-J wave-forecasting method. This marked the beginning of detailed explanations for wind-wave generation, swell propagation, and near-shore wave transformation using wave spectra. Although oceanographers accepted this concept, engineers found it too complex, delaying its application until the 1970s.

In the following chapters, wave elements and the forms and classifications of waves will be described first. Then, the main generating force of sea waves, the wind, and its statistical characteristics will be introduced. Subsequently, the observation methods of sea waves and the theory of random waves will follow. In the fourth section, wave spectra and the commonly used spectral theories in the engineering field will be introduced with emphasis. In the fifth and sixth sections, according to the different situations of the observation data mastered by researchers, methods for deriving design waves for coastal and ocean engineering will be introduced. In the seventh section, a series of propagation deformations of waves after they propagate to the near-shore area under the influence of topography and structures, as well as the corresponding derivation theories, will be further discussed.

5.2 Wave Elements and Classification

5.2.1 Wave Elements

Waves in the ocean are one of the forms of seawater movement, and they are generated as a result of a combination of external forces, gravity, and seawater surface tension. There are many external factors that cause seawater disturbance, such as wind, changes in atmospheric pressure, celestial gravitational forces, submarine earthquakes, and human-induced ship movements. The period of seawater fluctuations caused by these factors can vary over a very wide range, such as half a day to a day for tidal waves, tens of minutes for tsunamis, seconds for wind waves, and less than 1 s for surface tension waves. This chapter will focus on wind-induced gravity waves, which is a general term for wind waves, surge waves, and nearshore waves. Wind waves are mainly waves generated by the direct action of the wind. Surge waves are waves that stop, turn, or leave the windy zone and propagate into windless waters. Surge waves propagate into shallow waters and are deformed according to water depth and topography, resulting in refraction, diffraction, and breaking of waves and the formation of nearshore waves.

Although the wave profile is complex in shape, it is often idealized as a regular profile, as shown in Fig. 5.1 and characterized by various wave elements.

The individual wave elements are defined as follows:

(1) Wave crest: the part of the wave profile that rises above the surface of the still water, the highest point of which is called the top of the crest.
(2) Wave trough: the part of the wave profile below the surface of the still water, the lowest point of which is called the bottom of the trough.
(3) Wave crest line: the line connecting the tops of each wave crest, vertical to wave propagation.
(4) Wave direction line: the line orthogonal to the crest line, i.e., the direction of wave propagation.
(5) Wave height: the vertical distance between the top of the adjacent wave crest and the bottom of the wave trough, usually expressed as H in metres (m). Giant waves of up to 15 m have been recorded in the Taiwan Strait in China.
(6) Wavelength: the horizontal distance between the tops of two adjacent wave crests (or wave troughs), usually expressed in L and measured in metres (m). Each wave can have a wavelength of hundreds of metres, while tidal waves can have wavelengths of several kilometres.
(7) Wave period: the time it takes for two successive wave crests passing a fixed point in space, usually expressed in T, measured in seconds (s). The wave period along the coast of China is generally 4–8 s, and long waves with a period of 20 s have been recorded.
(8) Wave steepness: the ratio of wave height to wavelength, usually expressed as $\delta = H/L$. Wave steepness is commonly found in the ocean and ranges from 1/30 to 1/10.

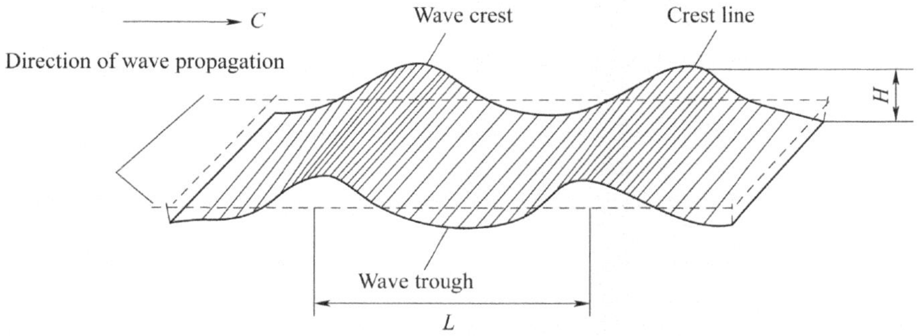

Fig. 5.1 Wave elements

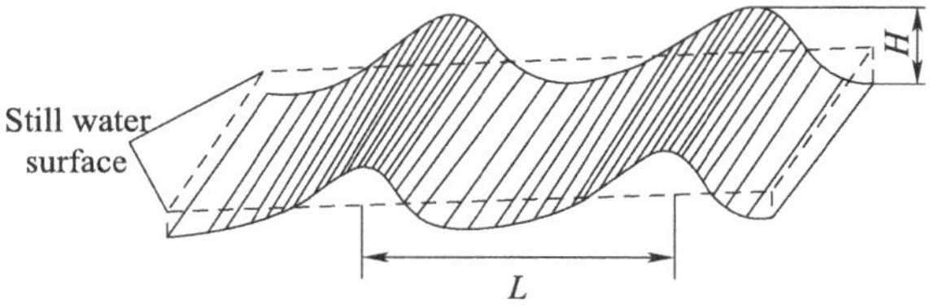

Fig. 5.2 The profile of wind waves

(9) Wave speed: the speed at which a waveform moves, usually expressed as C, which is equal to the wavelength divided by the period, i.e., $C = L/T$, in metres per second (m/s).

5.2.2 *Wave Classification*

Waves can be classified according to a variety of criteria.

5.2.2.1 Forced, Free, and Mixed Waves

Forced waves are waves in which the disturbance force causing the wave acts continuously on the water surface and the nature of the wave depends on the nature of the disturbance force. Wind waves generated by the direct action of the wind are forced waves whose shape is asymmetrical with respect to the vertical axis, as shown in Fig. 5.2, with the leeward side of the wave being steeper than the windward side. Free waves are waves that continue to propagate under gravity after the disturbance force has disappeared, and are no longer entirely dependent on the original disturbance force. If the wind stops or if the waves leave the windy area and propagate into calm waters, the waves are called free waves. The shape of the swell is relatively regular, and the wave surface is smooth. The sea surface is often subject to both wind waves and swell propagating from outside the windy area, and their superposition is known as mixed waves.

5.2.2.2 Capillary, Gravity, and Long-Period Waves

According to the nature of the restoring force acting on the water mass when it is disturbed from its equilibrium position, waves can be classified. When the restoring force is mainly surface tension, the resulting waves are known as capillary waves or surface-tension waves. For example, when the wind is very weak, tiny ripples that appear on the sea surface are capillary waves. Their period is often less than 1 s. When the wavelength is larger, the restoring force that brings water particles back to their equilibrium position is mainly gravity. Such waves are called gravity waves, such as wind waves, swells, seismic waves, and ship waves, etc. Long-period waves

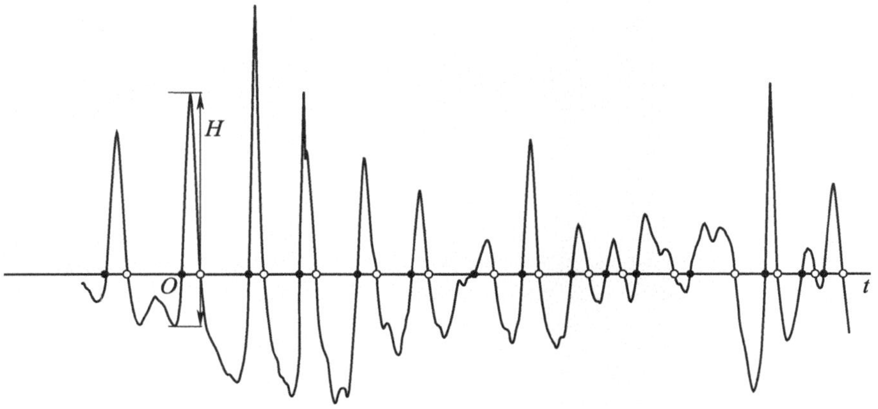

Fig. 5.3 Actual wave measurement record on the sea surface

mainly refer to tidal waves caused by the gravitational effects of the Sun and the Moon, but they also include ocean swells, wind-induced oscillations in bays, and other long-period fluctuations. For these waves, the restoring forces are gravity and the Coriolis force.

5.2.2.3 Irregular and Regular Waves

The wave elements of the individual waves that ensue on the sea surface are constantly changing. It is an irregular random phenomenon, and such waves are called irregular waves, as shown in Fig. 5.3 for a section of the actual wave measurement record on the sea surface. In order to facilitate the study of wave motion, an ideal series of identical wave elements is used to replace the actual irregular sea-wave series. This ideal wave is called the regular wave, such as the regular wave generated by artificial methods in the laboratory. Sea swells also approach regular waves.

5.2.2.4 Long- and Short-Crested Waves

On the sea surface, one can clearly observe a succession of crests and troughs. When the crest lines are very long and almost parallel to each other, the waves are called long-crested waves or two-dimensional waves. A swell is a long-crested wave. In the case of strong winds, the crest lines become indistinguishable, and the crests and troughs form a pattern resembling a chessboard. Such waves are called short-crested or three-dimensional waves. Wind waves are usually short-crested waves.

5.2.2.5 Forward and Standing Waves

Waves that propagate forward or toward the shore are called progressive waves. A standing wave (or stationary wave) is a wave that does not propagate forward, with the crests and troughs rising and falling periodically in place. It results from the interference between an incident wave and the reflected wave when the wave encounters a steep coastal cliff or a vertical wall.

5.2.2.6 Deep-Water and Shallow-Water Propagating Waves

Waves propagating in waters with a water depth d greater than half the wavelength are called deep-water propagating waves (deep-water waves for short). Deep-water waves are not affected by the bottom of the seabed, and the trajectories of water particles are nearly circular, and fluctuations are mainly concentrated in the water layer below the surface at a certain depth and are therefore referred to as short waves. When the water depth becomes less than half the wavelength, deep-water waves are transformed into shallow-water propagating waves (shallow-water waves for short). Shallow-water waves are affected by seabed friction, and water-particle orbits become flattened (nearly elliptical). Because the water depth is small relative to the wavelength, they are also referred to as long waves.

According to micro amplitude wave theory (linear theory), the interrelationship between wave speed, wavelength, and period of a regular deep-water wave is as follows:

$$C_0 = L_0/T_0 = \sqrt{gL_0/2\pi} \text{ or } L_0 = gT_0^2/(2\pi) \tag{5.1}$$

The interrelationship between wave speed, wavelength and period of a regular shallow-water wave is

$$C = \sqrt{\frac{gL}{2\pi}\tanh\frac{2\pi d}{L}} \text{ or } L = \frac{gT^2}{2\pi}\tanh\frac{2\pi d}{L} \tag{5.2}$$

where $2\pi d/L$ is the shallow-water factor expressed as a function of the hyperbolic tangent; the subscript "0" indicates deep-water wave parameters, whereas the quantities without a subscript refer to shallow-water wave parameters.

Observations show that the period of the wave is constant after it passes from deep water into shallow water, i.e.,

$$T = T_0 \tag{5.3}$$

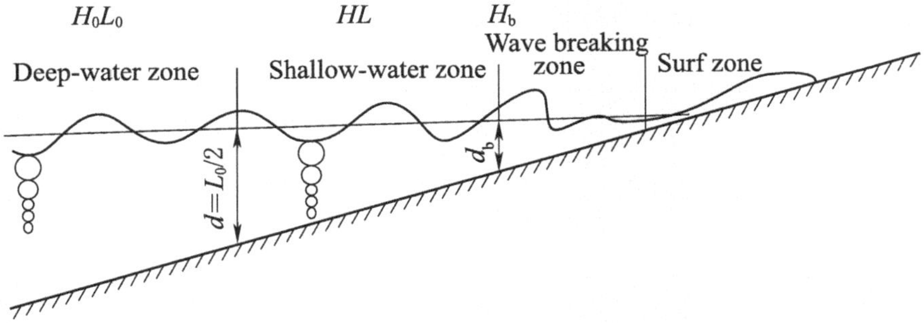

Fig. 5.4 Coastal zone delineation

5.2.2.7 Oscillating and Translatory Waves

In a wave cycle, the trajectory of water particle motion is closed or nearly closed, that is, the particles oscillate about fixed positions. Such waves are called oscillating waves. In contrast, when water particles undergo a net forward displacement during a wave cycle, the wave is called a translatory wave. As shallow-water waves approach the shore, they deform and steepen until breaking occurs. After breaking, the waves continue to propagate shoreward, forming surf waves that break repeatedly. The final stage of wave breaking contributes to the formation of nearshore currents.

For gently sloping beaches, inshore waters can be roughly divided into four zones according to the depth of water, i.e., deep-water zone, shallow-water zone, wave breaking zone, and surf zone, as shown in Fig. 5.4. In fact, due to the irregularity of waves, it is not possible to delineate the exact location of the boundaries of each zone, and Fig. 5.4 shows this only according to ideal regular waves. The depth of water at the point of wave breakage is indicated by d_b, and the corresponding wave height is called the breaking wave height, indicated by H_b.

5.3 Basic Knowledge of Wind Conditions

5.3.1 Formation and Types of Winds and Characteristics of China's Offshore Wind Conditions

5.3.1.1 Formation and Types of Winds

The temperature difference between the Earth's north and south poles and the equator provides the energy required to drive the Earth's atmospheric circulation. The conversion of heat energy to kinetic energy to produce atmospheric motion necessarily involves the horizontal, upward, and downward movement of the atmosphere. The main cause of atmospheric motion is the existence of differences in air pressure in the horizontal and vertical directions. This mutual balance between gravity and vertical pressure gradient force is called static equilibrium, which greatly limits the vertical movement of the atmosphere; therefore, except in special cases, the vertical movement of air in local areas can be neglected. It is mainly the horizontal movement of air formed by the difference in air pressure in the horizontal direction that is called wind. Therefore, the distribution of air pressure with time and space and its changes becomes the information that must be grasped for the analysis of wind speed, wind direction, and its changes, and thus the basis for the study of the formation of waves and their changes.

The main cause of air movement is the existence and maintenance of horizontal air pressure gradients, and there are four factors that control the horizontal movement of air near the ground (wind), namely the air pressure gradient forces acting on microclusters of air, the geodynamic deflection forces formed by the Earth's rotation (i.e., the Coriolis force), centrifugal forces, and frictional forces. Descriptions of various forces are detailed in ESM 5.1.

5.3.1.2 Characteristics of China's Offshore Wind Conditions

Wind conditions vary from zone to zone around the globe due to the different origins of air masses, and the characteristics of wind conditions offshore China are mainly characterized by monsoons, cold gales, and typhoons.

1. Monsoon

Due to the thermal differences between land and sea as the dominant factor, with the seasonal changes caused by the movement of high and low pressure centers and wind bands, the formation of winter and summer winds prevailing in almost opposite directions, from sea to land in summer, and from land to sea in winter, is known as the monsoon. China is a typical monsoon-affected country, with northerly winds prevailing from October to March and southerly winds prevailing from June to August. In winter, the Bohai Sea and the Yellow Sea have more west, northwest, and north winds, accounting for 60%; the East China Sea has more north and northeast winds; the South China Sea has more northeast and east winds, accounting for about 88%. In summer, there are mostly southeasterly, southerly, and southwesterly winds, with the Bohai Sea and Yellow Sea accounting for about 50% and the East China Sea and South China Sea accounting for about 56%.

2. Cold Wave

National Meteorological Centre (NMC): after the invasion of cold air, the temperature drops by more than 10 °C within 24 h, and the lowest temperature drops below 5 °C, it is called a cold wave, which is one of the main winter weather processes in China and usually lasts 3–5 days. The path of a cold wave is relatively stable. Cold waves usually originate in the polar regions, pass through Siberia, and move southward into China. Wind speeds can reach Beaufort force 8–9, with gusts up to force 10–11. The cold wave gale is likely to cause water gain (storm surge) on some windward coasts, such as Laizhou Bay in Shandong Province, and special attention should be paid to this phenomenon in engineering design.

3. Typhoon

A typhoon is a tropical cyclone originating over ocean surfaces between 5° and 20° north and south of the equator, mostly in the Pacific Ocean (about 66%), especially in the Western Pacific Ocean (more than 36%). NMC classifies tropical cyclones into four levels according to the maximum sustained wind speed (on the Beaufort scale): tropical depression (below force 8), tropical storm (force 8–9), strong tropical storm (force 10–11), and typhoon (force 12 and above). Typhoons have a diameter of 600–1000 km, the distribution of isobars in the typhoon area is approximately circular, the central pressure can drop below 90,000 Pa, the horizontal pressure gradient is generally 15–16 Pa/km, the maximum wind speed can reach 50–60 m/s, the typhoon center moves at a speed of 10–50 km/h. Typhoons mostly occur from July to October. Typhoon paths are complex, and it is difficult to find two typhoons with identical paths. For China's coast, they are mainly classified as westbound-moving typhoons, landfalling typhoons, and recurving typhoons.

Another feature of typhoons is that there is an eye area at the center of the typhoon, called the typhoon eye, where the wind suddenly decreases, or even becomes absent, the clouds disappear, and the pressure drops to its lowest point. The diameter of a typhoon's eye ranges from 10 to 40 km. Due to the presence of the typhoon eye, the maximum wind speed of a typhoon does not occur at the center, but on a circle with a radius of R at a certain distance from the center. Figure 5.5 shows the wind speed distribution within the typhoon area. The origin of the figure indicates the typhoon center, U is the wind speed at a distance r from the center, U_{max} is the maximum wind speed, R is the distance from the typhoon center to the place where U_{max} appears, called the maximum wind speed radius.

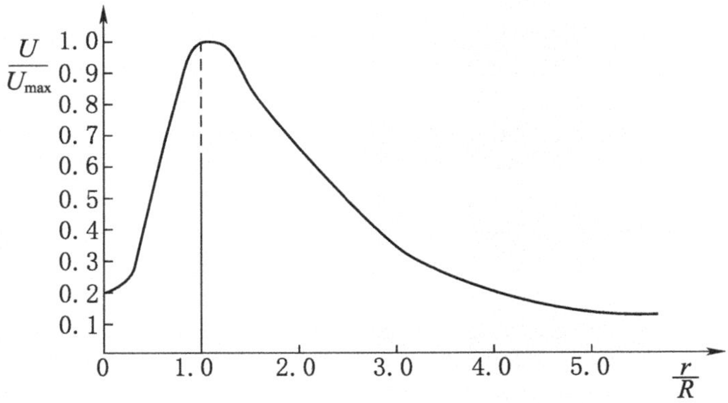

Fig. 5.5 Schematic diagram of wind speed distribution within the typhoon area

The cyclonic wind equation is

$$-\frac{1}{\rho}\left|\frac{\Delta p}{\Delta n}\right| + \frac{U^2}{r} = 0 \tag{5.4}$$

The general formula for maximum wind speed is obtained:

$$U_{\max} = k\sqrt{\frac{\Delta p}{100}}\ (m/s) \tag{5.5}$$

where $\Delta p = P_\infty - P_0$, Pa; P_0 is the pressure at the center of the typhoon; P_∞ is the pressure of the outermost closed isobars of the typhoon; k is an empirical coefficient, which varies with sea area and has been studied by scientists in China, the range of k values in the South China Sea area is 5.6–7.2, and it is found that k values have a tendency to decrease with increasing latitude, in Shanghai area $k = 5.7$, in Yingge Sea $k = 6.5$.

The radius of maximum wind speed R reflects the steepness of the typhoon pressure distribution; the larger the R, the greater the wind speed (at $r \geq R$). R varies among different typhoons and at different times. The value of R can be calculated according to the formula derived from the pressure field model as follows:

$$R = r\ln\frac{P_\infty - P_0}{P - P_0} = r\ln\frac{\Delta P}{P - P_0} \tag{5.6}$$

where P is the barometric pressure at any station in the typhoon area; r is the distance of the station from the typhoon center.

Equation (5.6) is derived from the Myers model of the barometric field in the United States, and its expression is

$$P = P_0 + \Delta p \exp(-R/r) \tag{5.7}$$

Others include the Japanese Fujita air pressure field model, which has a distribution form essentially identical to the above equation.

There are many factors affecting R. For a typhoon process, the value of R also varies greatly. In this case, the differences in the R values obtained for different r values at different moments, in different directions, and in the same direction are often very large. In order to avoid this phenomenon, different r values are generally selected at the same time in the direction in which the typhoon is moving, and R is calculated separately, and then the average value is taken as the maximum wind speed radius for that typhoon at a given moment. To do this, the barometric pressure values at the same moment at each station need to be consulted, which is a huge amount of work and requires repeated calculations.

To overcome this shortcoming, Graham and Nunn of the United States proposed an empirical formula, which takes into account not only the pressure at the center of the typhoon, but also the influence of geographical latitude and the speed of movement of the typhoon center on the value of R. More importantly, it avoids the trouble involved in calculating according to the barometric field model, with the following expression:

$$\begin{aligned} R = {} & 28.52\tanh\left[0.0873(\varphi - 28^\circ)\right] \\ & + 12.22\exp[(P_0 - 1013.2)/33.86] + 0.2V_F + 37.22 \end{aligned} \tag{5.8}$$

where R is the radius of maximum wind speed, km; V_F is the speed of movement of the typhoon center, km/h; φ is geographic latitude, (°); P_0 is the central typhoon pressure, hPa.

Typhoon is the main serious catastrophic weather along the coast of China, mainly manifested as a disaster caused by violent winds, heavy rainfall attacks and increased water. All coastal provinces in China have been hit by typhoons, causing serious damage, especially Fujian, Guangdong, Hainan, and Taiwan provinces, which are the most affected, and typhoons are catastrophic weather systems that need special attention in the design of harbor projects.

5.3.2 *Wind Observation*

The determination of wind speed fields over offshore waters from the distribution of isobars on ground-based weather maps is often not very reliable and can only be used for forecasting or hindcasting purposes, so in situ observations of wind conditions must be made.

The characteristics of the wind are often expressed in two quantities: wind speed and wind direction. Wind speed is the horizontal distance that air travels in a unit of time; its unit is generally expressed in metres per second (m/s), kilometres per hour (km/h), or nautical miles per hour (nm/h). The wind direction refers to the direction in which the wind blows and is generally expressed using the 16 bearings. The conversion relationship between the 16 bearings and degrees is shown in Table 5.1, with calm conditions denoted as C. As the wind field is a time-varying process, its instantaneous changes also exhibit significant pulsation characteristics, so the determination of wind conditions requires the observation of wind speed and direction over a period of time and the determination of their average and sometimes instantaneous maximum values.

In order to study the unstable and pulsating characteristics of wind speed to meet the special needs of certain engineering designs, it is sometimes necessary to carry out wind speed spectroscopy observations. With wind speed spectra, it is possible to analyze the dynamic response of structures or certain components under the action of the wind, so that the design can avoid resonance phenomena where the self-oscillation frequency of the structure is close to the frequency of the energy significant part of the wind speed spectrum.

Table 5.1 Conversion table of 16 bearings and corresponding degrees

Bearing	Degree	Bearing	Degree
N (North)	348.9°–11.3°	S (South)	168.9°–191.3°
NNE (North-northeast)	11.4°–33.8°	SSW (South-southwest)	191.4°–213.8°
NE (Northeast)	33.9°–56.3°	SW (Southwest)	213.9°–236.3°
ENE (East-northeast)	56.4°–78.8°	WSW (West-southwest)	236.4°–258.8°
E (East)	78.9°–101.3°	W (West)	258.9°–281.3°
ESE (East-southeast)	101.4°–123.8°	WNW (West-northwest)	281.4°–303.8°
SE (Southeast)	123.9°–146.3°	NW (Northwest)	303.9°–326.3°
SSE (South-southeast)	146.4°–168.8°	NNW (North-northwest)	326.4°–348.8°

On land, wind measurement stations should be set up in open, unobstructed locations free from buildings. In China, wind observations are generally made using automatic anemometers capable of measuring wind direction. The instrument consists of a wind speed sensor, a wind direction sensor, an indicator, and a recorder, with the sensor mounted on an outdoor mast. The average wind speed and direction for any 10 min interval can be obtained from the recorder's paper strip of the recorder. The indicator gives the instantaneous wind speed and direction. The wind measurements on the hydrometeorological station reports are mostly 2 min averages measured with the indicator.

To supplement the coastal stations, the national marine stations require our ships sailing along the coast to report hydrometeorological information from their waters to the shore stations four times a day at regular intervals. When there is no anemometer at sea, the wind speed is often measured visually using sea surface features, while the wind direction is determined by the compass. The visual wind speed is expressed in the Beaufort wind scale, as shown in Table 5.1 (ESM 5.2), which divides the wind into 12 levels according to wind speed and describes the corresponding sea surface features. Subsequent developments in anemometry revealed the existence of wind speeds exceeding Force 12 in nature, leading to the extension of the Beaufort wind scale to 17 levels.

On the surface weather map, the measured wind speed and direction at a given time at each station are indicated by symbols, see Fig. 5.6. The symbol "·" on the map indicates the station location; the vertical line pointing to the station indicates the wind direction; each long line segment perpendicular to the vertical line represents a wind speed of 4 m/s, each short line segment represents 2 m/s, e.g., "" indicates a wind speed of 6 m/s with a northerly wind direction, and a triangle represents 20 m/s, e.g., "" indicates a northerly wind with a wind speed of 24 m/s.

The records obtained from wind measurement are analyzed and collated, and then filled in various reports. The monthly reports are frequently consulted for engineering design purposes. It should be noted that the monthly reports from hydrometeorological stations under the China Meteorological Administration (CMA) are compiled four times daily at fixed intervals (02:00, 08:00, 14:00, and 20:00 Beijing Time). In

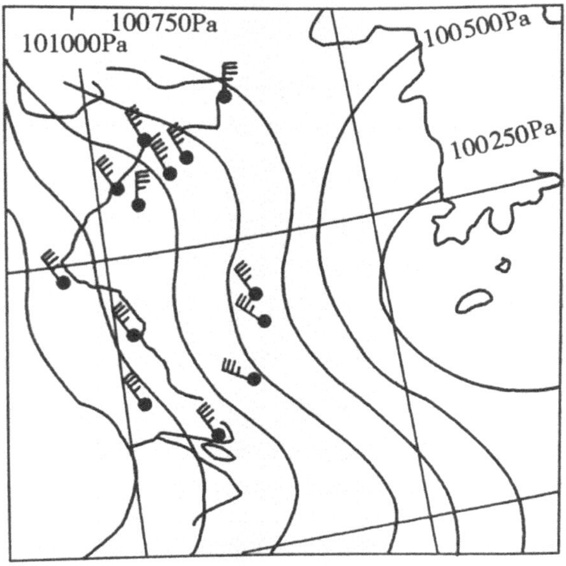

Fig. 5.6 Surface weather map

contrast, wind and wave measurements from hydrographic stations operated by the Ministry of Natural Resources are synchronized and also conducted four times daily, but at daytime hours of 08:00, 11:00, 14:00, and 17:00 Beijing Time. This difference is primarily attributed to the technical limitations of the wave gauges currently in operation.

5.3.3 Statistics of Wind Measurement Data

Upon collecting wind data from hydrometeorological stations or hydrological stations, for engineering planning and design purposes, it is necessary to compile statistics and generate various wind roses, which are also called wind rose diagrams because of their rose-like shape.

A wind rose is a graphical representation of the distribution of four quantities of wind: time period, wind speed, wind direction, and frequency of occurrence. Wind roses are generally drawn in 16 directions. There are various combinations of the four parameters, and it is often not possible to present all of them in a single wind rose, various forms of wind roses are often produced according to engineering requirements.

5.3.3.1 Rose Diagram of Wind Conditions

(1) Rose diagram of wind frequency. The wind data collected over many years (four times a day) are counted by direction, and the frequency of each wind direction is expressed as a percentage and plotted on polar coordinates on a specific scale. Calm wind (wind speed = 0) can be represented by a hollow circle with the frequency of no wind as the radius and the origin of the polar coordinates as the center of the circle, or by a direct numerical indication. A rose diagram of wind frequency can be drawn for the whole year, for a particular quarter or month, as required to meet the needs of statistical port operations or construction days. The meteorological seasons are divided into spring from March to May, summer from June to August, autumn from September to November, and winter from December to February.

In order to reflect the frequency of wind at all levels on the wind frequency rose diagram, the wind statistics are divided into wind levels. And then the number of times the wind is greater than or equal to a certain level (usually a level 6 wind) in each direction is added up and divided by the total number of observations, which is divided by a diagonal line on the diagram, as shown in Fig. 5.7.

To generate a more reliable wind frequency rose diagram, 1–3 years of continuous wind data is generally required (after rigorous validation and verification), or alternatively, a representative set of years with complete observational records may be selected.

From the wind frequency rose diagram it is possible to determine the wind direction that occurs most frequently in the area during a certain period of time, i.e., the direction of the prevailing wind as well as the direction of the gale. As shown in

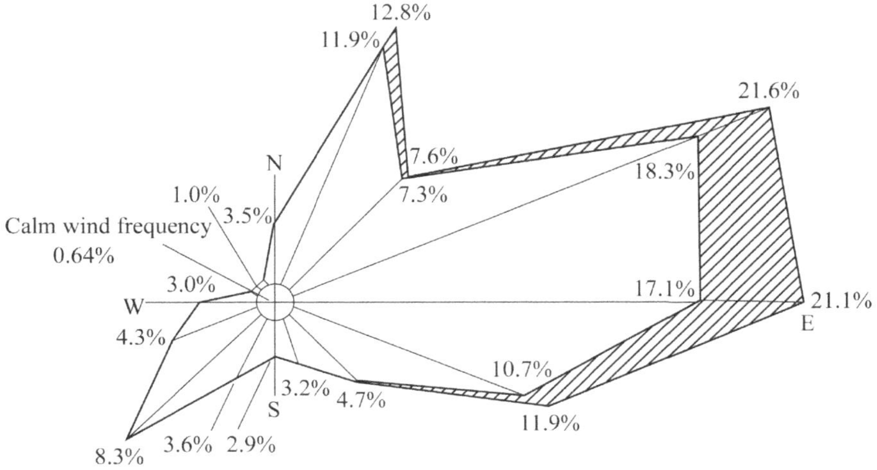

Fig. 5.7 Wind frequency rose chart

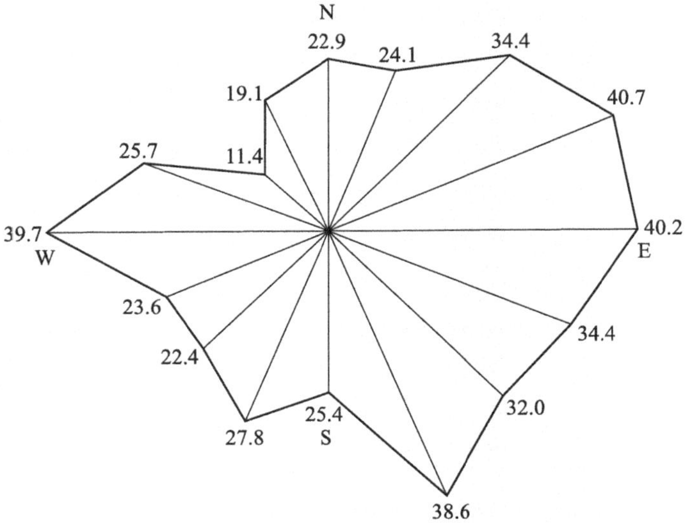

Fig. 5.8 Maximum wind speed rose diagram (unit: m/s)

Fig. 5.7, the prevailing wind directions in the area are ENE and E, as well as the direction of the gale.

(2) Maximum wind speed rose diagram. In order to ensure the reliability of the maximum wind speed rose diagram, it is generally necessary to consult the wind measurement records for 20–30 years, paying special attention to the wind information that has appeared in history but may have been missed, and trying to make up for it if necessary. The maximum wind speed rose diagram can be used to determine the strongest wind direction in the area. Figure 5.8 shows that the strongest wind directions in the area are ENE, E, and SSE, all with a maximum wind speed of 40 m/s.

Owing to the substantial spatiotemporal variations in wind speed, a unified standard must be established for wind speed determination in terms of time interval and measurement height. Technical specifications for harbour engineering in China stipulate that: the standard adopted for wave calculation is the 2 min mean wind speed at a height of 10 m above sea level; the standard applied to the design of port structures is the 10 min mean wind speed at a height of 10 m above sea level. Where the acquired data fail to meet these standards, it is necessary to conduct height conversion and time interval conversion of wind speed.

5.3.3.2 Conversion of Wind Speed

(1) Height conversion of wind speed. As mentioned above, air movement in the near-surface region of the friction layer is highly complex due to the effects of surface friction: wind direction gradually deviates from the direction of isobars, and even if the pressure field remains unchanged with height, the vertical distribution of wind speed still varies, that is, wind speed gradually decreases as the height above the surface reduces. It is generally recognized that the vertical variation law of wind speed with height near the ground (or sea surface) conforms approximately to the logarithmic law. For the height range from 100 m above the sea surface up to the top of the friction layer, if the pressure field exhibits no significant variation with height, the vertical variation law of wind speed with height approximates to the power law.

Expressed logarithmically, the relationship between the ratio of wind speeds at different elevations near the earth (sea) surface is

$$\frac{U_{z_2}}{U_{z_1}} = \frac{\lg z_2 - \lg z_1}{\lg z_1 - \lg z_0} \tag{5.9}$$

where z_1, z_2 are elevations; z_0 is the surface roughness length, 0.03 m for the ground and 0.003 m for the sea; U_{z1} and U_{z2} are the wind speeds at the location.

From Eq. (5.9), once z_0 is determined based on the type of land or sea surface where the anemometer is installed, the corresponding wind speed U_{z2} at any arbitrary height z_2 can be converted/calculated, given the measured wind speed U_{z1} at the anemometer's installation height z_1.

For ease of use, Eq. (5.9) gives the wind speed at 10 m above the ground (sea) U_{10} and the wind speed at any point z above the ground (sea) U_z. The relationship is

$$U_{10} = K_z U_z \text{ or } K_z = U_{10}/U_z \tag{5.10}$$

where K_z is the wind speed height conversion factor. For stations on land, the calculated K_z values are listed in Table 5.1 (ESM 5.3).

The pattern of wind speed along the height from 100 m above the sea surface to the top of the friction layer is

$$\frac{U_{z_2}}{U_{z_1}} = \left(\frac{z_2}{z_1}\right)^p \tag{5.11}$$

where p is an exponent related to the vertical variation of air temperature, with a value ranging from 0 to 1. For open sea areas and scenarios where the sea surface wind speed is relatively high, a value of p between 0.13 and 0.15 may be adopted; under this condition, Eq. (5.11) is essentially consistent with Eq. (5.9) at a height of 100 m.

As there are few stations for systematic wind speed observations at sea, wind and wave projections are often made using wind speed information from coastal land-based weather stations in lieu of sea surface wind speed, i.e., the known wind speed onshore at a height of 10 m above the ground must be converted to a wind speed at a height of 10 m above the sea surface.

With the same horizontal pressure gradient, the wind speed at sea is generally greater than the wind speed on land, so multiply the wind speed onshore at a height of 10 m above the ground by a factor greater than 1.0.

$$(U_{10})_{\text{sea}} = (U_{10})_{\text{land}} K \tag{5.12}$$

where K is the coefficient of increase in sea surface wind speed.

It can be seen from Eq. (5.12) that the elevation of the shore station is not taken into account, meaning that wind energy is assumed to travel unimpeded up or down slopes. This is particularly important for shore stations located very close to the coast, as neglecting this effect would otherwise lead to a significant underestimation of wind speed. The K value is primarily related to the distance between the onshore station and the meteorological station on the sea island, and is also dependent on wind speed and direction. The general variation pattern of the K value is as follows: it is larger over sea areas farther from the coastline and smaller closer to the coast; it is larger when onshore wind speeds are low and smaller when onshore wind speeds are high; and it is larger for offshore winds (blowing from land to sea) than for onshore winds (blowing from sea to land). In the absence of local measured comparative data on sea and land winds, K values may be adopted with reference to Table 5.2, which only accounts for the distance between sea and land stations and does not incorporate the effects of wind speed and direction.

(2) Time interval conversion of wind speed. Even under a highly stable pressure field, wind speed remains unstable and exhibits significant pulsation; therefore, wind speed values are typically represented by the average over a specified time interval, which is referred to as the time interval. In engineering design in China, the commonly used time intervals are 10 min and 2 min, with the corresponding wind speeds termed the 10 min time interval mean wind speed $\overline{U}_{10}$ and the

Table 5.2 Sea and land wind speed conversion factors

Distance from the sea or island to the coast/km	K (on shore wind)
≤ 2	< 1.10
2–30	1.10–1. 14
30–50	1.14–1.23
50–100	1.23–1.30
> 100	Determined from actual measurements or survey information

2 min time interval mean wind speed $\overline{U}_2$, respectively. The former is applied to wind load calculations, while the latter is used for wave prediction.

Under the same wind conditions, wind speed varies with the selected time interval: the shorter the time interval, the higher the corresponding wind speed value. Thus, when collecting wind speed data, it is essential to note the time interval adopted for the wind speed measurement. Conversion between wind speeds at different time intervals is derived through correlation analysis of timed and self-recording wind speed records from parallel observations, and has been validated through statistical analysis of partial datasets.

$$\overline{U}_2/\overline{U}_{10} \approx 1.103 \tag{5.13}$$

This value can be used as a reference when no comparative information is available from actual measurements.

Another noteworthy issue is that the majority of meteorological stations in China employed wind pressure plates for wind measurement prior to 1968. This type of instrument is subject not only to visual observation errors but also to inertial errors induced by wind gusts; consequently, the recorded wind speeds are overestimated, and a correction factor K (where $K < 1$) must be multiplied by the measured data when applied. This factor is related to the wind speed and the slope of the terrain where the station is located. The higher the wind speed, the lower the value of K. The steeper the terrain slope, the lower the value of K. For wind speed less than or equal to 40 m/s and terrain slope less than or equal to 30°, the range of K values is 0.55 to 1.0. In practice, if observational data after 1968 are sufficient, pre-1968 datasets should be avoided whenever possible.

In summary, when collecting data on wind conditions, special attention should be paid to the geographical location of the hydrometeorological stations, the types of anemometers employed, the height of the wind gauges above the ground, the time intervals for wind speed averaging, as well as their interannual changes. All necessary conversions must then be performed to ensure consistency within the statistical dataset.

5.4 Sea Wave Observation and Statistical Patterns of Wave Elements

5.4.1 Methods of Wave Observation

Since the 1960s, in order to meet the needs of construction, the State Oceanic Administration has established a series of marine hydro-meteorological observation stations in coastal regions to conduct systematic observations and accumulate long-term datasets. In addition, a number of temporary observation stations have been established in the port construction sites as needed to supplement the national stations.

Wave gauges are available in a wide variety of types, which can be broadly classified by their operating principles into shore-based optical wave gauges, pressure wave gauges, acoustic wave gauges, wave staffs, shipboard wave gauges, telemetric gravimetric wave gauges, as well as laser wave gauges that have not yet been standardized for mass production. This book focuses on three types of wave gauges: the visually observed shore-based optical wave gauge, which was the earliest type adopted by domestic stations and is still in service to date; the technologically advanced telemetric gravimetric wave gauge featuring continuous automatic recording; and the pressure wave gauge.

The national standard *Specifications for Oceanographic Survey—Part* 2: *Marine Hydrographic Observation* (GB/T 12763.2–2007) stipulates that wave observation items include sea state, wave type, wave direction, wave height, and wave period, with concurrent observations of wind speed, wind direction, and water depth required.

Sea state refers to the external characteristics of the sea surface under the action of the wind and is divided into 10 classes, which are determined from charts and tables against the sea surface, as shown in Table 5.1 (ESM 5.4).

Wave type refers to the morphological characteristics of waves, which are categorized into three classes: wind waves, swell waves, and mixed waves. In observation records, wind waves are denoted as F; swell waves as U; mixed waves dominated by wind waves as F/U; mixed waves dominated by swell waves as U/F; mixed waves where wind waves and swell waves coexist with comparable magnitudes as FU; and calm sea conditions (no waves) as C or a blank entry.

Wave direction, like wind direction, refers to the direction from which the waves are coming and is recorded in 16 bearings, see Table 5.1. When waves are present on the sea surface, but the direction of the waves is difficult to identify, mark the recorded result with an asterisk (*). When wind waves and swell waves coexist, the wave directions of both should be recorded separately.

Ocean waves exhibit irregularity, making it difficult for shore-based optical wave measurement instruments to continuously observe and record every wave in a wave train. To facilitate observations and satisfy practical application requirements, strict observation procedures are adopted (please refer to ESM 5.5).

5.4.2 Statistics of Wave Data Information

When collecting wave data, special attention should be paid to the geographical location of the wave measurement points and the types of wave meters.

The raw data from wave observations are entered into a daily report, analyzed and entered into a monthly report, and then statistically compiled into an annual report. The monthly reports are often consulted in the design of harbor construction projects.

For the monthly report recorded by shore-based optical wave meters, the contents include wind speed, wind direction, sea state (level), wave type, wind and wave direction, swell direction, period, wave height, maximum wave height, and water depth at the wave measurement point recorded four times a day for the month. It should be noted that the period in the report refers to the average period; wave height refers to the average of the first 10 large waves out of 100 consecutive waves, in descending order of wave height; and maximum wave height refers to the maximum of the 100 wave heights, which should be used with particular attention.

For telemetric wave data, various characteristic values of wave height and period are generally recorded, and if necessary, the shape of the wave profile or the frequency spectrum at the time can also be retrieved for study.

A diagram that depicts the occurrence frequency and magnitude of waves of different directions and grades in a specific sea area is defined as a wave rose diagram. To produce a wave rose diagram, wave observation data from a sea area over many years should be collated statistically, with wave heights or periods graded first, generally at intervals of 0.5–1.0 m for wave heights and 1 s for periods. To derive a reliable wave rose diagram, continuous observational data spanning 1–3 years is generally required, or alternatively, data from representative typical years may be selected.

There are various ways to draw a wave-height rose diagram, one of which is shown in Fig. 5.9, and a corresponding cycle frequency rose diagram can be drawn. Wave rose diagrams can also be drawn on a monthly or seasonal basis, depending on the construction and operation needs of the project. This wave-height rose diagram is read by direction and frequency. Each petal points to the wave approach direction, and its total radial length gives the occurrence frequency (read from the percentage scale). Where petals are subdivided by different hatching/line styles, the segments represent wave-height classes; longer segments indicate a higher share of that class, and the segments sum to the total frequency for that direction. The central value P_c denotes the proportion of calm (no-wave) conditions. This figure is schematic, and the wave-height legend is not fully shown; wave-height intervals should follow the legend in the original source.

5.4.3 *Statistical Distribution of Fixed Point Wave Heights and Periods*

If wave observations are conducted using a telemetric gravimetric wave gauge with continuous automatic recording functionality, it is possible to obtain a time-series record of the wave surface elevation at a fixed point on the sea surface. See Fig. 5.10 for an example.

This record is clearly different from a regular wave, and represents a complex irregular wave train, so the wave elements must be defined separately. The horizontal axis in Fig. 5.10 represents time and also the hydrostatic surface, while the vertical axis represents the vertical displacement of the wave surface relative to the hydrostatic surface. The intersection of the wave surface across the horizontal axis from top to bottom is called the lower span zero (e.g., points 3 and 6), while the intersection of the

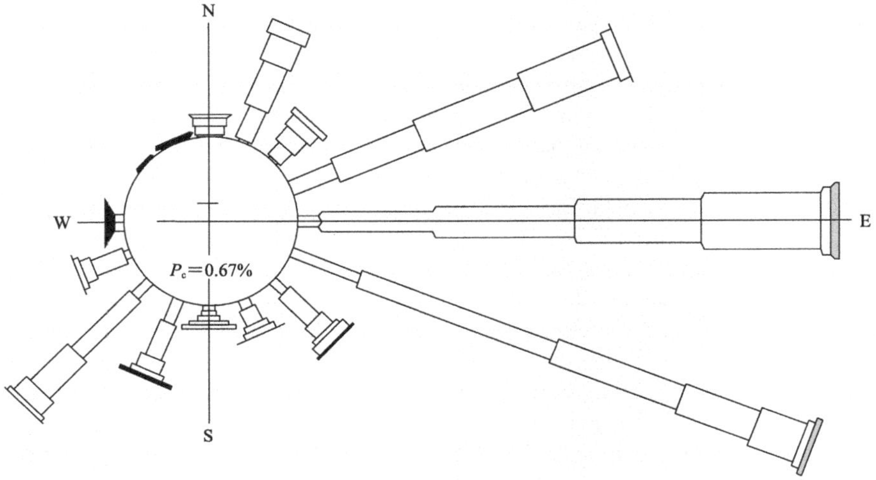

Fig. 5.9 Schematic wave-height rose diagram

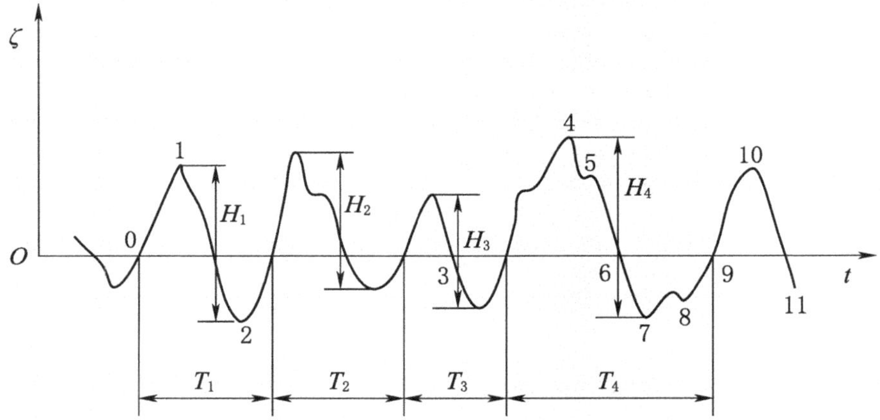

Fig. 5.10 Definition of irregular wave elements

wave surface across the horizontal axis from bottom to top is called the upper span zero (e.g., points 0 and 9). The time interval between two adjacent upper span zeroes (or lower span zeroes) is called the period, and the periods thus read in sequence are unequal, the average of which is called the mean period $\overline{T}$. The highest point of the wave surface is taken as the peak (e.g., point 4), and the lowest point is taken as the trough (e.g., point 7), while points 5 and 8 cannot be taken as the peak or trough. The vertical distance between the peak and trough is defined as the wave height. Statistics show that whether the wave height and period are defined by the upper or lower zero point, the average value is basically the same.

5.4.3.1 Statistical Distribution of Wave Heights and Periods

If a vertical profile of the wave surface is taken in the direction of wave propagation (wave direction line) at a fixed point in time, the curve of the wave surface with position is obtained. The same method can be used to read the wavelengths in sequence and calculate the average wavelength $\overline{L}$. The analysis of the measured data shows that the relationship between the mean period $\overline{T}$ and the mean wavelength $\overline{L}$ of irregular waves is different from that of regular waves, and can be approximated by the following equation:

$$\overline{L} = k\left(\frac{g\overline{T}^2}{2\pi}\right) \tag{5.14}$$

The coefficient k can be deduced from the fully developed Laumann spectrum, where $k = 2/3$.

Table 5.3 shows an example of a wave series of 100 consecutive waves obtained from actual observations at a station along the coast of China using the up-crossing zero method. It can be seen from this table that the individual waves in a wave series not only vary in size, but also in the order in which they appear. If another 100 consecutive waves are measured during this wind wave event (under the condition that the wind waves are in a steady state), the sequence and magnitude of the wave heights and periods of this wave train will differ from those of the previous one. Nevertheless, they both characterize the same sea state. Therefore, the distribution of wave magnitudes within a wave series is random and can thus be regarded as a random event.

The laws of random contention can be studied by means of mathematical statistics. In this case, the individual wave heights or periods of different sizes in the wave train are the random variables in the random event, while all the wave heights or periods during the steady state of the wind and waves are the total, and the 100 consecutive waves taken from them are the samples. We will use the samples to estimate the overall pattern of variation.

The first eigenvalue reflecting the pattern of random events is the sample mean, and the mean wave height $\overline{H}$ of the wave train in Table 5.3, or the mean period $\overline{T}$, is calculated using the following formula:

$$\overline{H} = \frac{1}{N}\sum_{i=1}^{N} H_i, \quad \overline{T} = \frac{1}{N}\sum_{i=1}^{N} T_i \tag{5.15}$$

where N is the total number of samples taken.

In this example, $N = 100$, $\overline{H} = 2.2$ m, the mean wave height reflects the overall size of the wave train. For example, the mean wave heights calculated from the records of two wind waves are different, so it is clear that the wave with the larger mean wave height is larger than the wave with the smaller mean wave height.

Table 5.3 A wave series of 100 consecutive waves

H/m	T/s	H/m	T/s	H/m	T/s	H/m	T/s	H/m	T/s
2.0	9.2	1.3	5.3	0.8	4.5	0.6	11.4	2.1	9.2
3.0	6.6	3.2	7.3	2.5	6.6	1.4	6.6	2.7	9.8
2.5	6.6	5.3	6.8	4.1	7.3	1.6	5.6	3.2	8.6
3.1	6.9	3.3	6.9	3.8	6.9	1.1	5.3	1.9	5.6
1.6	8.6	1.5	8.3	1.7	6.9	1.6	8.3	0.2	4.1
1.9	7.1	1.2	8.6	1.0	5.3	2.1	6.0	1.4	7.9
2.2	5.4	1.9	6.6	2.0	5.8	1.1	23	2.1	5.6
3.3	7.1	1.5	5.6	1.8	5.8	3.0	6.9	3.3	6.6
3.0	6.6	3.1	6.6	2.0	9.4	2.6	6.9	2.2	7.9
4.9	7.5	1.8	6.4	1.8	8.3	1.7	8.8	2.1	6.4
1.6	8.1	1.4	4.5	1.3	9.6	1.5	4.5	1.6	7.5
1.5	8.1	1.8	5.8	1.3	6.8	3.9	7.1	1.3	8.3
0.9	4.3	1.8	6.2	1.5	5.4	3.0	8.1	2.4	7.5
1.1	5.4	1.5	4.3	1.0	4.1	2.4	16.1	3.7	7.3
3.1	7.5	4.3	6.6	2.0	5.8	3.3	6.2	3.8	6.4
3.2	6.8	4.8	7.1	1.4	7.5	2.0	6.4	2.4	6.2
2.3	6.6	4.1	6.9	0.3	3.6	1.1	6.2	2.6	7.3
1.2	4.5	3.9	6.6	1.3	10.5	2.5	5.8	1.3	4.3
1.5	4.9	2.9	6.4	2.0	8.4	2.1	5.3	2.2	6.8
2.7	6.2	0.7	4.1	2.0	8.1	3.5	7.1	3.3	8.1

In order to study the distribution of wave heights or period sizes, frequency histograms must be drawn. For this purpose, the ratio of the wave heights $H_i/\overline{H}$ is calculated, which is defined as the wave height modulus ratio coefficient k_i and obviously, the modulus ratio coefficient of the mean wave height is equal to 1.0.

In order to draw a frequency histogram, divide the wave train into groups with appropriate spacing (i.e., class interval) $\Delta H/\overline{H}$ and calculate the wave height corresponding to the upper and lower limits of each spacing, as given in Column 1 and Column 2 of Table 5.4, in this case, take the class interval $\Delta H/\overline{H} = 0.2$. Count the number of occurrences of wave height in each group n_i, as given in Column 3 of Table 5.4, and divide n_i by N to obtain the class frequency of wave height in each group:

$$f_i = n_i/N \tag{5.16}$$

See Column 4 in Table 5.4, obviously $\Sigma f_i = 1.0$. The frequency of the wave heights varies from group to group, with more wave heights occurring around the modulus ratio coefficient equal to 1.0, i.e., the mean wave height, and less frequently at the ends.

Table 5.4 The wave height modulus ratio coefficient

$k_i = \frac{H_i}{\overline{H}}$	Wave height grouping H_i	Number of occurrences n_i	Class frequency f_i	Average frequency $f_i/\frac{\Delta H}{\overline{H}}$	Cumulative number of times $\sum n_i$	Cumulative frequency $F/\%$
1	2	3	4	5	6	7
2.4–2.2	$5.3 \geq H > 4.8$	2	0.02	0.10	2	2
2.2–2.0	$4.8 \geq H > 4.4$	1	0.01	0.05	3	3
2.0–1.8	$4.4 \geq H > 4.0$	3	0.03	0.15	6	6
1.8–1.6	$4.0 \geq H > 3.5$	5	0.05	0.25	11	11
1.6–1.4	$3.5 \geq H > 3.1$	9	0.09	0.45	20	20
1.4–1.2	$3.1 \geq H > 2.6$	10	0.10	0.50	30	30
1.2–1.0	$2.6 \geq H > 2.2$	9	0.09	0.45	39	39
1.0–0.8	$2.2 \geq H > 1.8$	18	0.18	0.90	57	57
0.8–0.6	$1.8 \geq H > 1.3$	23	0.23	1.15	80	80
0.6–0.4	$1.3 \geq H > 0.9$	14	0.14	0.70	94	94
0.4–0.2	$0.9 \geq H > 0.4$	4	0.04	0.20	98	98
0.2–0.0	$0.4 \geq H > 0$	2	0.02	0.10	100	100
		100	1.0			

In order to find the frequency at which one wave height may occur within each class interval, i.e., the average frequency, it is assumed that any wave height within a class interval has an equal chance of occurring, and that the total frequency of all wave heights within the class interval is equal to the class frequency. The average frequency equals the class frequency divided by the class interval, i.e., $f_i/\frac{\Delta H}{\overline{H}}$, as shown in Column 5 of Table 5.4.

The histogram of the average frequency of the wave height shown in Fig. 5.11a is plotted with the modulus ratio coefficient as the vertical coordinate and the average frequency as the horizontal coordinate. Obviously, the area of each rectangle on the graph is f_i of each group and the sum of each area is 1.0. When the class interval is close to 0, the histogram tends to curve, the area enclosed by the curve and the vertical axis is 1.0, then the horizontal coordinate is transformed into frequency density, and the curve is the frequency density curve. The wave height frequency density curve is characterized by small values at the ends, and large values in the middle, i.e., the wave height near the mean value has the most chances to appear.

In engineering design, it is often required to know the chance of a wave train being greater than or equal to a certain wave height, i.e., the cumulative frequency of a certain wave height, or it is required to know the value of the wave height for a given cumulative frequency. To do this, simply find the cumulative frequency F given in Column 7 of Table 5.4.

$$F = \frac{\sum n_i}{N} \times 100\% \tag{5.17}$$

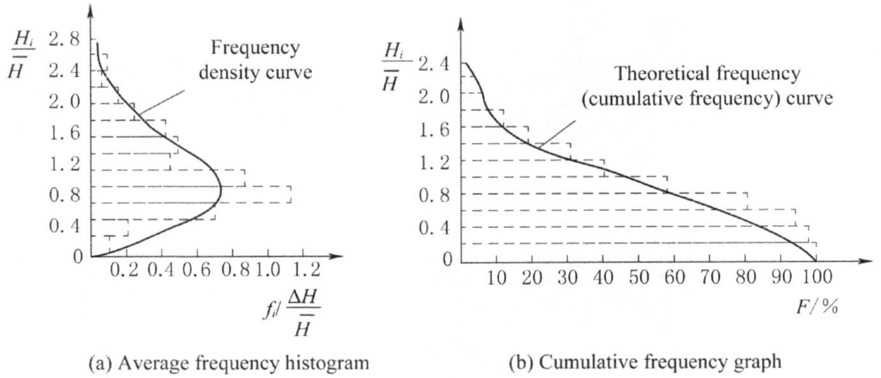

Fig. 5.11 Rayleigh distribution

The empirical cumulative frequency graph for wave height can be plotted according to Column 1 and Column 7 of Table 5.4, and the cumulative frequency curve is obtained when the class interval is close to 0, as shown in Fig. 5.11b.

We have now found the statistical distribution of the wave heights of a particular wave train using statistical methods. In the same way, the statistical distribution of periods and wavelengths can also be found. Figure 5.12 shows a histogram of the average frequency of the periods of the wave series in Table 5.3.

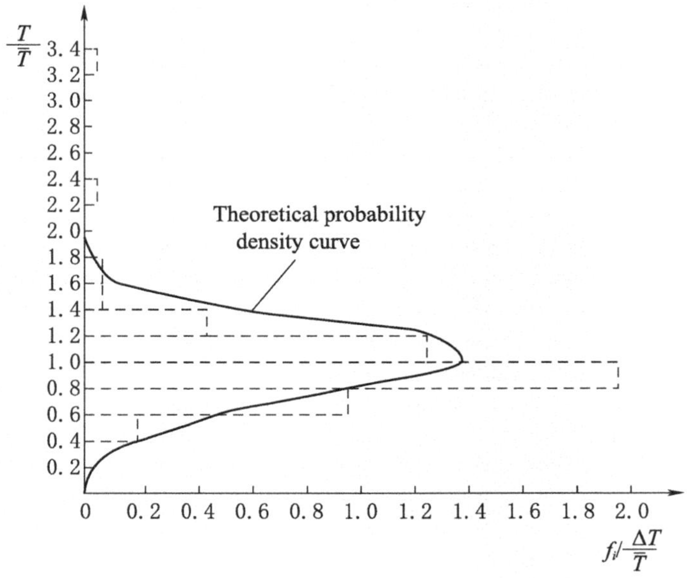

Fig. 5.12 Wave period frequency distribution histogram and theoretical probability density curve

If the number of waves taken is too small, the representativeness of the sample cannot be guaranteed, and the results will be unstable; conversely, if the number of waves taken is too large, the waves cannot be guaranteed to be in a stable and constant state, i.e., the consistency of the sampling cannot be guaranteed. This affects the reliability of the results.

5.4.3.2 Theoretical Distribution Functions for Fixed Point Wave Heights and Periods

The size of storms over the ocean varies greatly, and so do the scales of waves, which at first glance may not seem to follow a pattern. However, statistical analysis of a large number of wave observations has shown that the frequency distribution characteristics of the wave trains are very similar, regardless of the size of the waves. As a result, a theoretical study of the distribution of waves was undertaken to confirm the empirical statistics shown in Figs. 5.11 and 5.12.

1. Theoretical Distribution Function of Wave Height and Wavelength

The complex wave is assumed to be the result of the superposition of many simple harmonics of unequal amplitude, unequal frequency, and misplaced phase, as shown in Fig. 5.13. Curves 1 to 4 show four simple harmonics with different amplitudes, frequencies, and initial phases, and the result of their superposition is curve 5, which shows that the combined waves are much more complex than the original simple harmonics. It is therefore conceivable that the superposition of more simple harmonics will result in more complex fluctuations, each of which is called a constituent wave.

Based on these assumptions, the equation for the wavefront at a fixed point in the sea can be written as

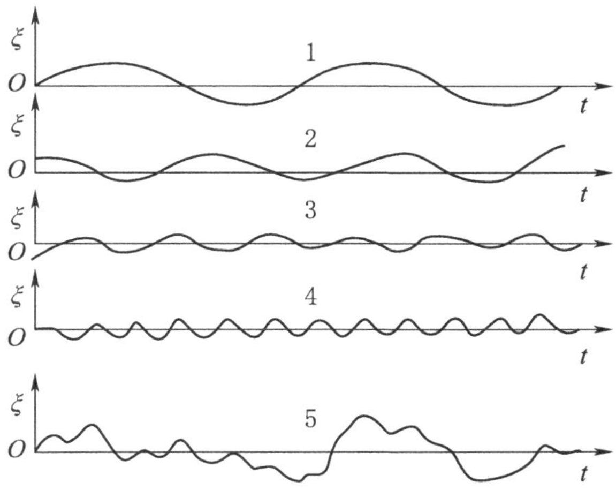

Fig. 5.13 Synthesis of simple harmonic waves. ζ—the height of the wave surface above the still water surface

$$\zeta(t) = \sum_{n=1}^{\infty} \zeta_n = \sum_{n=1}^{\infty} a_n \cos(\omega_n t + \varepsilon_n) \qquad (5.18)$$
$$\omega_n = 2\pi / T_n$$

where a_n is the amplitude of the constituent waves; ω_n is the circular frequency of the constituent wave; T_n is the period of the constituent wave; ε_n is the initial phase of the constituent wave; t is the time.

The initial phase of each constituent wave is arbitrary and is a random quantity that varies between 0 and 2π. Since the value of the cosine function at time t varies with the phase, if the phase is random, the value of the function is also a random quantity. ζ is thus the sum of an infinite number of random quantities, so it is also a random variable. Since the number of constituent waves can be considered to be nearly infinite, the amplitude of each constituent wave can be considered to be extremely small.

The Lyapunov theorem in probability theory states that if a random variable is the sum of a large number of independent and mutually uncorrelated random variables of very small magnitude, then the distribution of their sum tends to follow a normal distribution, regardless of their individual distribution laws. The probability density function of ζ can therefore be written as

$$f(\zeta) = \frac{1}{\sigma\sqrt{2\pi}} \exp\left(-\frac{\zeta^2}{2\sigma^2}\right) \qquad (5.19)$$

where σ is the standard deviation of the wave surface elevation. The other parameter of the normal distribution is the mean value, which for a standardized normal distribution is $\overline{\zeta} = 0$, i.e., the mean position of the wavefront is the still water level.

The empirical probability density curve is very similar to that described by Eq. (5.19), as shown by the data.

Longuet-Higgins used envelope theory to derive the probability density function for the wave amplitude a as

$$f(a) = \frac{a}{\sigma^2} \exp\left(-\frac{a^2}{2\sigma^2}\right) \qquad (5.20)$$

The relationship between the mean amplitude $\overline{a}$ and σ can be obtained using the method of moments about the origin.

$$\overline{a} = \int_0^{\infty} a f(a) \mathrm{d}a = \sqrt{\frac{\pi}{2}}\sigma \quad \text{or} \quad \sigma = \sqrt{\frac{2}{\pi}}\overline{a} \qquad (5.21)$$

Substituting $H = 2a, \overline{H} = 2\overline{a}$ into Eq. (5.21) yields the theoretical probability density function of the wave height:

$$f(H) = \frac{\pi H}{2\overline{H}^2} \exp\left[-\frac{\pi}{4}\left(\frac{H}{\overline{H}}\right)^2\right] \tag{5.22}$$

where $\overline{H}$ is the average wave height of the wave train.

This probability density function is known as the Rayleigh distribution, and the shape of its curve is shown in Fig. 5.11a. As can be seen from the figure, it fits well with the empirical frequency histogram, with a coefficient of variation of

$$C_v = \sigma_H/\overline{H} = M_2^{1/2}/\overline{H} = (4/\pi - 1)^{1/2} \approx 0.522 \tag{5.23}$$

It can also be seen from Fig. 5.11a that the wave height distribution is asymmetric, skewed, and positively skewed, with a skewness coefficient of

$$C_s = M_3/M_2^{3/2} = (2 - 6/\pi)/(4/\pi - 1)^{3/2} \approx 0.635 \tag{5.24}$$

where M_2, M_3 are the second and third central + + moments of $f(H)$, with the general formula:

$$M_n = \int_0^\infty (H - \overline{H})^n f(H)\mathrm{d}H \tag{5.25}$$

The wave height H_m corresponding to the maximum value of the probability density on the curve can be determined using the condition for extreme values $\mathrm{d}f(H)/\mathrm{d}H = 0$.

$$H_m = \sqrt{2/\pi}\overline{H} \approx 0.8\overline{H} \tag{5.26}$$

The above equation shows that the average wave height is not the wave height with the highest chance of occurrence, and that the wave height with the highest chance of occurrence is smaller than the average wave height. Integrating Eq. (5.22) from H to ∞ yields the cumulative frequency function $F(H)$, i.e., the distribution function, as shown in Fig. 5.11b.

$$F(H) = \int_H^\infty f(H)\mathrm{d}H = \exp\left[-\frac{\pi}{4}\left(\frac{H}{\overline{H}}\right)^2\right] \tag{5.27}$$

What is often required in engineering design is a wave height corresponding to a specified cumulative frequency $F(\%)$, for which Eq. (5.27) is rewritten as

$$H_F/\overline{H} = \left(\frac{4}{\pi}\ln\frac{1}{F}\right)^{1/2} \tag{5.28}$$

where H_F is the mean wave height of the highest fraction F of waves.

The test of the measured data proves that Eqs. (5.12) and (5.27) are only applicable to deep water areas. When the wave recession is gradually transmitted from deep water to shallow water, the distribution of wave height changes, and the water depth d starts to affect the distribution of wave height. If $H/d = H^*$, the wave height distribution can be expressed as

$$\left.\begin{array}{l} F(H) = \exp\left[-\frac{\pi}{4(1+H^*/\sqrt{2\pi})}\left(\frac{H}{\overline{H}}\right)^{\frac{2}{1-H^*}}\right] \\ H_F/\overline{H} = \left[\frac{4}{\pi}\left(1+\frac{H^*}{\sqrt{2\pi}}\right)\ln\frac{1}{F}\right]^{\frac{1-H^*}{2}} \end{array}\right\} \tag{5.29}$$

The distribution can be represented by the following distribution function. This distribution function is called the Gluhovsky distribution. When $H^* = 0$, i.e., when the water is very deep, Eq. (5.29) reduces to Eq. (5.27) in deep water.

Equation (5.29) is empirical and cannot be derived theoretically, but it is accepted because it agrees well with the interstitial data and converges with Eq. (5.27) at $H^* = 0$. It should be noted that the mean wave height $\overline{H}$ in Eqs. (5.5 and 5.29), is no longer the mean wave height in deep water, but the average wave height at the corresponding water depth d. Please refer to ESM 5.6 for more information.

In engineering practice, in addition to the wave height exceedance probability, it is sometimes necessary to determine the maximum wave height in a series of N waves. Obviously, this maximum wave height is also a random quantity due to the different sample sizes N taken. Based on probability theory, the mathematical expectation of the maximum wave height in deep water $H_{\max}$ can be approximated as a function of the number of waves N.

$$H_{\max}/\overline{H} = \frac{2}{\sqrt{\pi}}(\ln N)^{1/2} \tag{5.30}$$

In shallow water, $H_{\max}$ is also related to H^*.

$$H_{\max}/\overline{H} = \left[1 + \frac{H^*(1-H^*)}{2\sqrt{2\pi}}\right]\left[\frac{4}{\pi}\ln N\right]^{\frac{1-H^*}{2}} \tag{5.31}$$

Similarly, when $H^* = 0$, Eq. (5.31) reduces to Eq. (5.30).

The distribution of the wavelength L can be derived in the same way as the wave height distribution, which follows the same form as the wave height distribution and is consistent with the actual observations, i.e.,

$$f(L) = \frac{\pi}{2}\frac{L}{\overline{L}^2}\exp\left[-\frac{\pi}{4}\left(\frac{L}{\overline{L}}\right)^2\right] \tag{5.32}$$

2. Theoretical Distribution Function of the Cycle

With the distribution of wavelengths, the probability density function of the period can be derived using the relationship between period and wavelength in microwave Eq. (5.1) as follows:

$$f(T) = 4\Gamma^4\left(\frac{5}{4}\right)\frac{T^3}{\overline{T}^4}\exp\left[-\Gamma^4\left(\frac{5}{4}\right)\left(\frac{T}{\overline{T}}\right)^4\right] \tag{5.33}$$

where $\overline{T}$ is the average period; Γ is Gamma function, $\Gamma^4\left(\frac{5}{4}\right) = 0.675$, $4\Gamma^4\left(\frac{5}{4}\right) = 2.7$.

The distribution function of the period from Eq. (5.33) is

$$F(T) = \exp\left[-\Gamma^4\left(\frac{5}{4}\right)\left(\frac{T}{\overline{T}}\right)^4\right] \tag{5.34}$$

Similarly, the coefficient of variation $C_v \approx 0.283$ and the coefficient of skewness $C_s \approx 0$ of the period distribution can be derived from the second and third central moments of $f(T)$. The distribution of periods is more concentrated than the distribution of wave heights and is almost symmetrical, so the period with the highest chance of occurrence is the average period, i.e., $T_m \approx \overline{T}$. The curve shown in Fig. 5.12 is the theoretical probability density function plotted from Eq. (5.33), which is in general agreement with the measured histogram of the mean frequency of the periods.

Observations show that the mean period of waves is almost constant as they move from deep to shallow water. The distribution pattern of the shallow-water wave periods is independent of the water depth and varies very little, with the probability density function proposed by Grukhovski as

$$f(T) = \frac{\pi}{1.2}\frac{T^3}{\overline{T}^4}\exp\left[-\frac{\pi}{4.8}\left(\frac{T}{\overline{T}}\right)^4\right] \tag{5.35}$$

Its distribution function is

$$F(T) = \exp\left[-\frac{\pi}{4.8}\left(\frac{T}{\overline{T}}\right)^4\right] \tag{5.36}$$

Comparing Eq. (5.33) with Eq. (5.35), Eq. (5.34) with Eq. (5.36), it can be seen that the theoretical distributions of the deep water and shallow water cycles are also very close to each other, and the two sets of equations can be used interchangeably.

For convenience, Eq. (5.34) can be rewritten as

$$T/\overline{T} = \left[\frac{1}{\Gamma^4(5/4)} \ln \frac{1}{F}\right]^{1/4} \tag{5.37}$$

Table 5.5 shows the results of the calculation according to Eq. (5.37). Comparing Table 5.5 with Table 5.1 in ESM 5.6, it can be seen that the distribution of periods is much more concentrated than the distribution of wave heights.

3. Joint Probability Distribution of Wave Height and Period

In practice, it is often necessary to know the most probable average period corresponding to a certain cumulative rate wave height and the range of possible periods, which requires knowledge of the joint probability distribution of wave height and period.

It was previously thought that in random wave phenomena, wave height and period were mutually independent random variables and that the joint probability density could be expressed as $f(HT) = f(H)f(T)$ based on the assumption of independence of random variables. In practice, however, wave height and period are not completely independent, but are correlated to a certain extent, and this assumption is inconsistent with observations.

There are already some theoretically derived joint probability distribution functions for wave height and period, but they are not consistent with the joint probability distribution derived by the actual observations and are difficult to apply in practice. Therefore, it is desirable to derive the distribution law from the actual data.

Domestic and international measurements show that among the many parameters that affect the joint probability distribution of wave height and period, the correlation coefficient $R(H,T)$ between wave height and period has the most significant impact on their joint probability distribution. Figure 5.14 shows the joint probability distribution of wave height and period, plotted for different correlation coefficients $R(H,T)$ based on an extensive set of domestic and international field measurements, and further validated using domestic data.

As can be seen from Fig. 5.14, when the R value is small, the distribution of the periods corresponding to different wave heights is almost symmetrical with respect to the average period $\overline{T}$, and the period with the highest chance of occurrence is the average period.

Table 5.5 The results of the calculation according to Eq. (5.37)

F/%	$T/\overline{T}$	F/%	$T/\overline{T}$	F/%	$T/\overline{T}$
0.5	1.67	20	1.25	70	0.85
I	1.62	30	1.16	80	0.76
2	1.56	40	1.08	90	0.62
5	1.46	50	1.01	95	0.52
10	1.36	60	0.94		

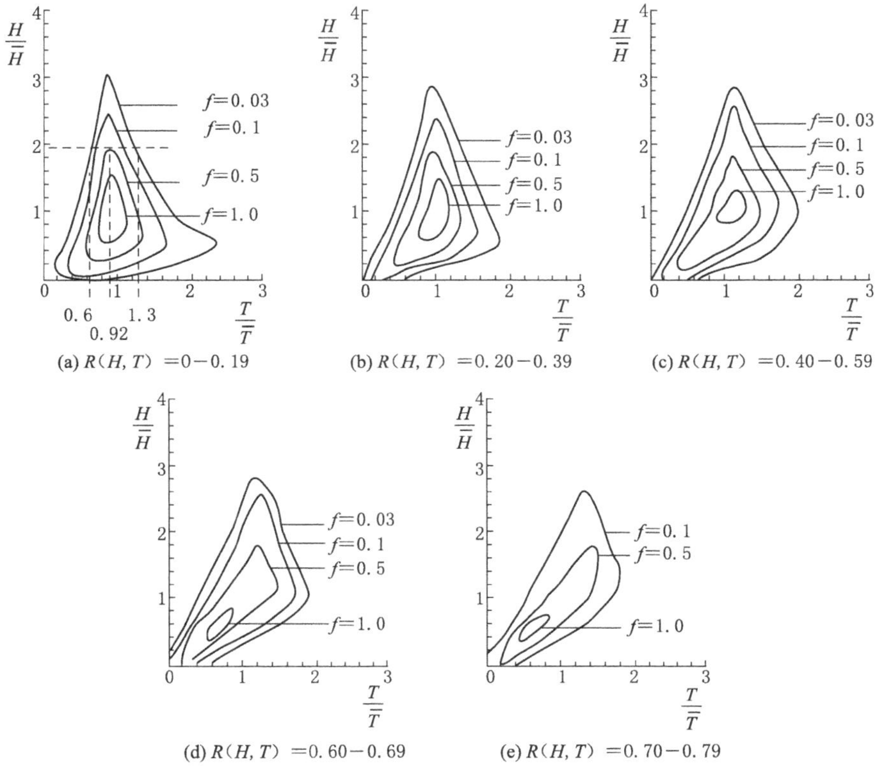

Fig. 5.14 Joint distribution of wave height and period

5.4.3.3 Two Characteristic Waves and Their Interrelationships

The state of a wave on the sea surface is usually described by the wave element as a characteristic quantity. The wave height is the most important of the wave elements, but as mentioned above, waves on the surface vary in size, so when we say what the height of a wave is, we should specify what the height means in a statistical sense, which is called the characteristic wave height. Another important wave element reflecting the size of a wave is the period, and again, when stating the size of the period, it is important to specify its statistical meaning, i.e., the characteristic period. The characteristic wave height and the characteristic period are collectively referred to as the characteristic wave.

1. Two ways of representing characteristic wave heights

There are currently two ways of representing characteristic wave heights.

(1) One is to select the wave height corresponding to a certain cumulative frequency in the wave train as the characteristic wave height, i.e., H_F (e.g., $H_{1\%}$, $H_{5\%}$, etc.).

The characteristic wave height reflects the probability of a given wave height occurring in the wave train, e.g., $H_{1\%}$ indicates a 1% probability of occurrence of a wave height greater than or equal to the wave height in the wave train, and so on. As mentioned above, wave heights obey certain distribution rules.

For example, the known wave height of any cumulative frequency in the wave train can be converted to a wave height of any required cumulative frequency. The average wave height $\overline{H}$ is the conversion medium between the various cumulative wave heights, so it is one of the most commonly used characteristic wave heights.

(2) Another method is to use the average wave height of the so-called partial large waves as the characteristic wave height, which is defined as follows: arrange the wave heights in the wave train in descending order of magnitude. The average value of the largest P fraction of these wave heights is called the average wave height of the P-partial large waves, denoted as H_P. For example, the average height of the largest 100 waves in a sequence of 1000 consecutive waves, or the largest 10 waves in a sequence of 100 consecutive waves, is referred to as the average wave height of 1/10 large waves ($H_{1/10}$), also known as the significant wave height. Similarly, the average height of the largest 1/3 of waves in a wave train, denoted as $H_{1/3}$, is also called the significant wave height. When expressed mathematically, the average wave height of partial large waves can be written as

$$H_P = \frac{1}{NP}\sum_{i=1}^{NP} H_i \tag{5.38}$$

where N is the total number of waves in the wave train.

If $P = 1/10$ or $1/3$, then the above equation yields

$$H_{1/10} = \frac{10}{N}\sum_{i=1}^{N/10} H_i \text{ and } H_{1/3} = \frac{3}{N}\sum_{i=1}^{N/3} H_i \tag{5.39}$$

2. Interrelationship of the two characteristic wave heights

Since wave heights obey a certain distribution law, there should be a relationship between the mean wave height of various partial large waves and between them and the wave height exceedance probability, which can be deduced theoretically. If $P = F$, for example, for deep-water waves, using Eq. (5.28), and $x = H_F/\overline{H}$, the modulus ratio coefficient can be written as

$$\frac{H_P}{\overline{H}} = \frac{1}{F}\int_0^F \left(\frac{4}{\pi}\ln\frac{1}{F}\right)^{1/2} \mathrm{d}F = \frac{1}{F}\int_0^F x\mathrm{d}F \tag{5.40}$$

Using the principle of integration by parts and substituting F with the expression given in Equation (5.27), the above equation eventually becomes

$$\frac{H_P}{\overline{H}} = \frac{H_F}{\overline{H}} + \frac{1}{F}\left[1 - \operatorname{erf}\left(\ln\frac{1}{F}\right)^{1/2}\right] \tag{5.41}$$

where erf(x) is the error function (or probability function), and its value varies between 0 and 1.

Since the second term in Eq. (5.41) is a positive decimal, if $P = F$, then $H_P/\overline{H}$ is greater than $H_F/\overline{H}$, i.e., $H_{1/10} > H_{10\%}$, $H_{1/3} > H_{33\%}$, etc.

The same principle can be used to derive the average wave height of some large waves in shallow water by substituting Eq. (5.29) for the cumulative. Since the equation is complex, the calculation results are listed in Table 5.1 (ESM 5.7) for convenience. Meanwhile, the conversion curves between the commonly used cumulative frequency wave heights at different water depths are also plotted for reference. In Fig. 5.15, the top horizontal coordinate is $\overline{H}/d$ and the bottom horizontal coordinates are $H_{4\%}/d$ and $H_{13\%}/d$.

The characteristic period corresponding to the wave height of a given cumulative frequency is based on a large amount of measured data, $T_{H_{4\%}} \approx T_{H_{1/10}} \approx 1.31\overline{T}, T_{H_{13\%}} \approx T_{H_{1/3}} \approx 1.15\overline{T}$.

The conditional distribution of the period at a given cumulative frequency wave height can also be theoretically derived from the respective distribution of wave height and period. This distribution also follows a normal distribution, which shows that for large wave heights the distribution of wave periods is more concentrated around the mean period, while for small wave heights the distribution of wave periods is more dispersed, i.e., for large wave heights there are generally no wave periods much larger than the mean period, while for small wave heights it is possible to encounter wave periods much larger than the mean period, which is particularly important in engineering design.

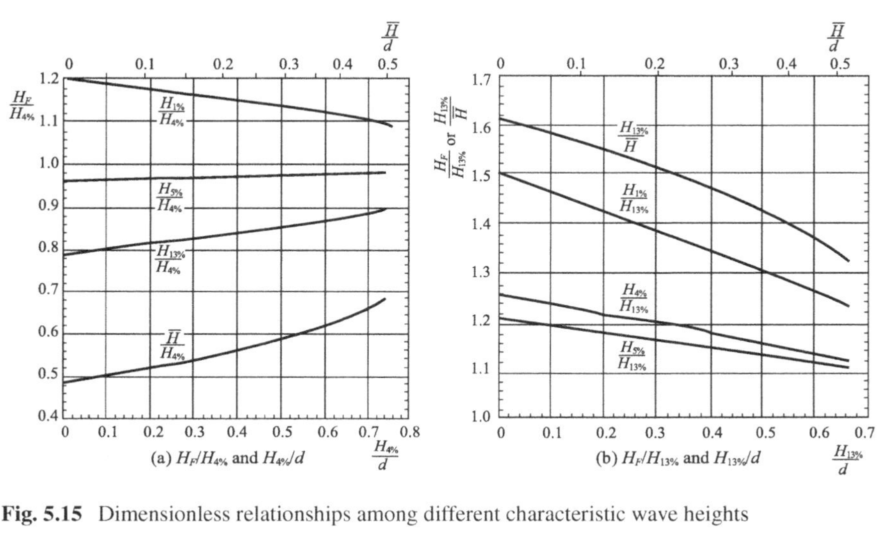

Fig. 5.15 Dimensionless relationships among different characteristic wave heights

5.4.4 Design Wave Criteria and Guidelines

5.4.4.1 Design Wave Criteria

The term design wave refers to the wave elements used in the design of various types of buildings and their components in harbor engineering. Different design wave criteria can be used for different design projects. Because of the high cost of harbor projects, design waves are chosen to take into account the adverse conditions that may arise during the reverse period of use, to ensure the safety of the building, and to satisfy the principle of economy.

The criteria for the design of waves are twofold: the design of the wave's return period and the design of the wave's cumulative frequency.

The return period of a design wave is the average number of years that a particular wave train occurs, and represents the long-term (in decades) statistical distribution of wave claims. The return period criterion for design waves mainly reflects the age and importance of the building: the method of deriving design waves for the specified return period criterion is described in Sect. 5.6.

The cumulative frequency criterion for design waves refers to the probability of occurrence of design wave elements in an irregular wave train on the actual sea surface, and represents the short-term (in minutes) statistical distribution of wave elements. During this statistical period, the sea surface is considered to be in a constant state, or the average state of the wave elements does not change over time. The cumulative frequency criterion for wave design mainly reflects the different nature of wave action on different types of buildings.

5.4.4.2 Wave Height Cumulative Frequency Criteria for Wave Design in China's Harbor Projects

In order to simplify practical work, it is common in design to replace the actual irregular waves on the sea surface with an equivalent regular wave train and to use the wave element of a representative characteristic wave in the irregular wave train as the wave element in this equivalent regular wave train. This representative characteristic wave height should be selected as the design wave height of a cumulative frequency from the irregular wave train according to a certain criterion. This criterion is the cumulative frequency criterion for the design wave.

The *Code of Hydrology for Harbour and Waterway* (JTS 145–2015) (2022) stipulates that the wave height of the design wave should be based on different building types using different wave height cumulative frequency criteria, as shown in Table 5.6.

The design wave height cumulative frequency criteria in the table are mainly based on practical experience from domestic and international projects and the results of model tests with irregular waves. This criterion reflects that different types of structures and different components of the same structure exhibit varying responses to

Table 5.6 Different wave height cumulative frequency criteria

Building type	Department location	Calculation content	Wave height cumulative frequency $F/\%$
Vertical wall and pile bent	Superstructure, wall or pile foundations	Strength and stability	1
	Bed and substrate blocks	Stability	5
Ramped	Chest wall or embankment top squares	Strength and stability components	1
	Face protection blocks or blocks	Stability	13
	Bottom protection blocks	Stability	13

Note $F = 5\%$ is preferred for low mean wave height to water depth ratios ($\overline{H}/d < 0.3$)

wave action. In general, vertical-wall and pier-and-column (including piled) structures are more sensitive to waves, and individual large waves in the wave train can significantly impact structural safety. For sloping structures, experience shows that damage occurs gradually. Local damage caused by individual large waves is not decisive and does not affect overall stability, and such damage is also easier to repair. The cumulative frequency criteria also differ for different components of the same structure. For critical components whose failure would compromise the safety of the entire structure, a higher design cumulative frequency criterion should be adopted; conversely, a lower criterion may be used for non-critical components.

If the wave height is greater than the ultimate wave height in shallow water, the ultimate wave height is used as the design wave height for all types of buildings.

For the design wave used to check the water surface berthing stability within the harbor area with shelter, the cumulative frequency of wave height shall be taken as 4%.

With regard to the design period and the design wavelength, as mentioned above, the periods corresponding to the wave heights of different cumulative frequencies are different. However, they are only statistical rules, and the theoretical relationships are not yet fully established. For the sake of simplicity, the design period is specified in the specification using the average period $\overline{T}$, the design wavelength is determined according to the regular wave formulas (5.1) and (5.2) by $\overline{T}$. In practical calculations, Eq. (5.2) has to be solved via iterative computation, which is rather inconvenient to use. Therefore, the value is usually obtained by referring to Schedule 6 based on the deep-water wavelength $L_0 = \frac{g\overline{T}_0^2}{2\pi}$ and local water depth d.

5.5 Basic Knowledge of Wave Spectra

5.5.1 The Concept of Wave Spectrum

A wave is a very complex fluid motion that cannot be adequately described by simple regular fluctuations, and the statistical laws described in the previous section only reflect the patterns of its external manifestations, not its internal structure. In order to further investigate waves, the concept of wave spectra is introduced here. The wave spectrum reveals the internal structure of waves and their energy distribution. The internal structure of waves and their external behavior are closely linked, as they describe the same phenomenon from two different perspectives. It is also because of the complexity of the internal structure of waves that their appearance is so varied.

As discussed above, complex waves can be understood as the superposition of many (theoretically infinite) simple harmonics of unequal amplitude and frequency, with misaligned initial phases. Longuet-Higgins therefore proposed the wave surface equation for a fixed point on the sea surface as Eq. (5.42):

$$\zeta(t) = \sum_{n=1}^{\infty} \zeta_n = \sum_{n=1}^{\infty} a_n \cos(\omega_n t + \varepsilon_n) \tag{5.42}$$

Since the initial phase of the constituent wave ε_n is a random quantity varying between 0 and 2π and uniformly distributed, its probability density function is

$$f(\varepsilon) = \frac{1}{2\pi} \tag{5.43}$$

According to the wave energy formula of the micro-amplitude wave theory, the energy of each component wave in the vertical water column (from the fluctuating water surface to the bottom of the water) can be determined as

$$E_n = \frac{1}{2}\rho g a_n^2 \tag{5.44}$$

where ρ is the density of water.

The wave energy in any frequency interval (ω-$\omega + \delta\omega$) is

$$E_n = \frac{1}{2}\rho g \sum_{\omega}^{\omega+\delta\omega} a_n^2 \tag{5.45}$$

This energy is clearly proportional to the frequency interval $\delta\omega$, and is related to the energy of the constituent waves of the frequencies in this interval, such that

$$S(\omega)\delta\omega = \frac{1}{2}\sum_{\omega}^{\omega+\delta\omega} a_n^2 \tag{5.46}$$

where $S(\omega)$ is a function of frequency.

In order to illustrate the physical meaning of the function $S(\omega)$, examine the quantity

$$S(\omega)\rho g = \frac{1}{\delta\omega}\sum_{\omega}^{\omega+\delta\omega} \frac{1}{2}\rho g a_n^2 \tag{5.47}$$

The summation symbol represents the sum of the energy of the constituent waves with frequencies in the range from ω to $\omega + \delta\omega$, divided by $\delta\omega$ to obtain the average energy in this interval. $S(\omega)$ is clearly proportional to Eq. (5.47) and $S(\omega)$ multiplied by the constant ρg is the energy provided by the component waves with frequencies in the interval between ω and $\omega+\delta\omega$, i.e., it represents the distribution of wave energy with respect to the component wave frequencies. If $\delta\omega = 1$, Eq. (5.47) represents the energy per unit frequency interval, thus $S(\omega)$ is proportional to the energy per unit frequency interval, i. e., the energy density. $S(\omega)$ is also called the energy spectrum, because it reflects the energy density, and it is also called the frequency spectrum, because it gives the distribution of energy relative to frequency.

Obviously, integrating the wave spectrum over the entire frequency range from 0 to ∞ results in the total wave energy, i.e.,

$$E = \rho g \int_0^{\infty} S(\omega)\mathrm{d}\omega \tag{5.48}$$

If the frequency ω is set as the horizontal coordinate and $S(\omega)$ is set as the vertical coordinate, the distribution of wave energy with respect to frequency can be plotted, see Fig. 5.16.

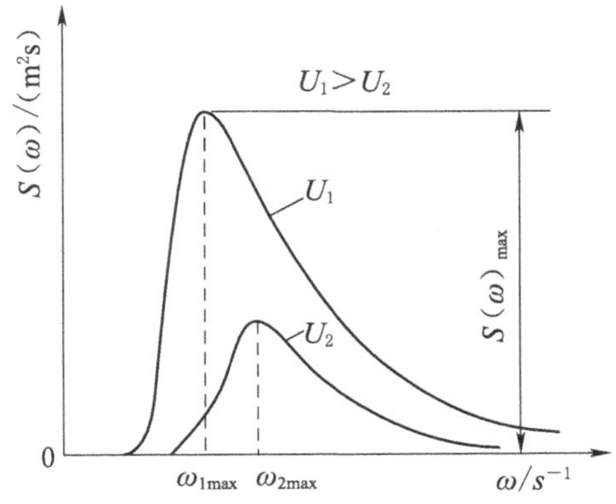

Fig. 5.16 Schematic of wave spectrum

As can be seen from Fig. 5.16, the spectrum around $\omega = 0$ is very small. As ω increases, the spectrum density increases sharply to a maximum value of $S(\omega)_{\max}$, which corresponds to the frequency called the peak frequency of the spectrum $\omega_{\max}$. As ω continues increasing, the spectrum density decreases rapidly and then decreases slowly, and finally tends to zero. Theoretically, $S(\omega)$ is distributed over the range $\omega = 0$ to ∞, but in practice, a significant portion is concentrated within a relatively narrow frequency band. In other words, among the component waves that make up ocean waves, those with very low or very high frequencies contribute little energy; the main part of the energy is concentrated within a certain frequency range.

The characteristics of this internal structure of waves are also reflected in the external properties of ocean waves: the waves we observe directly are irregular in height and length, yet those with very low frequencies (very long periods) or very high frequencies (very short periods) are rare. Most waves have frequencies within a relatively narrow band. Clearly, the area enclosed by the wave spectrum curve and the horizontal axis in Fig. 5.16 is proportional to the total wave energy E. As the wind speed U increases, the total area under the spectrum curve increases, leading to a rise in total energy. The frequency range covered by the significant portion of the spectrum also expands, and this significant portion shifts toward lower frequencies with increasing wind speed. Consequently, the wave periods of the significant portion in wind-generated waves increase accordingly.

From Eq. (5.44), if the units of length and time are taken as m and s, then the units of spectral density are $m^2 \cdot s$.

Theoretically, it can be shown that the variance of the wavefront vertical coordinate $\zeta(t)$, σ^2 is proportional to the total energy of the fluctuations, i.e.,

$$\sigma^2 = \int_0^{\infty} S(\omega)\mathrm{d}\omega = m_0, \ r = 0 \tag{5.49}$$

where m_0 is the zero-order moment of the spectrum; r is the number of order moments.

The general formula for each order moment of the spectrum is

$$m_r = \int_0^{\infty} \omega^r S(\omega)\mathrm{d}\omega \tag{5.50}$$

The variance of the wave surface σ^2 is a statistical eigenvalue statistically derived from the appearance of the wave, while m_0 is an eigenvalue of the wave's internal energy structure, thus, the external appearance of the wave is linked to its internal energy structure through Eq. (5.49).

Sometimes the horizontal coordinate of the spectrum ω (also called the circular frequency) is replaced by the frequency $f\,(f = 1/T)$, when $S(\omega)$ is converted to the spectrum $S(f)$ represented by f. Since $\omega = 2\pi/T = 2\pi f$, the ratio of the two frequencies $\omega/f = 2\pi$ is a constant. Let the interval of $\delta\omega$ plus the corresponding f be δf and the energy is equal in both intervals, i.e., $S(f)\delta f = S(\omega)\delta\omega$, so that $S(f) = S(\omega)\delta\omega/(\delta f)$, and therefore

$$S(f) = 2\pi S(\omega) \text{ or } S(\omega) = S(f)/(2\pi) \tag{5.51}$$

This concept is extremely important and must not be confused in calculations.

The emergence of wave spectra has opened up new avenues for wave research and has brought about a huge reform in the design of harbor projects in terms of wave and wave force projections, laying an important foundation for reliability studies, changing the old method of designing on the basis of a single ideal wave rule and further considering the dynamic effects of complex wave series on buildings to ensure safety and reduce costs. In addition, wave spectra are also important for the study of sediment movement and coastal evolution in the coastal zone.

5.5.2 Examples of Wave Spectra Forms

To date, many wind and wave spectra have been proposed both domestically and internationally, the most common general form of which is

$$S(\omega) = \frac{A}{\omega^p} \exp\left(-\frac{B}{\omega^q}\right) \tag{5.52}$$

where A, B are both covariates containing either the wind elements (wind speed, wind time, wind distance) or the wave elements (wave height, period); p, q are indices, $p = 4$–6, $q = 2$–4.

A number of wind and wave spectra, which are currently widely used globally in engineering practice are provided in ESM 5.8.

5.5.3 The Relationship Between the Spectrum and the Wave Elements

The representation of waves in terms of wave spectra and the representation of waves in terms of statistically characterized wave elements are two different perspectives on the same wave phenomenon and it should therefore be possible to establish an interrelationship between the two representations. As mentioned above, the characteristic values of the wave elements of a gauge wave, such as mean wave height $\overline{H}$, mean period $\overline{T}$, significant wave height H_s, effective period T_s, or other characteristic wave elements, can be substituted into the various wave spectrum formulas above to obtain a wave spectrum. On the other hand, if a measured empirical spectrum (derived from the measured wave surface process lines) is available, the corresponding characteristic wave heights and periods can also be derived. The calculation of the various characteristic wave elements from the spectra is important both in application and in theory.

Using the wave surface expression $\zeta(t) = \sum_{n=1}^{\infty} \zeta_n = \sum_{n=1}^{\infty} a_n \cos(\omega_n t + \varepsilon_n)$ and related formulas, the relationship between various characteristic wave heights and the zero-order moment m_0 of the spectrum can be determined.

$$\left.\begin{aligned} \overline{H} &= 2.506\sqrt{m_0} \\ H_{\mathrm{rms}} &= 2.828\sqrt{m_0} \\ H_{1/3} &= 4.004\sqrt{m_0} \\ H_{1/10} &= 5.091\sqrt{m_0} \\ H_{1\%} &= 6.069\sqrt{m_0} \end{aligned}\right\} \tag{5.53}$$

where H_{rms} is the root-mean-square (RMS) wave height, defined as the square root of the sum of squared wave heights over the record.

5.6 Calculation of Design Waves from Wave Observations

As pointed out in Sect. 5.4, the definition of a design wave consists of two aspects: the return period criterion of the design wave and the wave train cumulative frequency criterion of the design wave. This section also discusses the distribution of wave parameters within a wave train, that is, the short-term distribution of wave parameters, and introduces the wave train cumulative frequency criteria for design waves in China's harbor engineering, as shown in Table 5.6. The return period criterion of the design wave refers to selecting an appropriate wave train as the design basis in port engineering design; only after the design wave train is selected can the design wave parameters be finally determined according to the wave train cumulative frequency criterion, which is why it is necessary to study the long-term distribution of ocean waves, the task of this section.

A particular wave train (which is represented by one of the characteristic waves, e.g., $H_{1/10}$, $H_{s,}$ or $\overline{\mathrm{H}}$) comes from the wave field created by a particular gale. In a year, several gales may occur in the waters of the port construction area, each with different wind conditions, and therefore different wave trains, and the number and intensity of gales occurring each year are also different and unpredictable, i.e., the occurrence of a wave train is a random event. This is the long-term (multi-year) distribution of waves. At this stage, the characteristic wave parameter used to represent each wave train is treated as a random variable. The collection of all wave trains that have occurred in the study area in the past, together with those that may occur in the future, constitutes the population, which may be regarded as infinite. The wave trains recorded over a finite number of years form a sample. On the basis of this sample, the occurrence frequency and cumulative frequency of a given wave train can be estimated; this constitutes the wave return-period problem.

There are two basic methods of deriving design waves for a certain return period in port engineering design. The first method is applicable when there is an oceanographic observation station in or near the harbor construction area, with accumulated long-term wave observation data. Based on these data, a series of wave sequences represented by characteristic parameters, such as $H_{1/10}$ and T, can be obtained. These sequences form a statistical series, and their distribution patterns are analyzed using mathematical statistics. Based on the derived distribution patterns, the characteristic waves likely to occur over many future years can be inferred. In the second method, if there is no marine hydrographic observation station in the port area or its vicinity, the local or neighboring meteorological stations' wind observations or weather charts can be used to estimate the wave parameters at that time based on the relationship between wind and wave parameters. The first method is then used to determine the design waves for a certain return period. Obviously, the first method is a direct projection method, which is more reliable, while the second method is an indirect projection method, which has a larger error, and therefore, a temporary wave observation station should be set up in the port sea area to accumulate information to test the design wave parameters projected by meteorological data. This section first introduces the first method of design wave projection for a certain return period.

5.6.1 *Criteria for the Return Period of Waves for the Design of China's Harbor Projects*

According to the practical experience of China's port construction over the years and foreign design standards, the technical specifications of China's port engineering specify different design wave return period standards for various types of buildings, see Table 5.7.

According to Table 5.7, when carrying out strength and stability calculations for vertical-wall, pile bent, and ramped structures, the wave return period for Class I and Class II structures should be taken as 50 years. Differences in safety requirements among building classes are reflected in the structural importance factors used for strength and stability calculations. For non-critical structures such as sloping berms, where failure would not result in significant consequences, a design wave return period of 25 years may be adopted.

Table 5.7 Design wave return periods for different types of structures

Building type	Building class	Return period	Building type	Building class	Return period
Vertical wall	I, II, III	Once in 50 years	Ramped	I, II	Once in 50 years
Pile bent	I, II, III	Once in 50 years		III	Once in 25 years

Additionally, the code states that when the maximum wave height observed historically exceeds the 50-year event, the design standard may be appropriately elevated for important structures, such as lighthouses or other buildings whose failure would lead to particularly severe consequences. Where necessary, verification based on the observed maximum wave height may also be considered.

For the design wave height to check the berthing stability in port waters, the return period may be determined based on operational requirements, but should not exceed a 2-year return period.

The return period for design waves represents the long-term (in decades) statistical distribution pattern of wave elements, and therefore, the return period criterion for design waves primarily reflects the service life and importance of the building. It should be pointed out that the return period is an average concept: adopting a 50-year return period design wave does not guarantee that no larger waves will occur within 50 years after the structure is completed. In order to investigate this issue, a probabilistic approach can be used to derive the probability that a building designed for a certain return period design wave may suffer damage during its service life, called the hazard rate (or encounter probability), and expressed as q.

Assuming that the return period of a structure's design wave is T years, the annual cumulative frequency is $P = 1/T$. If $T = 50$ years, then $P = 2\%$, corresponding to a design wave height of $H_{P=2\%}$. This means there is a 2% annual probability that a wave height greater than or equal to $H_{P=2\%}$ will occur, i.e., a 2% annual probability that the structure will sustain damage. Conversely, the annual probability that no such wave event will occur is, by the complementary probability theorem, $1-P = 98\%$.

Assuming that the design life of the building is m years, the probability of a wave being less than H_P in m years is f, called the safety rate, according to the probability multiplication theorem.

$$f = (1 - P)^m \tag{5.54}$$

Thus, by the inverse event theorem, the hazard rate q should be

$$q = 1 - (1 - P)^m \tag{5.55}$$

As can be seen from the above equation, the hazard rate q is not only related to the return period, but also to the age of the building. The longer the service life, the higher the probability of hazard under a specified return period; conversely, the longer the return period adopted, the safer the structure will be for a given service life.

For example, if the return period is one in 50 years, i.e., $P = 2\%$, and the life of the building, m, is also 50 years, the hazard rate of the building during its life, calculated by Eq. (5.55), is 63%, which can also be deduced from the theoretical approximation:

$$q = \lim_{m \to \infty} \left[1 - \left(1 - \frac{1}{m} \right)^m \right] \approx 1 - e^{-1} = 0.632 = 63.2\% \tag{5.56}$$

The above equation shows that a building designed only for the design waves derived from the return periods shown in Table 5.7, with a service life equal to the return period, has a constant 63% probability of damage during its service life; in other words, the safety probability is only 37%.

If we aim to account for both the service life of the project (project life) m and the risk of damage to the project during the service life, we can use Eq. (5.55) to back-calculate the required design wave return period T.

$$T = 1/P = \left[1 - (1 - q)^{1/m} \right]^{-1} \tag{5.57}$$

Using the above equation, the design wave return period required for the specified risk of damage to occur during the life of the project can then be calculated, as shown in Table 5.8.

As can be seen from Table 5.8, for a building with a desired project life of 50 years, the design wave return period is only 11 years if the hazard rate is taken to be 99%, i.e., basic damage to the building, whereas if safety is required to be guaranteed with 90% certainty for 50 years, a wave event of 475 years should be taken for the design, so the difference can be seen to be significant.

Clearly, it is reasonable to consider both the service life and the potential risk of damage to the building when designing, but the choice of a reasonable mix of the two is a complex issue that will involve the economic benefits of the project, damage losses, and other factors. Table 5.7, given in the current Port Engineering Code, only considers the return period as a criterion for engineering design, and it can be seen that the results thus obtained are compromised.

Finally, it should be noted that if, in the course of its use, a building encounters waves far larger than the design waves for the specified return period, it should be considered reasonable for the building to be damaged as a result, since it is not economic to require absolute safety for the building. However, in order to prepare for exceptional catastrophic weather, wave forecasting should be used, and all measures should be taken to protect buildings from damage or minimize damage as far as possible in the case of large waves that are to come.

Table 5.8 Wave return period required for specified risks to arise during the project's service life

Tolerable risk rate of failure	Desired project life/ years				Tolerable hazard rate of failure	Desired project life/ years			
	10	25	50	100		10	25	50	100
0.10	95	238	475	950	0.75	8	19	37	73
0.25	35	87	174	348	0.99	2.7	6	11	22
0.50	15	37	73	145					

5.6.2 *Calculating Design Waves Using Long-Term Wave Measurement Data*

The technical specifications of China's port engineering stipulate that when there is longer-term (more than 20 years) wave measurement data in the location of the port construction project or its adjacent sea area, a series of annual maximum wave heights (expressed as a certain characteristic wave) in different directions can be used for analysis to determine the design waves of different return periods in each direction, and the specific steps are as follows.

5.6.2.1 Identifying the Main Wave Directions

Before determining the design waves, long-term wind observations from local meteorological stations should first be analyzed. Wind-direction, wind-frequency, and maximum wind-speed rose diagrams should then be constructed following the method described in Sect. 5.3, in order to determine the prevailing local wind direction.

In order to study the main incoming wave directions in the port construction area, it is necessary to analyze the wind rose map in conjunction with the geographical location of the port site, as shown in the example in Fig. 5.17.

Figure 5.17 plots the wind rose for port site *A*. The solid line is the maximum wind speed rose, and the dashed line indicates the wind frequency rose. It can be seen from this diagram that the local north direction is a strong and constant wind direction, and the north direction of the harbor site is open to the sea, and high winds can create large waves. Although the headland is slightly sheltered, waves can still bypass the headland and reach the harbor area, so the southeast direction is also a major wave direction; the other east and northeast directions, although the maximum

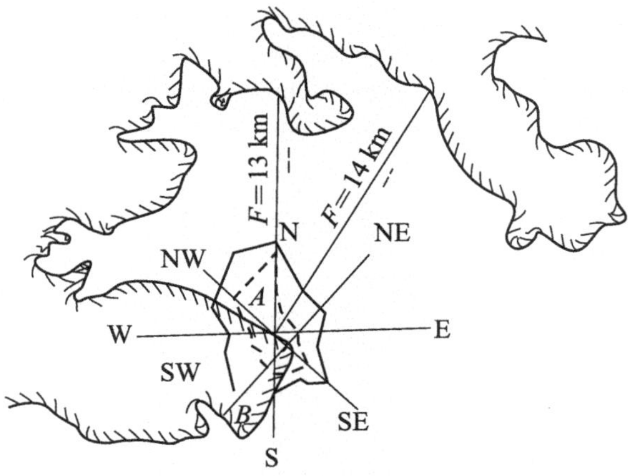

Fig. 5.17 Determination of the main incoming wave directions in the sites

wind speed and frequency are not as high as the north or southeast directions, but the sea is also more open in these two directions. If the site is changed to point *B*, and the wind rose chart remains largely unchanged, the main wave directions for site *B* should be east, southeast, south, and south west.

5.6.2.2 Collecting and Collating Wave Observations

When compiling wave data from adjacent oceanographic stations, priority should be given to examining the geographical conditions of the stations and comparing them with those of the harbor site, i.e., evaluating the directional applicability of the wave data from the stations. As shown in Fig. 5.17, the harbor site is at point *A*, the oceanographic station is at point *B*, and the wave buoy is deployed on the sea surface approximately 0.5 km southward offshore, directly in front of the observation station. It is clear that for east and southeast swells, the site is subjected to waves similar to those at the float, so the information from the station can be used to extrapolate waves from these two directions at the site. However, the waves at the float are not representative of the north and northeast swells, as they are limited by the topography to very small offshore swells and cannot be used to analyze the north and northeast swells at site *A*. For site *A*, the north and northeast swell can only be indirectly derived from meteorological data, see the next section.

In addition, the difference in water depth between the harbor site and the wave measurement float needs to be considered. It is known that the sounding buoy is located at the -30 m isobath, while the harbor breakwater is located at the -10 m isobath. Therefore, the east and southeast facing waves derived from the measured data at the wave measurement float are only representative of the waves at a depth of -30 m outside the harbor site, and these waves have to be deformed in shallow water (see Sect. 5.8) to obtain the waves at -10 m at the harbor site.

For sampling, to ensure the independence of samples, the annual maximum sampling method is employed to construct the dataset. Specifically, the largest wave event occurring each year in a given direction is selected as a sample from the observational records. This is justified because, for wave data, the correlation between annual maxima in consecutive years is extremely weak, and they can thus be treated as statistically independent.

In selecting the sample series, the wave direction should also be taken into account. In wave observation records, wave directions are recorded in 16 directions. However, when analyzing waves in a particular direction statistically, waves within one azimuth (i.e., 22.5°) of that direction are often counted as waves in that direction, because wave direction observations are not very accurate. However, if statistical analysis is required at 45-degree azimuth intervals, each wave direction must be aggregated into a single directional bin separated by 22.5°, based on the geographical location of the site, and each directional bin must be used uniquely without duplication.

Finally, it is critical to ensure the independence of the sample data. For instance, if a single storm wave event begins in late December of a given year and persists into early January of the following year, only the maximum wave height from this event can be selected as the extreme value sample for one year. No other large wave heights from the same event may be used as a sample for the other year, as these values originate from the same wave event and are therefore statistically dependent rather than independent.

It is important to emphasize that the purpose of selecting a sample is to deduce the nature of the general population, and that the correct selection of the sample is crucial in determining the reliability of the general pattern sought. For this reason, when consulting the observation records, attention should also be paid to the consistency of the data over the years, i.e., whether there have been any changes in the natural conditions that formed the data, e.g., whether any artificial structures have been constructed near the wave measurement floats during these years; whether the data have been measured using the same instrument, the same method at the same location; whether there have been any changes in the observation norms; whether the floats have been moved in position, etc. If any of the aforementioned conditions have changed significantly and cannot be corrected or converted, the affected data cannot be aggregated into a unified dataset. Instead, the dataset should be subdivided into segments or the affected years should be excluded before compiling the frequency analysis series.

Only through the above series of verification exercises will the compiled annual extreme series be more reliable and ready for the next step in the frequency analysis process.

5.6.2.3 Calculation of Design Waves Using the Frequency Analysis Method

The basic principles of frequency analysis have been described in detail in Chap. 3, and this section will only further illustrate them with examples in relation to the characteristics of the projected design wave.

(1) Calculation of empirical cumulative frequencies. Table 5.9 presents the series of annual maximum significant wave heights $H_{1/10}$ from the south direction (including merged data from SSW and SSE directions) over a 23-year period at an observation station, sorted in descending order. The third column in the table shows the modulus ratio coefficient, and the seventh column shows the empirical cumulative frequency P of the annual maximum wave heights, which is calculated using the mathematical expectation formula for cumulative frequency.

The empirical cumulative frequencies (solid dots) of wave heights from Table 5.9 are plotted on the normal probability grid shown in Fig. 5.18. To avoid confusion between the calculated cumulative frequencies and the cumulative frequencies of wave highs within the wave train described in Sect. 5.4, the return period is usually used instead of the cumulative frequency of annual extreme wave heights that may occur over a number of years, i.e., $T(\text{year}) = 1/P$ (see the upper horizontal coordinate in Fig. 5.18).

Table 5.9 Calculation results of the relationship between wave height and empirical cumulative frequency (waves from SSW and SSE directions)

m	$H_{1/10}$/m	k_i	k_i-1	$(k_i-1)^2$	$(k_i-1)^3$	$P=m/(n+1)\times 100\%/\%$
1	2	3	4	5	6	7
1	3.5	1.926	0.926	0.857	0.794	4.2
2	3.5	1.926	0.926	0.857	0.794	8.3
3	3.1	1.706	0.706	0.498	0.352	12.5
4	3.0	1.651	0.651	0.424	0.276	16.7
5	2.4	1.321	0.321	0.103	0.033	20.8
6	2.3	1.266	0.266	0.071	0.019	25.0
7	2.1	1.156	0.156	0.024	0.004	29.2
8	2.0	1.101	0.101	0.010	0.001	33.3
9	1.8	0.991	–0.009	0	0	37.5
10	1.6	0.881	–0.119	0.014	–0.002	41.7
11	1.6	0.881	–0.119	0.014	–0.002	45.8
12	1.6	0.881	–0.119	0.014	–0.002	50.0
13	1.5	0.826	–0.174	0.030	–0.005	54.2
14	1.5	0.826	–0.174	0.030	–0.005	58.3
15	1.4	0.771	–0.229	0.052	–0.012	62.5
16	1.3	0.715	–0.285	0.081	–0.023	66.7
17	1.3	0.715	–0.285	0.081	–0.023	70.8
18	1.3	0.715	–0.285	0.081	–0.023	75.0
19	1.3	0.715	–0.285	0.081	–0.023	79.2
20	1.1	0.605	–0.395	0.156	–0.062	83.3
21	1.1	0.605	–0.395	0.156	–0.062	87.5
22	0.9	0.495	–0.505	0.255	–0.129	91.7
23	0.6	0.330	–0.670	0.449	–0.301	95.8
$\sum$	41.79			4.348	1.599	

Note $\overline{H_{1/10}} = 1.817$ m

As can be seen from the graph, it is difficult to sketch a smooth distribution curve from these empirical frequency points, and the short age of the data makes it even more difficult to extrapolate the 50-year or 25-year wave that is required for engineering design, for which theoretical cumulative frequency curves are required.

(2) Selection of theoretical cumulative frequency curves. The purpose of selecting the theoretical cumulative frequency curve is to fit the empirical cumulative frequency points and thus achieve extrapolation.

The *Code of Hydrology for Harbour and Waterway* (JTS 145-2015) (2022) states that for the theoretical frequency curve of the annual extreme wave height and its corresponding period, the Pearson Type III distribution is generally used. This is

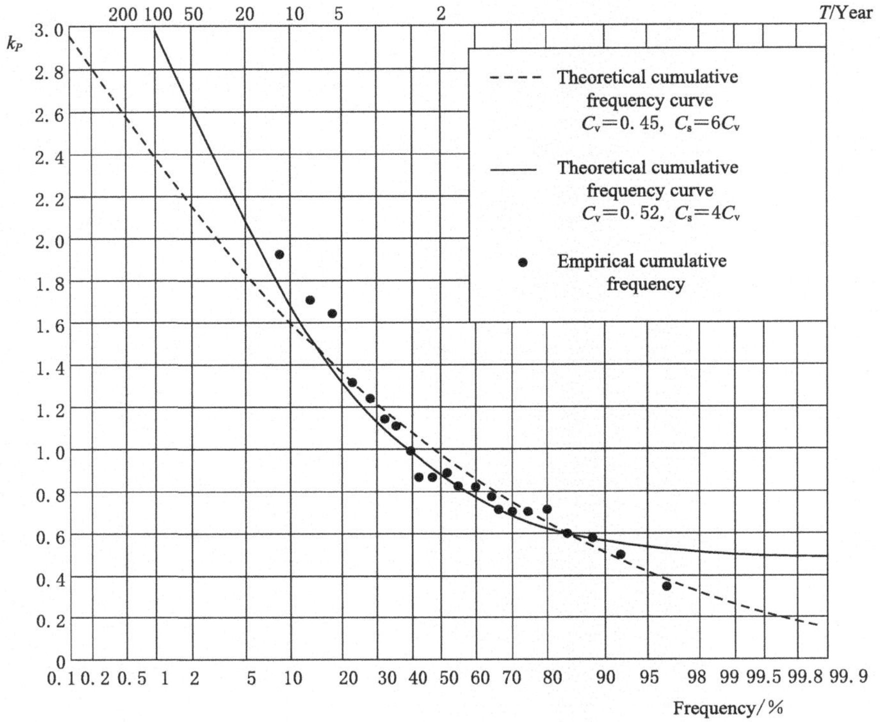

Fig. 5.18 Frequency distribution of S-directional annual extremes of $H_{1/10}$ at an observatory (Pearson Type III distribution)

because domestic experience shows that Pearson Type III curves generally fit well with the empirical frequency points of wave heights and periods, and secondly, such lines are usually used in the analysis of annual maximum wind speeds and flood frequencies of rivers, for which considerable experience has been accumulated. However, given that the statistical parameters derived from the actual sample data are subject to a certain margin of error, these parameters can be adjusted by means of the curve-fitting method during the calculation process. This approach thus entails a degree of arbitrariness, which becomes particularly pronounced when a small number of extreme values are present within the dataset. Therefore, it is necessary to explore more rational distribution models. The specification suggests that, when available, the principle of the best fit to the measured empirical cumulative frequency points can be used in parallel with other theoretical cumulative frequency curves, such as the Type I extreme value distribution, the log-normal distribution and the Weibull distribution, for comparison in many ways, which is conducive to the final determination of the design waves for different return periods.

1) Pearson Type III distribution. The principles and line characteristics of the Pearson Type III cumulative frequency curve have been described in Chap. 3

and are not repeated in this section. When the curve line type is used for the curve-fitting calculation, the three parameters included in the curve must first be found, namely the mean $\overline{H}$, $C_{v,}$ and C_s.

The mean value of the series shown in Table 5.9 $\overline{H_{1/10}} = 1.817$ m, which represents the average characteristic value of the annual maximum wave heights $H_{1/10}$ over the multi-year observation period. To investigate the degree of dispersion of the random variable relative to $\overline{H_{1/10}}$, the coefficient of variation should be calculated, which is the ratio of the standard deviation of the series to its mean, according to Eq. (5.58):

$$C_v = \sqrt{\frac{\sum_{i=1}^{n} (k_i - 1)^2}{n - 1}} \tag{5.58}$$

In this case, $C_v = 0.445$. To indicate whether the distribution of the random variable is symmetrical around the mean, calculate the skewness coefficient C_s, according to the following equation:

$$C_s = \frac{\sum_{i=1}^{n} (k_i - 1)^3}{(n - 1)C_v^3} \tag{5.59}$$

In this case, $C_s = 0.907$, which is positively skewed. Since the skewness coefficient is derived from the cube of deviations, and deviations are subject to sampling errors, the reliability of the C_s value is generally poor. Therefore, it is typically adopted as a multiple of C_v in engineering practice.

Due to the short age of the measured data, not only C_s, but also C_v, and even $\overline{H}$ are subject considerable errors. According to the error formula of Pearson Type III curves, the absolute error of the mean value $\sigma_H = 0.17$ m; the absolute error of the deviation coefficient $\sigma_{Cv} = 0.072$, so the value range of each parameter is $\overline{H} = (1.817 \pm 0.17)$ m, $C_v = 0.445 \pm 0.072$, and C_s is taken to be 2 to 6 times that of $C_{v.}$

Take $\overline{H} = 1.817$ m, $C_v = 0.45$, $C_s = 2C_v$, check the integral table of Pearson Type III in Schedule 2, and get k_P for the cumulative frequency, listed in Table 5.10. The data in the table are plotted in Fig. 5.17, and are linked into a smooth curve (dashed line) for comparison with the empirical cumulative frequency point. For this reason, the three parameters must be adjusted appropriately. As the average error is small, no adjustment is generally made. Take $\overline{H} = 1.817$ m, $C_v = 0.52$, $C_s = 4C_v$, k_P are looked up and also listed in the lower part of Table 5.10, which are then used to plot the theoretical frequency curve. Clearly, the adjusted theoretical cumulative frequency curve matches the actual observed data better and can represent the overall distribution of the sample, while serving as an extrapolation.

Thus, according to Table 5.10, the significant wave heights $H_{1/10}$ for the 50-year and 25-year events in the S direction at the observatory are calculated as follows:

Return period of 50 years: $H_{1/10} = 2.53 \times 1.817 \approx 4.6$ m, noted as $(H_{1/10})_{2\%}$.

Return period of 25 years: $H_{1/10} = 2.18 \times 1.817 \approx 4.0$ m, noted as $(H_{1/10})_{4\%}$.

Table 5.10 Calculation of k_P values for Pearson Type III distribution under different parameters

$\overline{H} = 1.817$ m, $C_v = 0.45$, $C_s = 2C_v$													
P/%	0.1	1	2	3	5	10	25	50	75	90	95	97	99
k_P	2.98	2.33	2.13	2.00	1.84	1.60	1.26	0.93	0.67	0.48	0.39	0.34	0.25
$\overline{H} = 1.817$ m, $C_v = 0.52$, $C_s = 4C_v$													
k_P	4.13	2.90	2.53	2.32	2.04	1.67	1.20	0.84	0.63	0.55	0.53	0.51	0.50

Once the design wave height $H_{1/10}$ for a given return period is available, the design wave heights for the other required cumulative frequencies can be obtained from Fig. 5.15 or from Table 5.1 in ESM 5.6. At this point it should be noted that in Table 5.1 of ESM 5.6 $\overline{H}$ refers to the mean value of the irregular wave series observed at one time (i.e., the mean value of the short-term distribution of wave heights), which is currently unknown, whereas the mean value in the above frequency calculation refers to the mean value of the annual maximum wave heights (expressed as $H_{1/10}$) in the series, i.e., the mean value of the long-term distribution of wave heights, which are very different and must not be confused.

The wave heights of the various cumulative frequencies for the two return periods required for the S-direction design of a station are thus obtained, as shown in Table 5.11.

The treatment of very large values is also described in detail in Chap. 3, and this section will only analyze it further in relation to the example of waves.

A station was hit by a typhoon in 1985, and there was a very large wave. As the wave occurred at night, and the wave-measuring float was swallowed by the wave, the sea state was not recorded. A post-event wind and wave hindcast using weather charts and a field survey resulted in a maximum wave height of $H_{1/10} \approx 6.8$ m. A survey of historical data also showed that this typhoon had not been seen for 120 to 130 years. For this reason, the return period N of the mega-value was assumed to be 100 and 150 years, respectively, as shown in Table 5.12.

The parameters are calculated according to the following formulas:

$$\overline{H_N} = \frac{1}{N}\left(H_N + \frac{N-1}{n-1}\sum_{i=1}^{n} H_i\right) \tag{5.60}$$

Table 5.11 Design wave heights for different cumulative frequencies under two return periods

Return period	$H_{1\%}$/m	$H_{4\%}$/m	$H_{5\%}$/m	$H_{13\%}$/m	$\overline{H}$/m
50 years	5.5	4.6	4.4	3.6	2.3
25 years	4.8	4.0	3.8	3.2	2.0

Table 5.12 Frequency analysis of wave heights with historical extreme event for 100-year and 150-year return periods

m	$H_{1/10}$ /m	$N = 100$				$N = 150$			
		k_i	$k_i - 1$	$(k_i - 1)^2$	$P = m/(n+1) \times 100\%$	k_i	$k_i - 1$	$(k_i - 1)^2$	$P = m/(n+1) \times 100\%$
2	5.0	1.963	0.963	0.927	8.3	1.973	0.973	0.947	8.3
3	3.7	1.452	0.452	0.204	12.5	1.461	0.461	0.213	12.5
4	3.6	1.413	0.413	0.171	16.7	1.421	0.421	0.177	16.7
5	3.0	1.178	0.178	0.032	20.8	1.184	0.184	0.034	20.8
6	3.0	1.178	0.178	0.032	25.0	1.184	0.184	0.034	25.0
7	2.8	1.099	0.099	0.010	29.2	1.105	0.105	0.011	29.2
8	2.8	1.099	0.099	0.010	33.3	1.105	0.105	0.011	33.3
9	2.6	1.021	0.021	0	37.5	1.026	0.026	0.001	37.5
10	2.5	0.981	−0.019	0	41.7	0.987	−0.013	0	41.7
11	2.4	0.942	−0.058	0.003	45.8	0.947	−0.053	0.003	45.8
12	2.3	0.903	−0.097	0.009	50.0	0.908	−0.092	0.008	50.0
13	2.3	0.903	−0.097	0.009	54.2	0.908	−0.092	0.008	54.2
14	2.1	0.824	−0.176	0.031	58.3	0.829	−0.171	0.029	58.3
15	2.1	0.824	−0.176	0.031	62.5	0.829	−0.171	0.029	62.5
16	2.1	0.824	−0.176	0.031	66.7	0.829	−0.171	0.029	66.7
17	2.1	0.824	−0.176	0.031	70.8	0.829	−0.171	0.029	70.8
18	2.0	0.785	−0.215	0.046	75.0	0.789	−0.211	0.045	75.0

(continued)

Table 5.12 (continued)

m	$H_{1/10}$ /m	$N = 100$				$N = 150$			
		k_i	$k_i - 1$	$(k_i - 1)^2$	$P = m/(n + 1) \times 100\%$	k_i	$k_i - 1$	$(k_i - 1)^2$	$P = m/(n + 1) \times 100\%$
19	2.0	0.785	–0.215	0.046	79.2	0.789	–0.211	0.045	79.2
20	2.0	0.785	–0.215	0.046	83.3	0.789	–0.211	0.045	83.3
21	1.8	0.707	–0.293	0.086	87.5	0.711	–0.289	0.084	87.5
22	1.6	0.628	–0.372	0.138	91.7	0.632	-0.368	0.135	91.7
23	1.3	0.510	–0.49	0.24	95.8	0.513	-0.487	0.237	95.8
$\sum_{i=2}^{23}$	55.1			2.133				2.154	
	$H_{N/10}$/m	k_N	$k_N - 1$	$(k_N - 1)^2$	$P = 1/(N + 1) \times 100\%$	k_N	$k_N - 1$	$(k_N - 1)^2$	$P = 1/(N + 1) \times 100\%$
1	6.8	2.669	1.669	2.786	0.99	2.685	1.685	2.839	0.66

$$C_{vN} = \sqrt{\frac{1}{N-1}\left[(k_N - 1)^2 + \frac{N-1}{n-1}\sum_{i=1}^{n}(k_i - 1)^2\right]} \quad (5.61)$$

The empirical frequency of mega values is calculated according to the following formula:

$$P_N = \frac{1}{N+1} \times 100\% \quad (5.62)$$

Then, $\overline{H_{100}} = 2.548$ m, $C_{v100} = 0.354$, and $\overline{H_{150}} = 2.533\,m$, $C_{v150} = 0.342$.

The above calculations show that whether N is taken for 100 or 150 years, the $\overline{H_N}$ and C_{vN} values are very small and within the margin of error, which means that the distribution parameters do not vary much despite the large difference in the selection of N. This suggests that historical investigations of extreme wave heights, although they do not provide precise return periods, are sufficient for frequency analysis as long as a reasonably appropriate range is available.

With the two parameters, the theoretical cumulative frequency curve is still obtained using the curve-fitting method, as shown in Fig. 5.19.

From Fig. 5.19, we can obtain design wave parameters for this station in the E direction:

50-year return period wave height: $H_{1/10} = 2.548 \times 2.39 = 6.1$ m.

25-year return period wave height: $H_{1/10} = 2.548 \times 2.02 = 5.1$ m.

It can be seen that a large wave of 50 years is smaller than a very large wave of $H_{1/10} = 6.8$ m that has historically occurred. In this case, this very large wave should be considered for calibration calculations for particularly important buildings.

If the annual maximum wave height series does not fit well with Pearson Type III curves, consider fitting by other line types, which are recommended by the code as follows.

2) Log-normal distribution. The basic principle of the log-normal distribution is that when a random variable is taken logarithmically, its distribution obeys a normal distribution. At this point, if the empirical cumulative frequency points are plotted on log-normal probability grid paper, the distribution is approximately linear, so the linear equation can be calculated using the least squares regression. Using the linear extrapolation, the design wave can be found for any return period, as shown in Chap. 3.

Table 5.13 presents the series of annual maximum wave heights $H_{1/10}$ in the SW direction at a certain observation station. This series contains an extraordinary value of $H_{1/10} = 5.8$ m, which is treated as an extreme wave height with an assigned return period of 150 years. The $H_{1/10}$ in the table is plotted on the log-normal probability grid in Fig. 5.20 according to the corresponding empirical cumulative frequencies, where the horizontal coordinates are the sub-grids of the normal distribution and the vertical coordinates are the logarithmic coordinates. From the graph, it can be seen that the empirical points exhibit an approximately linear distribution, which

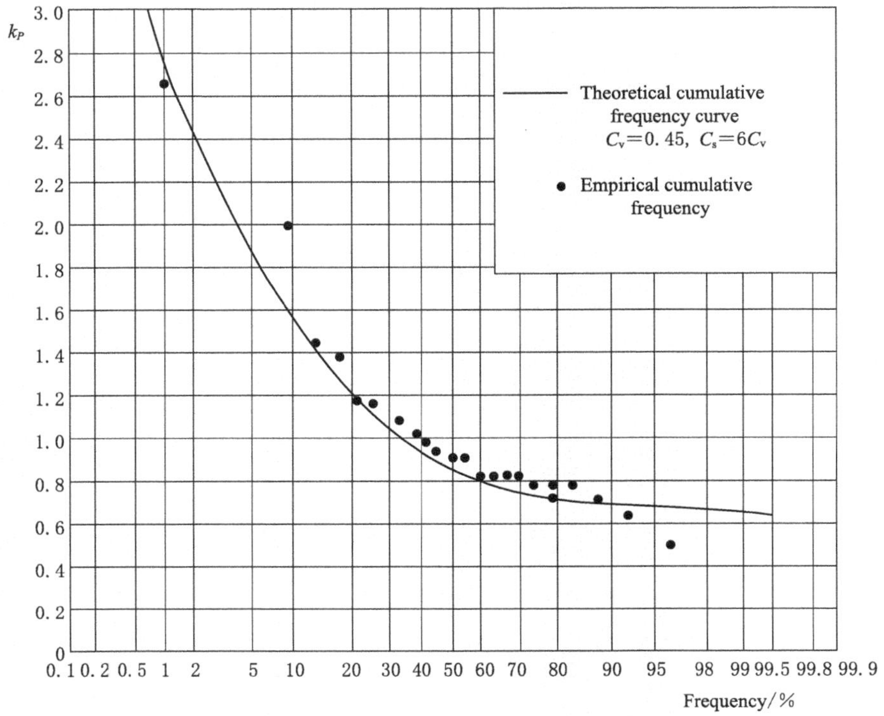

Fig. 5.19 Frequency distribution of the maximum annual wave height $H_{1/10}$ (E-direction) at a station (Pearson Type III distribution)

can be fitted by the law of log-normal distribution, i.e., the two parameters A and B determined using the least squares method in the format $y = A + Bx$. According to Chap. 4, the linear equation should be

$$\lg H_i = \overline{\lg H} + \sigma_{\lg H} t_i \tag{5.63}$$

where $\overline{\lg H}$ is the mean value of the wave height value after taking the logarithm, i.e., $\overline{\lg H} = \frac{1}{n}\sum_{i=1}^{n} \lg H_i$, which is the intercept of the line; $\sigma_{\lg H}$ is the standard deviation of the wave height when taken logarithmically, which is the slope of the line; t_i is the lower limit of integration of the normal distribution, since the empirical cumulative frequency P is known and the corresponding value of t_P can be obtained using the normal distribution integration table, as shown in Column 5 of Table 5.13.

Taking Column 3 and Column 5 of Table 5.13 for correlation analysis, we get two parameters $\overline{\lg H_N} = 0.4462$, i.e., $\overline{H_N} = 2.79$ m, and $\sigma_{\lg H} = 0.1176$, so the equation of the line is

Table 5.13 Log-normal distribution parameters for wave height frequency analysis

m	$H_{1/10}$/m	lg$H_{1/10}$	$P = m/(n+1) \times 100\%/\%$	t_m
1	2	3	4	5
2	4.0	0.602	8.3	1.385
3	3.7	0.568	12.5	1.150
4	3.5	0.544	16.7	0.966
5	3.5	0.544	20.8	0.813
6	3.3	0.519	25.0	0.675
7	3.2	0.505	29.2	0.548
8	3.1	0.491	33.3	0.432
9	3.1	0.491	37.5	0.319
10	3.0	0.477	41.7	0.209
11	2.9	0.462	45.8	0.106
12	2.9	0.462	50.0	0
13	2.7	0.431	54.2	−0.106
14	2.6	0.415	58.3	−0.209
15	2.6	0.415	62.5	−0.319
16	2.5	0.398	66.7	−0.432
17	2.3	0.362	70.8	−0.548
18	2.3	0.362	75.0	−0.675
19	2.2	0.342	79.2	−0.813
20	2.2	0.342	83.3	−0.966
21	2.1	0.322	87.5	−1.150
22	1.9	0.279	91.7	−1.385
23	1.8	0.255	95.8	−1.728
	H_N	lg$H_{N/10}$	$P = 1/(N+1) \times 100\%$	t_N
1	5.8	0.763	0.66	2.478

$$\lg H_{1/10} = 0.4462 + 0.1176t \tag{5.64}$$

The correlation coefficient for the straight line, $R = 0.99$, is extremely close. The straight line is plotted in Fig. 5.20 and extrapolated to show that it does fit well with the empirical points.

Using the straight lines or related equations in Fig. 5.20, the 50-year and 25-year design wave heights $H_{1/10}$ can be found, as shown in Table 5.14.

In the absence of this dedicated probability grid paper, it is equally possible to explore fitting empirical frequency points to a lognormal distribution using ordinary coordinate paper. In this case, it is sufficient to take the vertical coordinate y as lg$H$ and the horizontal coordinate x as the lower limit of integration t of the normal distribution.

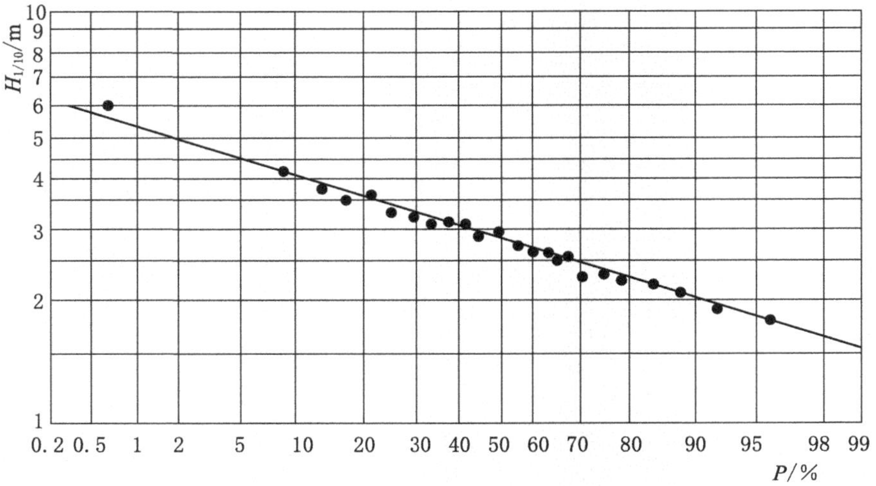

Fig. 5.20 Frequency distribution (log-normal) of the maximum annual wave height $H_{1/10}$ in the SW direction at a station

Table 5.14 Design wave heights $H_{1/10}$ for 50-year and 25-year return periods derived from a log-normal distribution

Return period T	$P = 1/T$	t_P	$\lg H_{1/10}$	$H_{1/10}$/m
50 years	2% (0.02)	2.054	0.69	4.9
25 years	4% (0.04)	1.751	0.65	4.5

The log-normal distribution is easy to calculate and avoids the human error caused when fitting the line compared to the Pearson Type III curve. It has been tested for many years with a series of examples of extreme wave heights and works well enough to be tried out.

3) Type I extreme value distribution (Gumbel distribution). The basic principle of the first type of extreme value distribution is that the original distribution of a random variable is known to be exponential, and multiple trials are made under the same conditions to obtain n groups of random variables, and the maximum or minimum value of each group (called the extreme value) is taken to form a series whose empirical cumulative frequency points are distributed in a straight line on the Gumbel probability grid paper. Because the new series of random variables is made up of extreme values, it is called the extreme value distribution, which is derived entirely from theory and is the true theoretical distribution, as detailed in Chap. 3.

According to the Type I extreme value distribution, the cumulative frequency P of the random variable x is

$$P(x \geq x_\alpha) = 1 - \exp[-\exp(-y)] \tag{5.65}$$

where $y = \alpha(x\text{-}u)$, α and u are two parameters.

This leads to the equation of the line:

$$x = u + \frac{1}{\alpha}y \tag{5.66}$$

The number of extremes is known, and a certain extremum and its corresponding empirical cumulative frequency P are given, the corresponding y_m can be calculated, which depends only on n and the order m of the extremes from largest to smallest:

$$y_m = -\ln\left[-\ln\left(1 - \frac{m}{n+1}\right)\right] \tag{5.67}$$

The resulting series is formed, and the least squares method can be used to solve for α and μ, or the approach in Chap. 3 can be used; the equation of the line is written as

$$x = \overline{x} + \lambda_{Pn}S_x \tag{5.68}$$

where $\overline{x}$ is the mean of the extremes; S_x is the mean squared deviation of the extreme values; λ_{Pn} is a function related to P and n only, listed in Schedule 3.

Table 5.15 shows the series consisting of the SE-oriented annual maximum $H_{1/10}$ at one station. The empirical cumulative frequency points were plotted on a Gumbel probability grid paper and are shown in Fig. 5.21. As can be seen from the figure, the empirical points are approximately linearly distributed, indicating the possibility of fitting with a Gumbel distribution.

Using the least squares method, we obtained the intercept $\alpha = 1.1515$, the slope $u = 4.823$, and the correlation coefficient $R = 0.991$. Alternatively, using Schedule 3, we calculated the sum of squared deviations $\sum x^2 = 896.12\ \text{m}^2$ and the goodness-of-fit statistic $S_x = 0.9607$. Based on these results, the final theoretical cumulative frequency curve was derived, with the theoretical cumulative frequency values P_0 listed in Table 5.15 and the curve plotted in Fig. 5.21. As shown in the figure, the theoretical cumulative frequency curve exhibits a good fit to the empirical points. Here, Q in the table represents the sum of squared differences between the theoretical and empirical cumulative frequencies, defined as

$$Q = \sum_{i=1}^{n} (P_0 - P)^2 \tag{5.69}$$

It is called the sum of squared residuals and is an indicator of how well the theoretical cumulative frequency curve fits the empirical cumulative frequency points, with smaller values indicating a better fit for the same n values.

Table 5.15 Frequency analysis of the annual maximum value of $H_{1/10}$ in the SE direction (Gumbel distribution)

m	$H_{1/10}$/m	P	P_0	$P_0 - P$	$H^2_{1/10}/\text{m}^2$
1	7.5	0.0313	0.0449	0.01360	56.25
2	7.2	0.0625	0.0628	0.00027	51.84
3	6.5	0.0938	0.1351	0.04135	42.25
4	6.5	0.1250	0.1351	0.01010	42.25
5	6.5	0.1563	0.1351	–0.02115	42.25
6	6.4	0.1875	0.1503	–0.03722	40.96
7	6.1	0.2188	0.2055	–0.01325	37.21
8	5.9	0.2500	0.2515	0.00146	34.81
9	5.8	0.2813	0.2775	–0.0038	33.64
10	5.8	0.3125	0.2775	–0.03505	33.64
11	5.7	0.3438	0.3055	–0.03820	32.49
12	5.6	0.3750	0.3358	–0.03923	31.36
13	5.4	0.4063	0.4026	–0.00370	29.16
14	5.4	0.4375	0.4026	–0.03495	29.16
15	5.2	0.4688	0.4772	0.00841	27.04
16	5.2	0.5000	0.4772	–0.02284	27.04
17	5.0	0.5313	0.5580	0.02674	25.0
18	4.9	0.5625	0.5999	0.03741	24.01
19	4.9	0.5938	0.5999	0.00616	24.01
20	4.9	0.6250	0.5999	–0.02509	24.01
21	4.8	0.6563	0.6422	–0.01402	23.04
22	4.7	0.6875	0.6844	–0.00309	22.09
23	4.6	0.7188	0.7258	0.00709	21.16
24	4.6	0.7500	0.7258	–0.02416	21.16
25	4.5	0.7813	0.7659	–0.01536	20.25
26	4.4	0.8125	0.8039	–0.00859	19.36
27	4.3	0.8438	0.8393	–0.00448	18.49
28	4.3	0.8750	0.8393	–0.03573	18.49
29	4.3	0.9063	0.8393	–0.06698	18.49
30	3.6	0.9375	0.9833	0.04581	12.96
31	3.5	0.9688	0.9899	0.02112	12.25

Note: $u = 4.824$, $\alpha = 1.1515$, $R = 0.991$, $Q = 0.0223$, $\sum H^2{}_{1/10} = 896.12\ \text{m}^2$

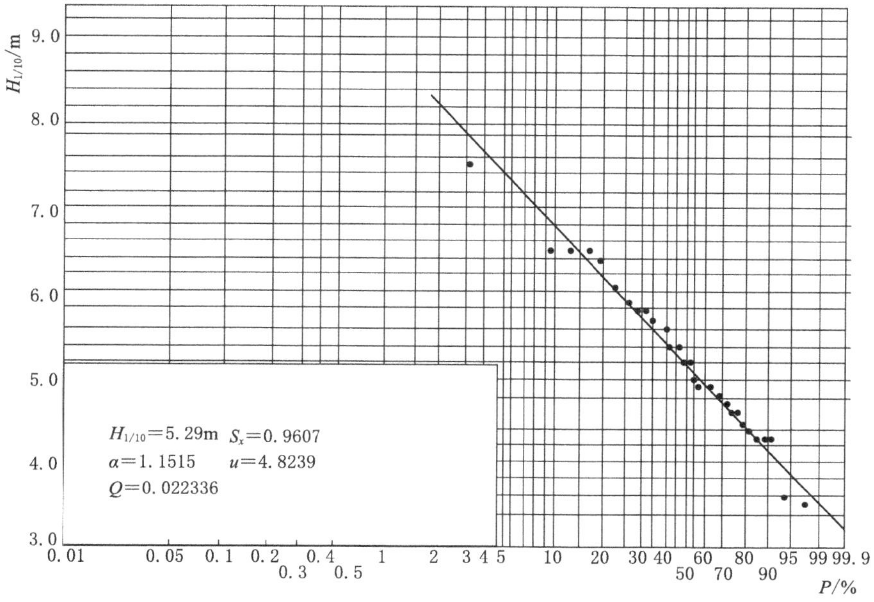

Fig. 5.21 Frequency analysis (Gumbel distribution) of the SE-oriented annual maximum $H_{1/10}$ at a station

In the absence of special paper for the Gumbel probability grid, the Gumbel distribution can also be plotted in a straight line using ordinary coordinate paper, except that the horizontal coordinates are taken as y values and the vertical coordinates are still uniform.

The wave heights $H_{1/10}$ for the 50-year and 25-year return periods in the SE direction at this station can be obtained either from Fig. 5.20 or through direct calculation, which are respectively:

50-year return period wave height: $H_{1/10} = 8.2\,\text{m}$.

25-year return period wave height: $H_{1/10} = 7.6\,\text{m}$.

4) The Weibull distribution. The Weibull distribution is classified into three-parameter and two-parameter forms. When researchers study the distribution of extreme waves, the two-parameter Weibull distribution is usually used, and its probability distribution function is

$$P(x) = 1 - \exp\left(-x^{\beta/\alpha}\right) \tag{5.70}$$

While $k = 1/\alpha$, its linearized form is

$$y = \ln\ln\left(\frac{1}{1-p}\right) = \ln k + \beta \ln x \tag{5.71}$$

The Weibull probability paper is constructed based on the aforementioned equation, as illustrated in Fig. 5.22. It features a logarithmic scale on the vertical axis and a scale of y-values on the horizontal axis. If the empirical cumulative frequency points for the annual extreme wave height $H_{1/10}$ are distributed in a straight line on the Weibull probability grid paper, one can consider fitting the Weibull distribution and solving for the two parameters using the least squares method to plot the theoretical cumulative frequency curve.

$$x = \exp\left[\frac{\ln\ln\left(\frac{1}{1-p}\right) - \ln k}{\beta}\right] \tag{5.72}$$

Table 5.16 shows the results of fitting the W-directional annual extreme wave height $H_{1/10}$ to the Weibull distribution at one station. The two parameters in this case are $k = 8.3696$, $\beta = -3.133$, the linear correlation coefficient $R = 0.989$, and the residual sum of squares $Q = 0.02782$, which shows a good fit.

In the absence of Weibull probability grid paper, the frequency distribution can also be plotted on ordinary coordinate paper.

The design wave height in the W direction for this station is calculated from Fig. 5.22 or by Eq. (5.72), as follows:

50-year return period wave height: $H_{1/10} = 6.8$ m.

25-year return period wave height: $H_{1/10} = 5.5$ m.

The log-normal distribution, the first type of extreme value distribution and the Weibull distribution all have only two parameters to be determined, so after converting the distribution function into a linearized form, it is easy to solve the two parameters by the least squares method on the computer, and then draw a theoretical

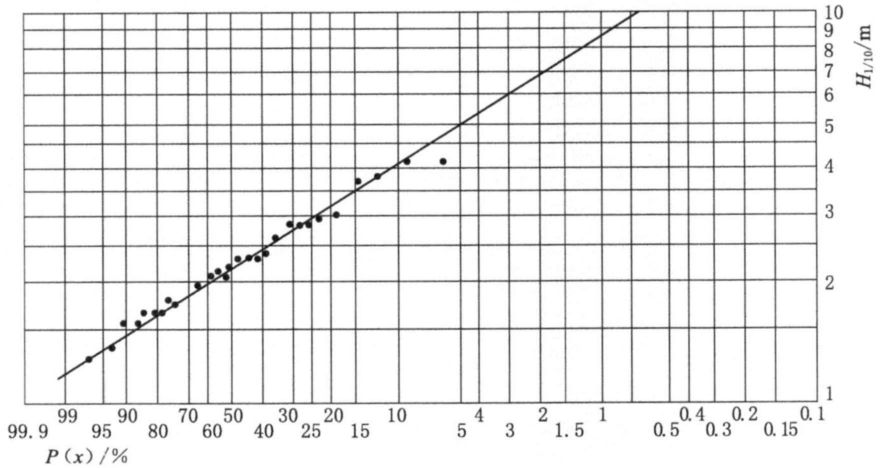

Fig. 5.22 Frequency distribution (Weibull distribution) of $H_{1/10}$ in the W direction at a station

Table 5.16 Frequency analysis of the annual maximum value of $H_{1/10}$ in the W direction (Gumbel distribution)

m	$H_{1/10}$/m	P	P_0	$P_0 - P$
1	6.8	0.0313	0.0204	–0.0109
2	4.2	0.0625	0.0891	0.0266
3	4.1	0.0938	0.0957	0.0019
4	3.7	0.1250	0.1296	0.0046
5	3.6	0.1563	0.1403	–0.0160
6	3.0	0.1875	0.2349	0.0474
7	2.9	0.2188	0.2575	0.0387
8	2.8	0.2500	0.2828	0.0328
9	2.8	0.2813	0.2828	0.0015
10	2.8	0.3125	0.2828	–0.0297
11	2.6	0.3438	0.3424	–0.0014
12	2.4	0.3750	0.4165	0.0415
13	2.3	0.1063	0.4597	0.3534
14	2.3	0.4375	0.4597	0.0222
15	2.3	0.4688	0.4597	–0.0091
16	2.2	0.5000	0.5072	0.0072
17	2.1	0.5313	0.5590	0.0277
18	2.1	0.5625	0.5590	–0.0035
19	2.1	0.5938	0.5590	–0.0348
20	2.0	0.6250	0.6147	–0.0103
21	2.0	0.6563	0.6147	–0.0416
22	2.0	0.6875	0.6147	–0.0728
23	1.8	0.7188	0.7347	0.0159
24	1.8	0.7500	0.7347	–0.0153
25	1.7	0.7813	0.7955	0.0142
26	1.7	0.8125	0.7955	–0.0170
27	1.7	0.8438	0.7955	–0.0483
28	1.6	0.8750	0.8533	–0.0217
29	1.6	0.9063	0.8533	–0.0530
30	1.4	0.9375	0.9459	0.0084
31	1.3	0.9688	0.9747	0.0059

Note: $k = 8.3696$, $\beta = -3.133$, $R = 0.989$, $Q = 0.0278$

Table 5.17 Parameters y_n and σ_n of the Gumbel distribution for different sample sizes n

n	y_n	σ_n	n	y_n	σ_n	n	y_n	σ_n
8	0.4843	0.9043	19	0.5220	1.0566	60	0.5521	1.1747
9	0.4902	0.9288	20	0.5236	1.0628	70	0.5548	1.1854
10	0.4952	0.9497	22	0.5268	1.0754	80	0.5569	1.1938
11	0.4996	0.9676	24	0.5296	1.0864	90	0.5586	1.2007
12	0.5035	0.9833	26	0.5320	1.0961	100	0.5600	1.2065
13	0.5070	0.9972	28	0.5343	1.1047	200	0.5672	1.2360
14	0.5100	1.0095	30	0.5362	1.1124	500	0.5724	1.2588
15	0.5128	1.0206	35	0.5403	1.1285	1000	0.5745	1.2685
16	0.5157	1.0316	40	0.5436	1.1413	∞	0.5772	1.2826
17	0.5181	1.0411	45	0.5463	1.1519			
18	0.5202	1.0493	50	0.5485	1.1607			

cumulative frequency curve, which is less convenient and faster than drawing the Pearson curve and eliminates the human factor. In practical applications, the residual squared sum Q is used as an indicator to select the optimal distribution model. This model is then compared with the Pearson curve to determine the final choice of distribution.

5) Poisson-Gumbel compound extreme value distribution. Large waves in the south-eastern and southern waters of China are usually caused by typhoons, and several typhoons occur each year, resulting in multiple large waves. In addition to the annual maximum sampling method—where design wave heights are derived by fitting theoretical cumulative frequency curves—the Poisson-Gumbel distribution can also be used to estimate design waves. This distribution takes into account not only the distribution of annual extremes but also the number of typhoon occurrences per year and is therefore more reasonable, and its basic principles are as follows.

For a given sea area, the number of times a typhoon passes near it each year is n. Obviously n is a random variable whose distribution should be some kind of discrete distribution taking values in the range of integers between 0 and k, and whose corresponding probability is P_k. The maximum wave height formed by each typhoon in the sea area each year is denoted as ζ_i, $i = 1$–n. One extreme value of n typhoon waves each year is denoted as $\zeta_n = \max(\zeta_i)$. Let $G(x)$ be the cumulative distribution function of the individual extreme wave heights ζ_i.

It is required to derive the distribution function $F(x)$ for the annual extreme value of typhoon waves ζ_n when the probability distribution of the number of typhoon occurrences n is P_k.

According to the addition and multiplication theorems of probability, $F(x)$ can be written as follows:

$$
\begin{aligned}
F(x) &= P(\varsigma < x) = \sum_{k=0}^{\infty} P(\varsigma < x, n = k) \\
&= \sum_{k=0}^{\infty} P(\varsigma < x|_{n=k})P(n = k) = \sum_{k=0}^{\infty} P[\max(\varsigma_i) < x]P_k \\
&= \sum_{k=0}^{\infty} P_k[G(x)]^k
\end{aligned} \tag{5.73}
$$

The above equation shows that $F(x)$ is made up of two distributions, the probability distribution of n and the probability distribution of ζ, so it is called the composite extreme value distribution.

After examination of some stations along the coast of China, the number of typhoon occurrences n per year conforms to the Poisson distribution, i.e.

$$
P_k = \exp(-\lambda)\frac{\lambda^k}{k!} \tag{5.74}
$$

where λ is the average number of typhoon occurrences per year, which can be calculated by the following formula:

$$
\lambda = \text{total typhoon numbers/total years}
$$

And the empirical cumulative frequency points of typhoon waves are approximately distributed in a straight line on the Gumbel probability grid paper, indicating that the distribution of typhoon waves conforms to the Gumbel distribution, i.e.,

$$
G(x) = \exp\big[-\exp(-x)\big] = \exp\{-\exp[-\alpha(H - u)]\} \tag{5.75}
$$

where α and u are the two parameters of the Gumbel distribution, and so

$$
F(x) = \sum_{k=0}^{\infty} \exp(-\lambda)\frac{\lambda^k}{k!}[G(x)]^k = \exp\{-\lambda[1 - G(x)]\} = 1 - P \tag{5.76}
$$

where P is the cumulative frequency of excess values, i.e., the inverse of the return period.

Substituting the expression for the $G(x)$ phantom into Eq. (5.76) yields

$$
x_P = -\ln\left\{-\ln\left[1 + \frac{\ln(1 - P)}{\lambda}\right]\right\} = \alpha(H_P - u) \tag{5.77}
$$

α and u can be obtained from the mean of a series consisting of the maximum wave height of all typhoon waves $\overline{H}$ and the mean squared deviation S, or solved by the least squares method, and their relationship with $\overline{H}$ and S are given by

$$\left.\begin{aligned} \alpha &= \sigma_n / S \\ u &= \overline{H} - \frac{y_n}{\alpha} \end{aligned}\right\} \tag{5.78}$$

where y_n and σ_n are parameters related to n only, as shown in Fig. 3.6.

The design wave height for a multi-year event under the influence of a typhoon can then be calculated by the following equation:

$$H_P = u + \frac{x_P}{\alpha} = -\ln\left\{-\ln\left[1 + \frac{\ln(1-P)}{\lambda}\right]\right\}\frac{1}{\alpha} + u \tag{5.79}$$

Let there be n typhoons in N years, and substitute the expressions for α and u into Eq. (5.79), we get

$$H_P = \overline{H} + \beta S \tag{5.80}$$

where

$$\beta = -\frac{1}{\sigma_n}\left(y_n + \ln\left\{-\ln\left[1 + \frac{N\ln(1-P)}{n}\right]\right\}\right) \tag{5.81}$$

As can be seen from Eq. (5.81), the value of β is only related to N, n, and P. For cases that are commonly encountered, this value has been calculated and is listed in Schedule 4.

There is an example available, please refer to ESM 5.9.

In summary, the frequency analysis of waves is a highly detailed and complex task. In particular, when the duration of wave observation data is short, the calculation results often exhibit discrepancies. Therefore, it is essential to conduct repeated analyses and calculations using various distribution models and select the one with the best fit. Furthermore, the results of any single calculation should never be treated as an unchanging truth. Instead, the findings of the frequency analysis must be continuously revised as more observation data are accumulated over time.

5.6.3 Calculating Design Waves Using Short-Term (1 to 3 years) Wave Measurement Data

When there is no marine hydrographic station in the local area of the port of construction or its adjacent waters, or when the actual data from the adjacent hydrographic stations cannot be applied due to different geographical locations, temporary local stations are often set up for wave observation, and when the data has accumulated for several years or at least a full year, the data can be used for frequency analysis and estimation of design waves.

When using short-term wave measurement data for frequency analysis, it is common to apply undifferentiated, all or daily maximum wave height values as samples, using different coordinate conversions to approximate a line distribution of empirical frequency points on various specialized coordinate papers, and then to adopt least squares method to find the straight line equation, the extrapolation method can be used to estimate the design wave height for a multi-year event. It should be noted that the various methods described below can only be used for reference.

The technical specifications for port engineering recommend a distribution model in which the wave height is expressed in uniform coordinates and the cumulative frequency P greater than or equal to a certain wave height is expressed in logarithmic coordinates.

When using short-term wave measurement data to deduce the design wave height, the cumulative frequency of the multi-year wave height can be considered as follows: if a total of n waves are observed in year a, the cumulative frequency of the maximum in year a is $P_a = 1/n$, and it can be deduced that the expected number of wave heights in year b will be $b \times (n/a)$ times, then the cumulative frequency of the maximum in year b will be

$$P_b = \frac{a}{bn} = \frac{a}{b} P_a \tag{5.82}$$

where P_a is the cumulative frequency of maximum values in a year of observations; P_b is the cumulative frequency of the design wave height corresponding to a b-year return period.

Please refer to ESM 5.10 for two examples.

5.6.4 Method of Deriving the Design Period Corresponding to the Design Wave Height

In engineering practice, the design wave parameters must include both the design wave height and the design wave period, as neither can be omitted. The magnitude of wave height and period do not correspond, i.e., the average period corresponding to the annual maximum wave height may not be the extreme value of the year, and it may be smaller than the annual extreme value. Therefore, it is not possible to take the annual maximum value of the mean period in a certain direction as a sample, but rather the mean period $\overline{T}$ corresponding to the annual maximum value of the wave height $H_{1/10}$. The frequency analysis is carried out in the same way as the frequency analysis of wave height. In this case, the theoretical frequency curve is still a Pearson Type III curve or other curve type, and the best fit to the empirical cumulative frequency points is used as the criterion for selecting the curve type. It should be noted that this method is generally applicable when the local large waves are mainly swells or mixed waves. If the extreme waves are dominated by wind waves, the resulting estimates may exhibit significant overprediction.

Figure 5.23 presents the results of the frequency analysis performed on the series of $\overline{T}$ corresponding to the annual extreme wave heights $H_{1/10}$ in the S direction, based on 31 years of data from an observational station. The empirical cumulative frequency points exhibit an approximately linear distribution on the Gumbel probability paper. Therefore, the Type I extreme value distribution is adopted for fitting. The fitting yields a residual sum of squares $Q = 0.0201$ and a correlation coefficient $R = 0.992$, indicating an excellent fit between the model and the observed data.

From Fig. 5.23, we can read or calculate the average period of the 50-year and 25-year events corresponding to the design wave height. The average periods for the 50-year and 25-year events are 8.0 s and 7.4 s, respectively.

It should be noted that the above approach is not the most reasonable. As discussed in Sect. 5.4, the period corresponding to a multi-year wave height cannot be a definite value either; it is also a random variable and has a certain distribution pattern. It is, therefore, more reasonable to use the conditional distribution of periods for known wave heights to find a possible range of design periods and their most likely values. There is no accepted method for this subject, and it is not presented here.

This section has thus provided a comprehensive introduction to the whole process of applying frequency analysis methods of mathematical statistics to derive design waves. These methods reveal the laws of wave variation to a certain extent, are quite scientific, can provide the quantitative data needed when carrying out engineering design, reflect years of practical experience, and are relatively mature. However,

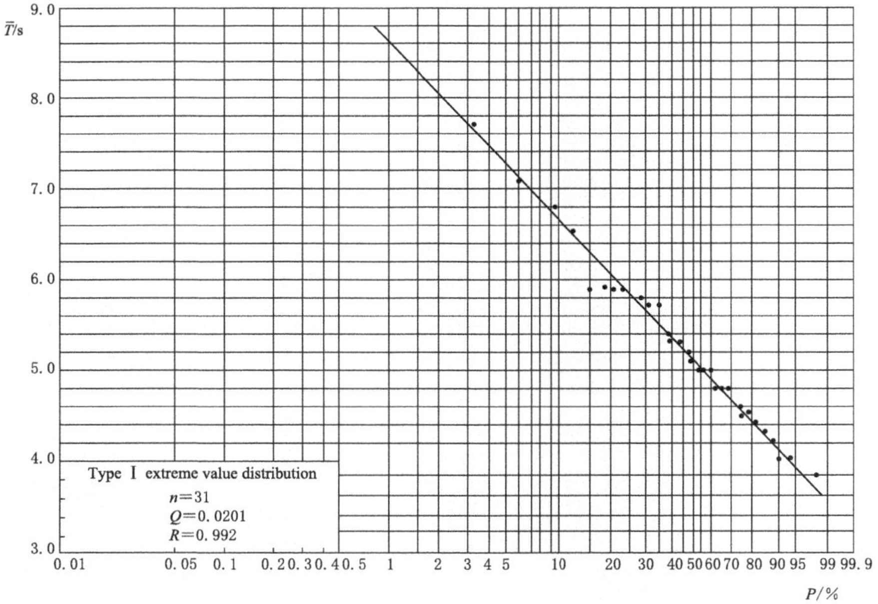

Fig. 5.23 Frequency analysis of the average period corresponding to the S-direction and the annual wave height at a certain station

there are also limitations: the representativeness of results is insufficient due to the limited duration of wave observation data, and there is occasionally poor agreement between empirical points and theoretical distribution curves. In particular, when annual extreme values are identical across multiple years in the dataset, the fitting process becomes more challenging. In addition, the physical concept of the frequency analysis method is not very clear, and it is difficult to explain the multi-year distribution pattern of waves in terms of the causal mechanism.

For these reasons, historically recorded extreme wave events represent the most valuable first-hand data. Their conceptual basis is more intuitive because they reflect real-world occurrences, and they therefore deserve careful analysis.

5.7 Extrapolation of Wind Wave Scales from Meteorological Data

The previous section discussed how to use measured data to project design waves, which is a more reliable method. However, when there are no marine hydrographic stations in the port area or its adjacent waters, the wind wave scales can only be extrapolated from local historical meteorological data. For this reason, it is important to have an understanding of wind wave growth theory.

5.7.1 Mechanisms for the Generation, Development, and Decay of Wind Waves

5.7.1.1 Wind Wave Formation Theory

According to fluid dynamics, when two different media, air and seawater, come into contact with each other and move relatively to each other, an interface wave is formed on their dividing surface as a result of their interaction. When air movement (wind) is predominant, mainly when wind acts on the surface of seawater, the interface wave is a wind wave.

The mechanism of wind wave growth has been studied since the nineteenth century, but these early theories differed greatly from actual observations and were not developed or applied. It was not until the middle of the twentieth century that scholars proposed the resonance and shear flow theories of wind wave growth, which gradually developed into the basis of modern wave formation theory. Their basic explanation is to divide the wind force on the water surface into two components: the wind shear stress τ between the wind and the water surface and the normal positive pressure N acting on the windward side of the wave by the wind, as shown in Fig. 5.24.

Fig. 5.24 Wind forces on the water surface

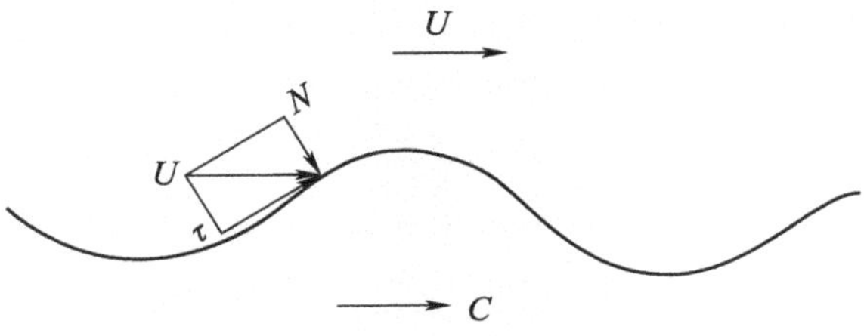

The shear stress τ between the wind and the water surface is proportional to the wind speed U. Since water particles primarily oscillate in place and the wave speed is the phase velocity, τ is independent of the wave speed C. The normal pressure N exerted by the wind on the windward face of the wave creates a pressure difference between the windward and leeward faces (in the absence of wind, the pressure on both sides of the wave surface is equal). Therefore, its magnitude is proportional to the difference between the wind speed U and the wave speed C, i.e., $U–C$. In other words, the wind works through the friction between the water point and the pressure difference between the windward and leeward sides of the wave to the water point, transferring the wind energy to the seawater and constituting the driving force for the growth of the wind waves.

At the beginning of the wind, the waves on the sea surface are small, i.e., U is greater than C during the whole period. Due to the simultaneous action of the two forces of τ and N, the wind will continue to transfer energy to the water body, so that the waves continue to develop, and the wave height and wavelength increase. As the wave scale grows, the wave speed C increases accordingly, leading to greater energy dissipation within the water due to friction. When the wave speed approaches the wind speed, the wind continues to transfer energy only through shear stress and no longer contributes energy via normal pressure. When $C > U$, the air impedes the wave's propagation and instead dissipates the wave's energy. Thus, the total energy input decreases as the wave speed increases. When the energy input is equal to the energy consumption, the wave stops developing and becomes stable, forming the largest wave that can be formed at a given wind speed.

When the wind stops, there is no new energy input to the seawater, and part of the wave energy propagates outward in all directions, while another part of the wave energy is constantly consumed by the molecular viscosity and turbulent viscous effects within the water column. In addition, air resistance, seabed friction, and infiltration also consume part of the wave energy, contributing to the gradual decay and eventual extinction of the waves.

This explanation of wind wave growth is a simplified approximation. For example, the relationship between τ on the water surface and U differs between the initial stage (when the water surface is calm) and after waves have formed (when the water surface is oscillating). Similarly, the relationship between N and the wind-wave speed difference changes with wave scale. Furthermore, this explanation completely ignores the feedback effect of the oscillating water surface on the airflow. Despite these limitations, it remains widely used due to its simplicity.

In conclusion, the generation, development, and decay of waves depend on the quantitative relationship between the energy intake and consumption of the water body; when the energy input is greater than the expenditure, the wind waves will grow and develop; conversely, the waves will tend to decay until they die out.

5.7.1.2 The Three Elements of a Wind Field

The growth and development of wind waves depend on three main factors: U, measured in meters per second (m/s); the wind duration t, the length of time the wind acts on the water surface, measured in hours (h); and the fetch (wind distance) F, the main horizontal extent of the sea area over which the wind blows, measured in kilometers (km), or meters (m).

The growth and propagation of wind waves are also influenced by local factors such as seafloor topography, water depth, and ocean currents, which further complicate the wind wave field; however, the three aforementioned factors are the most critical and are therefore referred to as the three elements of the wind field, and the following sections explain the impact of each wind field element on wave growth.

(1) Wind speed. The magnitude of wind speed U has the most obvious effect on wind wave growth. As the wind acts on the water surface, τ is proportional to U and N is proportional to U–C. Therefore, generally speaking, the greater the wind speed of the wind field, the greater the waves generated. When determining the wind speed and direction, we should try to rely on the wind information from shore stations and ship reports, and pay attention to their observation characteristics and the weather situation at that time. In the absence of actual data, calculations can be made on the basis of the distribution of isobars on ground weather maps, see Sect. 5.3.
(2) Wind duration. Wind duration refers to the continuous period during which wind acts on the sea surface under conditions of relatively constant wind speed and direction. Since wind transfers energy to seawater to generate wind waves, generally speaking, under constant wind speed, the longer the wind acts on the water surface, the more energy the water body acquires, and the larger the waves become.
(3) Wind distance. Wind distance F refers to the distance the wind blows across the water surface. To determine the fetch, it is first necessary to define the wind field or wind zone, where a wind zone is a water region with relatively uniform wind speed and direction. As shown in the air pressure distribution in Fig. 5.25, according to the definition of wind zones, three main wind zones can be roughly identified. The wind direction and wave direction in these zones are indicated by the arrows in the figure.

For the port site A near the shore, only the waves in the BC segment are propagated toward point A, so this is considered the effective wind zone for point A, abbreviated as the wind zone.

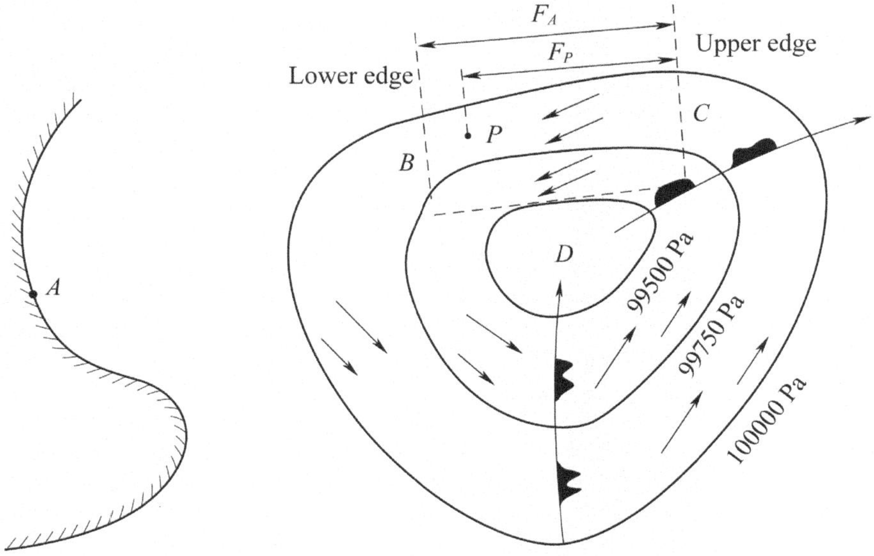

Fig. 5.25 Wind zone division

The wind zone on a weather chart can be represented by a rectangle. Along the direction of the wind, the distance from the upper boundary to the lower boundary of the wind zone is called the wind zone length, or wind distance F. For point A, the wind distance is F_A. If we want to forecast the waves at point P within the wind zone, the distance from the upper boundary of the wind zone to point P is F_P, which is the wind distance for this point. Therefore, wind distance refers to the effective wind zone extent and water length that actually contribute to wave generation at a specific point under given wind conditions. Since the development of wind waves requires a certain range of water area, it can be generally said that, under constant wind speed, the larger the wind distance, the larger the waves. In small reservoirs, no matter how strong the wind is, large waves like those in the open sea cannot be generated, because the wind distance is too small.

There are fixed and moving wind zones, as well as three states of wind wave development. These will not be elaborated here; please refer to ESM 5.11 for details.

5.7.2 Determination of Wind Field Elements

In the process of using meteorological data to project wind waves, two aspects need to be addressed. The first is how to collate and determine the wind speed, wind duration, and wind distance used in the calculations, as well as the depth of the water, i.e., the above-mentioned wind field elements; the second is how to establish the relationship between the wind field elements and the wave gist. These two aspects constitute the so-called wind wave forecasting (or hindcasting) method. The following discussion addresses the first issue.

It is known that wind wave growth depends on wind speed, wind duration, wind distance, and other wind field elements. When the water depth is shallow, wave growth is also influenced by water depth. Therefore, the correct determination of the wind field and the average depth of the water is the basis for calculating the wind wave parameters and must be treated with caution.

In the practice of engineering design, there are two general methods to determine the elements of the wind field. The first method is to use long-term wind observations from onshore meteorological stations in the vicinity of the port construction area or offshore wind measurements from more reliable ships in the sea as a basis for analyzing the wind speed, direction, and duration at sea. The second method is to use ground weather maps to determine the location of the wind field and its elements. This method is more suitable for the projection of wind waves in waters far from shore and those caused by typhoons.

The second method is also an effective approach for wind wave forecasting and hindcasting. While this method is relatively comprehensive and reasonable, it faces several challenges: the small scale of weather maps, sparse observation points over the sea, low accuracy of isobar drawing, and a 6 h time interval between consecutive weather maps. It is difficult to accurately determine the elements of the wind field. Therefore, it is still necessary to make reference to the wind measurement data from the neighboring shore stations as far as possible and carry out a test of coordination.

5.7.2.1 Determination of the Wind Speed

Wind speed has the greatest impact on wind wave elements and should be determined with great caution. Since engineering designs often require wind wave elements with a specific return period, such as 50 years or 25 years. When calculating wind waves using meteorological data, we have to assume that the return period of the wind wave elements is simply equal to the return period of the wind speed, i.e., a 50-year high wind speed produces a 50-year high wave. At this point, the maximum annual wind speed in the same direction as a certain wave direction can be found to form a series of annual maximum wind speeds, and statistical analysis is carried out according to the frequency analysis method described in Sect. 5.6, using Pearson Type III curves or other curve types as theoretical distribution curves to fit the line, to find the wind speed value for a certain return period in that wind direction as the calculated wind speed for the wind waves in the same return period of the projection, and then to find the wind speed for that return period according to the relationship between the wind field elements and the wave elements.

5.7.2.2 Determination of the Wind Distance

Wherever a selected wind field experiences significant changes in wind speed or wind direction, or is located at the boundary of a smaller water body, the location can be considered the boundary of the wind area.

When estimating wind waves in smaller water bodies such as bays, straits, or lakes, where the water area is relatively small, wind fields usually cover the entire water body. In such cases, the distance to the opposite shore in the wind direction can be used as the length of the wind area. If the water body is more open, the wind area can be determined based on the weather chart corresponding to the selected wind field. In this case, the intersection angle between the wind direction and the isobar can generally be ignored. It is assumed that as long as the angle between the tangent direction of the isobar at a certain point and the line connecting that point to the calculation point does not exceed 30° (when the isobar is relatively flat) or 45° (when the isobar curvature is large), the point can be included in the wind area. The tangent direction of the isobar at that point then represents the wind direction.

For slowly moving pressure systems, multiple calculation points can be selected within the system, and their corresponding wind areas may overlap. In other words, different wind areas can be selected for different calculation points.

On surface weather charts, if there is little change in wind direction between two time intervals, the length of the wind area should be measured from the later weather chart. Conversely, if there are significant changes in weather conditions during this time interval, the timing of wind direction shifts should be analyzed, and wind areas should be determined in two or more stages.

Once the wind area is delineated, the wind distance can be measured. If the proposed engineering site is outside the wind area, the wind distance is the distance between the upper and lower boundaries of the wind area. If the proposed engineering site is within the wind area, the wind distance is measured from the proposed engineering site in the upwind direction to the upper boundary of the wind area. Clearly, if the proposed engineering site is located within a bay, the wind area typically covers the entire water body, and the wind distance is the distance to the opposite shore in the upwind direction (Fig. 5.26). In practice, when using meteorological data to calculate design waves, it is usually based on observation data from meteorological stations, and the range of the wind area cannot be determined. Therefore, it is generally assumed that the wind area is sufficiently large (covering the entire water body). For convenience, the wind distance for each wind direction, i.e., the distance to the opposite shore in each wind direction, can be measured in advance. In cases where the wind distance for certain directions (facing the open sea) is particularly large, e.g., when the wind distance extends to infinity in the upwind direction, the wind distance can be determined based on empirical relationships, such as the formula proposed by Akitoshi Yoshiba: $F = 5 \times 105/U$.

In coastal areas, estuaries, bays, and even lakes and reservoirs, the growth of wind waves is often affected by boundaries on the one hand and by obstacles (such as small islands or reefs) along certain directions on the other hand. In such cases, the wind area length (wind distance) should be determined based on the effective wind area

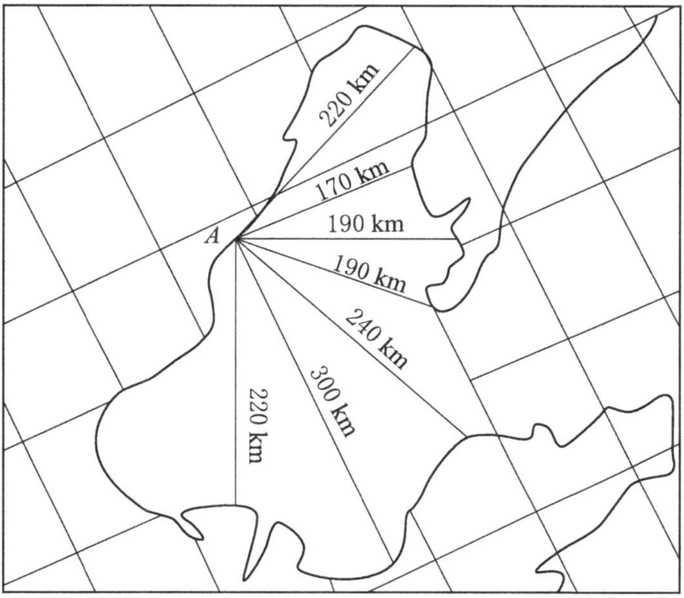

Fig. 5.26 Determination of the wind distance in a bay

length or effective wind distance, which can be expressed as follows:

$$F_e = \frac{\sum_{i=0}^{n} F_i \cos^2 \alpha_i}{\sum_{i=0}^{n} \cos \alpha_i} (i = 0, \pm 1, \pm 2 \ldots) \quad (5.83)$$

In the formula, each F_i represents the wind distance to the opposite shore in a specific direction (if it extends to infinity, it is calculated using empirical relationships). A straight line drawn along the predominant wind direction is defined as the main ray (wind distance in the main wind direction), where $i = 0$, $\alpha_0 = 0$, and the wind distance is F_0 from the proposed engineering site, within a 45° range on each side of the main ray. Lines are drawn at regular intervals. The wind distance along each line represents F_i, and the angle between each line and the main ray is denoted as α_i. The spacing of these lines (angular interval) is generally 22.5°, but can also be 11.25°, 7.5°, etc., as illustrated in Fig. 5.27.

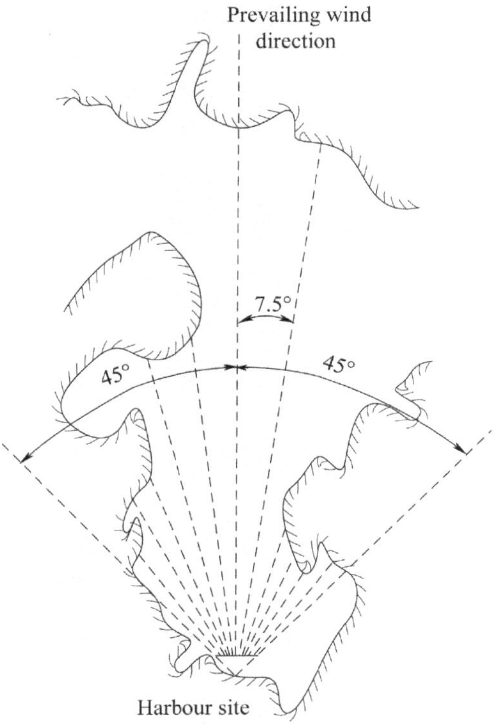

Fig. 5.27 Schematic diagram for setting up rays to calculate effective wind area

5.7.3 Introduction to Commonly Used Methods for Wind Wave Calculation

After selecting the wind field elements, the relationship between the wind field elements and wave elements can be utilized to determine the variation of wave elements over time or location.

Currently, there are many wave forecasting methods proposed in China and internationally, most of which are empirical or semi-empirical. The former is limited to the range of measured data and specific on-site conditions, rendering its application highly restricted and gradually obsolete. The latter, based on measured data, identifies certain empirical relationships, followed by corresponding theoretical generalizations. It may derive energy methods from studying the energy transfer and consumption processes during wave formation, or spectrum methods from studying the distribution of energy in formed waves, or combine energy and spectrum methods to calculate wave elements. These methods have been simplified to propose calculation formulas and diagrams for engineering applications. The calculation methods continue to be validated and refined through measured data. The following are several representative methods, including widely used methods worldwide and those recommended in the *Code of Hydrology for Harbour and Waterway* (JTS 145-2015) (2022).

Multiple representative methods are available for determining offshore wind wave parameters, including Wilson's wave forecasting approach (1965), Bretschneider's wave prediction method (1970), and the JONSWAP-based wind-wave forecasting technique. For detailed specifications, refer to the supplementary materials accessible via ESM code 5.12.

5.8 Shallow Water Deformation of Waves

The wave elements calculated above are those of deep-water waves at a distance from the harbor site, or at depths where the wave measurement buoys are still some distance away from the harbor site. These waves are taken as incident waves, which continue to propagate towards the harbor site and change significantly as they enter shallow water, and calculations that take this change into account are often referred to as shallow water deformation calculations.

As waves pass from deep water to shallow water, they change in height, wavelength, speed, and profile. The main reasons for the changes in shallow water are the effects of shallow water depths, complex topography, submarine friction, current action, and obstacles (islands, buildings, etc.). Wave deformation due to shallow water also includes refraction, diffraction, reflection, and breaking, and is further complicated by wind action on the nearshore surface. These phenomena are important characteristics of shallow nearshore waves. In other words, the scale and structure of the wave elements in shallow inshore waters are the result of a combination of these phenomena.

In practice, it is very difficult to consider the results of all these phenomena simultaneously in the analysis of shallow water deformation of waves. Usually, the deformation phenomena are treated separately, i.e., the wave deformation due to a certain influence is analyzed and discussed independently, and the results are expressed in terms of a certain coefficient, e.g., the shallow water coefficient, the refraction coefficient, the diffraction coefficient, the friction attenuation coefficient, etc. It is assumed that the linear superposition of all these coefficients gives a combined result. In addition, in the deformation analysis, only the most influential factors can be considered on a case-by-case basis, and minor factors are ignored. Recently, the methods used to determine the wave elements near the shore have included wave transformation due to shallow water depths, wave refraction due to irregular isobath distribution, wave diffraction due to obstacles, wave deformation due to stronger currents (including currents), the effect of friction on gently sloping beds, and wave breaking in shallow nearshore waters. The problem is simplified by ignoring the continuing effect of wind on the nearshore surface, i.e., the nearshore is mainly swell, and ignoring underwater penetration, etc.

There are two ways to analyze wave deformation in shallow water: the first is to select some kind of characteristic wave (corresponding wave height and period) representing nearshore waves to replace the actual waves as regular waves, and use the energy method to analyze the deformation under the effect of shallow water depth, etc., called the regular wave calculation method; the second is to use the concept of wave spectrum to describe the structure of nearshore shallow-water waves and calculate their statistical properties, called the irregular wave calculation method. Obviously, the second method is more in line with the actual situation at sea and is more difficult. The regular wave calculation method can be further divided into two calculation approaches based on the wave theory applied: one based on linear wave theory, and the other on nonlinear wave theory. The latter is more consistent with the actual sea waves in nearshore shallow waters, but involves more complex calculations. Accordingly, this section is confined to the linear (small-amplitude) wave theory approach.

5.8.1 Wave Deformation Due to Shallowing Water Depth

Deep-water waves are transmitted in a forward direction into a gently sloping shallow area relative to the water depth $d/L_0 < 1/2$. When the wave crest line is parallel to the isobath or the wave direction line is perpendicular to the isobath (see Fig. 5.28), the magnitude and distribution of wave energy change as the water depth becomes shallower and the wave elements change accordingly.

5.8.1.1 Variation of Wavelength and Wave Speed

Actual wave observations show that the period of the wave as it travels from deep water to the coast is essentially constant or varies very little, i.e., it is considered to be $T = T_0$ (Any wave element below with a subscript 0 indicates a deep-water wave element, while those without a subscript are shallow-water wave elements).

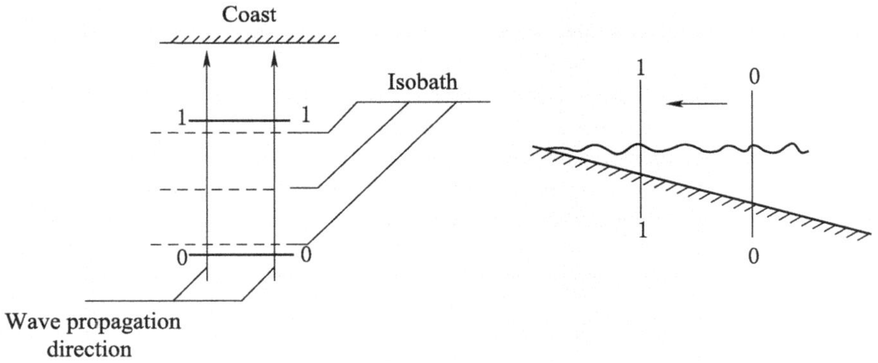

Fig. 5.28 Deep-water waves are transmitted positively into shallow waters

Figure 5.28 shows two sections along the wave direction, Section 0-0 in deep water and Section 5.1-5.1 in shallow water. For stable waves, the number of wave crests per unit time across Section 0-0 and across Section 1-1 should be equal, the waves will be discontinuous between the two sections, which indicates that the wave period does not change during wave propagation from deep water to shallow water.

For a constant wave period, the wavelength and wave speed will decrease as the water becomes shallower, according to linear wave theory.

From Eqs. (5.1 and 5.2), and $T = T_0$, we can get

$$\left.\begin{aligned} L/L_0 &= \tanh \tfrac{2\pi d}{L} \le 1.0 \\ C/C_0 &= \tfrac{L/T}{L_0/T_0} = L/L_0 = \tanh \tfrac{2\pi d}{L} \end{aligned}\right\} \tag{5.84}$$

Equation (5.84), despite its simplicity, is implicit and requires an iterative solution for its variables. For ease of use, Eq. (5.84) has been plotted against the relative water depth in a curve (Fig. 5.29) and in a table (Schedule 6).

The horizontal coordinates in Fig. 5.29 are the relative water depths d/L_0, $d/L_0 = 0.5$ is the boundary between deep and shallow water. The graph $\frac{C}{C_0}\frac{L}{L_0}$ or d/L is the change curve of wavelength and wave speed in shallow water, which shows that the wavelength and wave speed decrease rapidly as the water depth becomes shallower.

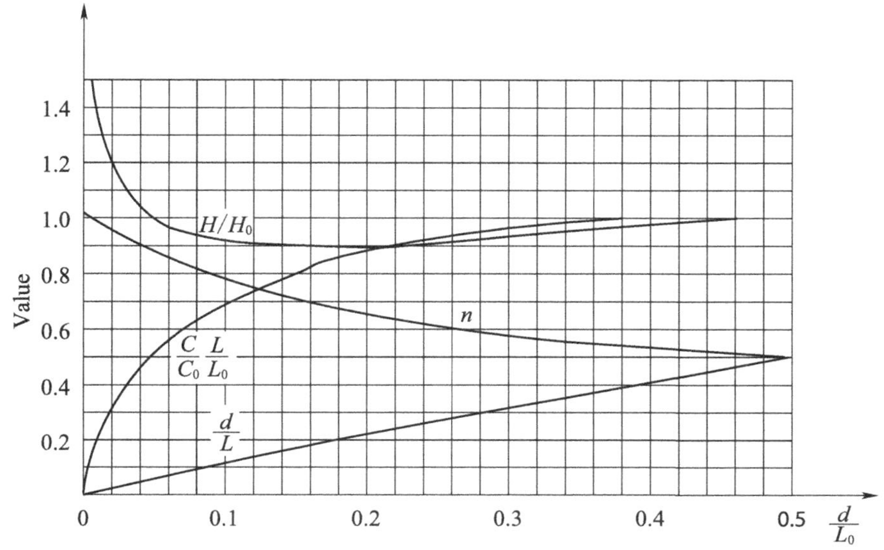

Fig. 5.29 Wave element deformation curves due to shallowing water depth

5.8.1.2 Change in Wave Height

When deep-water waves propagate normally into shallow water, the wave rays remain parallel, as shown in Fig. 5.28. It can be assumed that the energy between two wave rays remains essentially unchanged, with no lateral exchange of wave energy across the wave rays. In other words, the energy input per unit time into the deep-water Section 0-0 and the energy output from the shallow-water Section 1-1 between two wave rays should be equal. Neglecting wave energy losses such as bottom friction and seepage, the wave energy flux between two wave rays is conserved, leading to the equation:

$$EnCb = E_0 n_0 C_0 b_0 \tag{5.85}$$

where E_0, E are the wave energy per unit area of the water column in deep and shallow water; C_0, C are wave speeds in deep and shallow water; b_0, b are the wave ray spacing in deep and shallow water, when the two wave rays are parallel, $b = b_0$; n_0, n are wave energy transfer rates in deep and shallow water.

Wave energy transfer rate n or n_0 denotes the fraction of the wave's energy that travels forward in a wave cycle E_n to the total energy E of a wave cycle, i.e., $n = E_n/E$. According to linear wave theory, the expression for the wave energy transfer rate in deep and shallow water is Eq. (5.86) shows that the wave energy transfer rate varies between 0.5 and 1.0 in shallow water, and at very shallow depths, $\sinh x \approx x$.

$$\left.\begin{aligned} n_0 &= \tfrac{1}{2} \\ n &= \tfrac{1}{2}\left(1 + \frac{\frac{4\pi d}{L}}{\sinh\frac{4\pi d}{L}}\right) \end{aligned}\right\} \tag{5.86}$$

According to linear wave theory, the wave energy per unit area of the water column is proportional to the square of the wave height, i.e.,

$$\left.\begin{aligned} E_0 &= \tfrac{1}{8}\gamma H_0^2 \\ E &= \tfrac{1}{8}\gamma H^2 \end{aligned}\right\} \tag{5.87}$$

where γ is the capacity of water.

Thus, from Eqs. (5.85) and (5.87), we obtain

$$\frac{H}{H_0} = \sqrt{\frac{n_0 C_0}{nC}} = \sqrt{\frac{n_0 L_0}{nL}} = K_s \tag{5.88}$$

where K_s is the shallow-water coefficient, which represents the change in wave height in deep and shallow water when there is no refraction.

As can be seen from Eq. (5.88), the shallow water coefficient is only related to wavelength and water depth, and its curve with relative water depth has been plotted in Fig. 5.29. As can be seen from the figure, during the initial stage of wave propagation into the shallow water zone, the wave height decreases slightly. However, once the wave enters the region where $d/L_0 < 0.163$, the wave height gradually increases and exceeds the deep-water wave height until the wave is too steep to maintain and breaks. For practical convenience, a table based on Eq. (5.88) has been prepared and is shown in Schedule 6.

5.8.2 *Wave Refraction*

Waves propagate from deep water towards the shore into shallow water ($d/L_0 < 1/2$). The wave propagation direction or wave direction line will be deflected in the plane and cause a change in wave height; meanwhile, the wave crest line will also become curved with the submarine topography, and eventually tend to be compatible with the coastline or close to parallel.

5.8.2.1 Wave Refraction Principle

The reason for wave refraction in shallow water is that the wave travels towards the shore at an angle between the crest line and the isobath, and the speed of the wave is different at different points on the same crest line because the depth of the water is different. The wave speed decreases as the water depth decreases, in deeper water, the wave speed is greater and the wave propagates faster; in shallower water, the wave speed is smaller and the wave propagates slower, resulting in the wave crest in deeper water propagating faster than the wave crest in shallower water, which reduces the angle between the crest line and the isobath, i.e., the crest line gradually tends to be parallel to the isobath, as shown in Fig. 5.30.

According to Snell's law, the refraction angle of a wave is related to its propagation speed. In Fig. 5.31, *PN* is an isobath (a line of constant depth) separating two regions with water depths d_1 and d_2. As a wave propagates from the deeper region (d_1) into the shallower region (d_2), its speed changes from C_1 to C_2, and its direction is refracted. At depth d_1, the angle between the wave crest line and the isobath is α_1, and the wave speed is C_1. Take two closely spaced wave rays. During a time interval Δt, the segment at the *M* end advances to point *P*, covering a distance $MP = C_1\Delta t$. Meanwhile, the segment at the *N* end enters the d_2 region and advances to point *Q*, covering a distance $NQ = C_2\Delta t$. The corresponding angle after refraction is α_2. Based on the geometric relationships in triangles *PNM* and *PNQ*, the following relation holds:

In triangles *PNM* and *PNQ*:

$$\frac{\mathrm{MP}}{\sin\alpha_1} = \frac{\mathrm{NQ}}{\sin\alpha_2} \text{ or } \frac{C_1\Delta t}{\sin\alpha_1} = \frac{C_2\Delta t}{\sin\alpha_2} \tag{5.89}$$

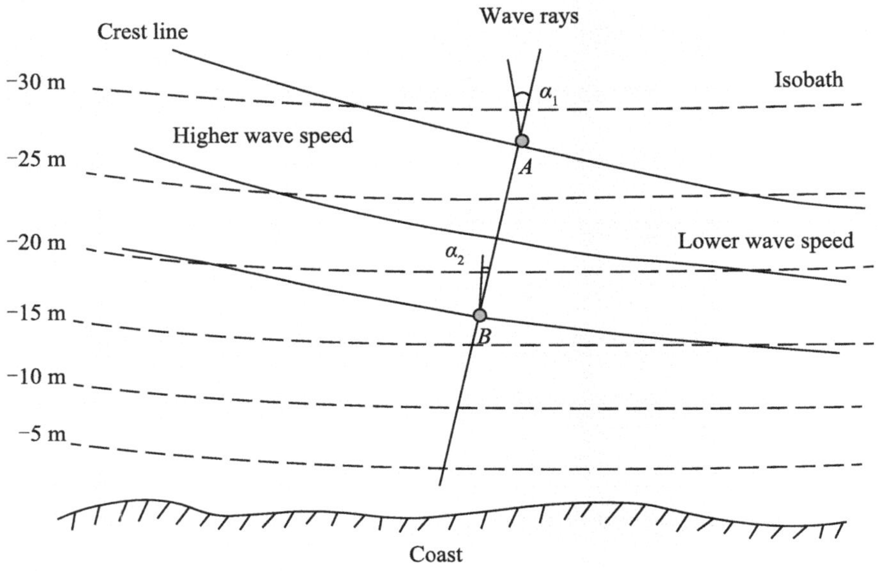

Fig. 5.30 Refraction of waves

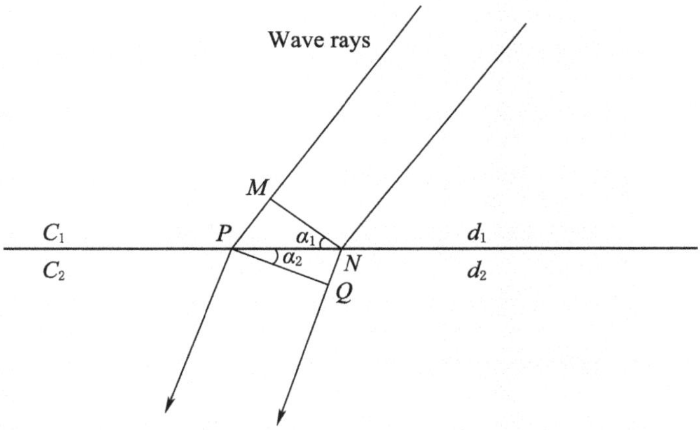

Fig. 5.31 Snell's law

Then

$$\frac{\sin \alpha_1}{C_1} = \frac{\sin \alpha_2}{C_2} \tag{5.90}$$

Due to the effect of refraction, wave rays often no longer remain parallel after entering shallow water. However, according to the principle of wave energy flux

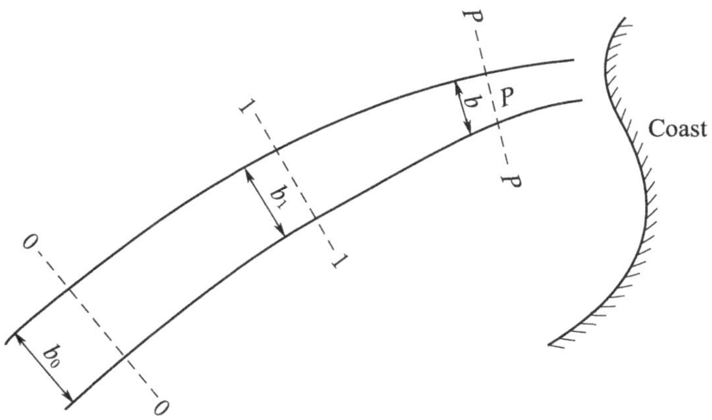

Fig. 5.32 Wave to line refraction deformation

conservation, the energy input per unit time through a deep-water Section 0-0 and the energy output through any shallow-water Section 1-1 between two wave rays must still be equal, as illustrated in Fig. 5.32. Then, from Eq. (5.85), we get.

$$\left.\begin{aligned} \frac{H}{H_0} &= \sqrt{\frac{n_0 L_0}{nL}}\sqrt{\frac{b_0}{b}} = K_s K_r \\ H &= H_0 K_s K_r \end{aligned}\right\} \tag{5.91}$$

where K_r is called the refraction coefficient, $K_r = \sqrt{\frac{b_0}{b}}$.

Equation (5.91) shows that when the wave is travelling diagonally close to the shore, in addition to the change in wave height due to the shallow water depth, the change in wave height will also occur due to the change in the spacing between the two wave lines as a result of refraction, which can be reflected by K_r.

In the shallow water zone, if the wave rays diverge, i.e., $b_0 > b$, then $K_r < 1$, and the wave height will decrease due to refraction. If the wave rays converge, i.e., $b_0 < b$, then $K_r > 1$, and the wave height will increase due to refraction. When $K_r = 1$, there is no refraction effect, which corresponds to the case of normal wave incidence. As a result of refraction, the wave rays of nearshore waves often no longer remain parallel but undergo complex variations in response to irregular underwater topography. Figure 5.33 illustrates wave refraction induced by bays and headlands.

Equation (5.91) indicates that to derive the refraction coefficient K_r, it is necessary to first plot the wave refraction diagram (i.e., the wave ray refraction pattern), before the wave height at a specific location near the harbor site can be calculated.

If the wave refraction diagram is constructed starting from the deep-water zone, with the deep-water wave height known, the wave height at an arbitrary point P near the harbor site (Fig. 5.32) can be determined as follows: measure the corresponding widths b (at point P) and b_0 (in the deep-water zone) between two wave rays, calculate

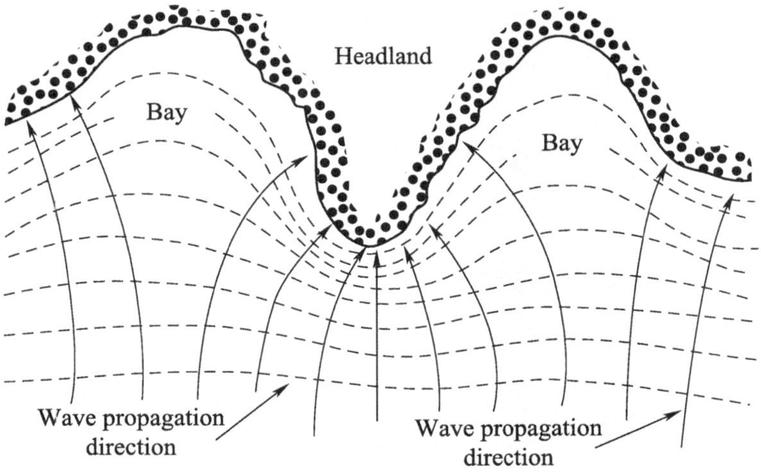

Fig. 5.33 Wave ward line refraction deformation in shallow coastal waters

the value of K_r, use the relative water depth d/L_0 at point P to look up K_s in Schedule 6, and finally obtain the wave height at point P using Eq. (5.91).

If the wave refraction diagram is constructed starting from a point in the shallow water zone with water depth d_1 (e.g., the depth of a wave buoy), where the local wave height H_1 (which may be a shallow-water wave height) is known (see Fig. 5.32), the wave height at point P shall be calculated using the following equation:

$$H = \sqrt{\frac{b_1}{b}\frac{K_{sP}}{K_{s1}}}H_1 \tag{5.92}$$

where b_1 denotes the width between two wave rays at water depth d_1; K_{sP} is the shallow-water coefficient at point P (obtained by looking up Schedule 6 using d_P/L_0); K_{s1} is the shallow-water coefficient at water depth d_1 (obtained by looking up Schedule 6 using d_1/L_0).

Equation (5.92) essentially describes wave transformation from deep water to the depth d_1, where refraction effects are neglected due to the relatively large water depth. The wave direction after refraction can be directly measured from the wave refraction diagram.

It should be noted that when calculating according to Eq. (5.91) or Eq. (5.92), the average wave height $\overline{H}$ is generally adopted. If the known wave height in deep water is a exceedance wave height, e.g., $H_{1/10}$, the wave height can be converted to $\overline{H}$ according to Fig. 5.15. The wave period is still taken as the average period $\overline{T}$.

Please refer to ESM 5.13 to get a calculation example.

5.8.2.2 Introduction to Numerical Methods for Plotting Refraction

The graphical method of refraction mapping has the following drawbacks. Firstly, in complex topographic situations, the mapping process is somewhat arbitrary; for example, the processing of the average isobath, the processing of the wave direction lines and isobath lines when they are close to parallel, etc., depends on the judgement and experience of the mapper, and is prone to considerable errors. Secondly, due to the large workload, the wave direction lines are often not spaced too closely, so it is difficult to obtain comprehensive and accurate results. Thirdly, it is not easy to accurately display and deal with problems such as intersecting lines. Currently, numerical calculations are increasingly being used to map refraction in complex terrain. For more information on numerical methods, please refer to ESM 5.14.

5.8.2.3 Frictional Attenuation of Waves

If the wave propagation distance in shallow water is long or the bottom slope is very gentle (1/2,000–1/1,000), and the water bottom is relatively rough, in addition to the above-mentioned effects of shallow water depth and refraction, the effect of friction on the water bottom on wave height should also be considered.

The frictional attenuation is commonly evaluated using the initial wave height H_0, the wave period T_0, the wave propagation distance in the shallow-water zone Δx, the mean water depth d, and the bottom-friction coefficient f (typically 0.01–0.02). First, set $T = T_0$. Then, from Fig. 5.34, determine the frictional attenuation coefficient K_{f} corresponding to a propagation distance Δx. As indicated by the figure, when the horizontal propagation distance is long and the water depth is small, frictional attenuation becomes pronounced and generally cannot be neglected.

In summary, the wave height at the end of this distance is

$$H = H_0 K_{\mathrm{s}} K_{\mathrm{r}} K_{\mathrm{f}} \tag{5.93}$$

where H_0, H are the average wave height in deep and shallow water.

5.8.3 *Breakage of Inshore Waves*

5.8.3.1 Nearshore Wave Breaking Mechanism

The breaking mechanism of nearshore waves can be explained as follows: according to Fig. 5.28, the wavelength of a wave becomes shorter when it enters shallow water, and the wave height decreases slightly at the beginning, but then gradually increases, so when the wave propagates to a certain shallow water, the wave steepness increases rapidly. On the other hand, because the water depth at the trough is smaller than at the crest, the wave trough is more affected by the friction of the seabed, so the speed of

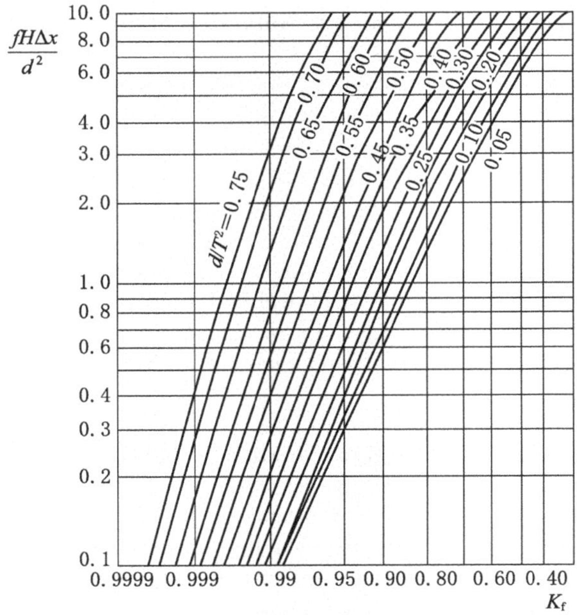

Fig. 5.34 Frictional attenuation coefficient of wave height over a constant-depth seabed

wave propagation at the trough is lower than at the crest, the wave crest gradually goes forward to the trough, the waveform is distorted and the front slope becomes steeper. After a certain water depth, the wave may reach the limit of steepness (theoretical limit $H/L \approx 1/7$) and lose stability, or the wave breaks because the front slope dips steeply or the crest breaks (theoretical limit of the wave crest apex angle is 120°). These are all wave-breaking phenomena.

The water depth at the point of wave breaking is referred to as the critical water depth (also called the breaking water depth), denoted as d_b. The corresponding wave height is called the breaking wave height, denoted as H_b. Since this is the maximum possible wave height under the condition of water depth d_b, it is also known as the limiting wave height. The nearshore area from the outer edge where waves first begin to break up to the shoreline is termed the breaker zone or surf zone. Within this area, waves may undergo multiple breaking events until they eventually form up-rushing swash flows near the shore. Inside the surf zone, a significant portion of wave energy is dissipated through friction, turbulence, and sediment transport, while also generating onshore-directed currents. This effect of wave breaking requires full consideration in engineering practice.

5.8.3.2 Wave Breaking Patterns

Wave breaking patterns vary, depending mainly on the steepness of the waves in deep water and the slope of the nearshore water bottom, and can be broadly classified into four types.

(1) Spilling breaking wave. This is a wave type in which white foam first appears on the wave crest and gradually spreads forward over the wave front until the wave collapses. Spilling breakers exhibit a relatively symmetric profile between their front and rear phases. The breaking process develops gradually and is limited to a portion of the water near the wave crest. This type of breaker typically occurs in relatively deep water when waves are steep, and the seabed slope is gentle, as shown in Fig. 5.35a.
(2) Plunging breaking wave. This is a wave that gradually steepens in front of the crest and finally plunges forward, creating a breaking wave that splashes forward and is accompanied by the entrainment of air. This form of breaking wave is one in which the body of water breaks at the top of the crest in a single instant. This type of breaking wave will occur in deep water when the wave is moderately steep, and the bottom slope is large, as shown in Fig. 35b.
(3) Slumping breaking wave. Its characteristics are somewhere between those of a roll-breaking wave and the dispersive breaking wave described below. The bottom of the slope before the wave becomes straight, with a breaking wave, and the climbing of the slope is similar to that of a dispersive breaking wave,

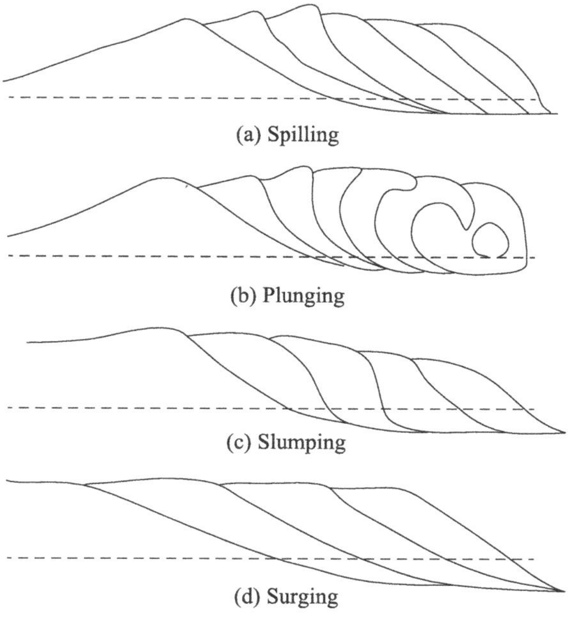

Fig. 5.35 Wave breaking patterns

which is a type of breaker from the plunging breaker to the surging breaker, as shown in Fig. 35c.

(4) Surging breaking wave. This is a type of wave that gradually steepens in front of the wave and starts to break from below as it travels, but the wave crest remains largely unbroken, with a very disorganized state in front of the wave, and creeps up the seabed slope. This type of breaking wave often occurs in deep water when the wave is gentle, and the bottom seabed slope is large, as shown in Fig. 35d.

There are no strict boundaries between the above types of breaking waves, and they often appear interspersed with one another. In general, when the deep-water wave steepness exceeds 0.06, spilling breakers often form as the wave propagates into shallow water; when the deep-water wave steepness is between 0.06 and 0.03, spilling breakers also tend to form on relatively flat seabeds whereas plunging breakers often occur on steeper seabeds; and when the deep-water wave steepness is less than 0.009, surging breakers appear on steeper seabeds while plunging breakers form on gentler seabeds. Figure 5.36 illustrates the approximate boundaries for these breaking wave types, with the horizontal coordinate representing the seabed slope and the vertical coordinate representing the deep-water wave steepness.

Of course, Fig. 5.36 is based on laboratory experiments with regular waves; actual waves are irregular and have a much more complex breaking pattern.

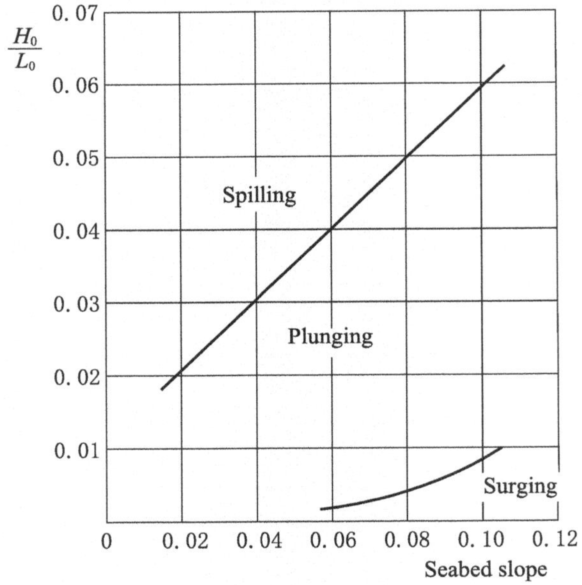

Fig. 5.36 Boundaries of various forms of breaking waves

5.8.3.3 Calculation of Breaking Wave Height and Breaking Water Depth

From the perspective of practical engineering needs, the limiting wave height when the wave is on the verge of breaking in shallow water (breaking wave height H_b), and the water depth at breaking (breaking water depth d_b) are important. Currently, there are two main theoretical approaches to study the relationship between H_b and d_b: one is the energy method, and the other is based on solving the governing equations under sloping seabed boundary conditions. However, regardless of the above methods, accurately calculating breaking wave height and breaking depth remains challenging. The majority of research findings are derived from laboratory simulations and in situ field observations. Figure 5.37 synthesizes the results of previous model tests on regular wave conditions at home and abroad.

The horizontal coordinate is the ratio of the breaking depth to the deep water wavelength, and the vertical coordinate is the ratio of the breaking wave height (i.e., the limiting wave height) to the breaking depth.

For waters with a very gentle seabed slope ($i < 1/100$), Fig. 5.37 does not provide additional results. Table 5.18 presents experimental results under regular wave conditions for seabed slopes i in the range 0 to 0.01. This table reflects that the maximum ratio of breaking wave height to breaking water depth is not less than 0.55 for a flat bottom. For slopes of $1/120 \leq i \leq 1/50$ of the sea, experimental data are lacking in both Fig. 5.37 and Table 5.18. In this case, based on solitary wave theory, the maximum value of the ratio H_b/d_b can be taken as 0.78. For safety reasons, the value of less than 0.6 in Table 5.18 is adjusted to 0.60.

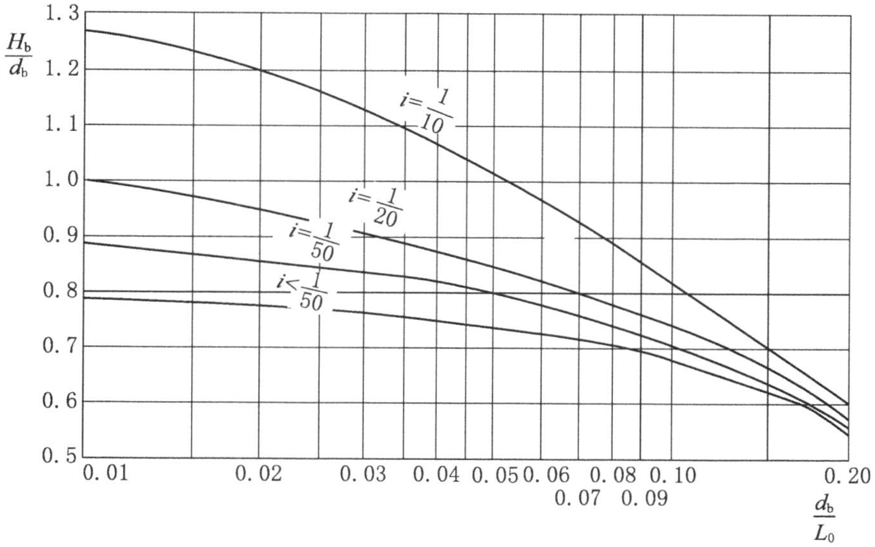

Fig. 5.37 Ratio of breaking wave height to breaking water depth i-Seabed slope

Table 5.18 Maximum ratio of breaking wave height to breaking water depth on gentle slopes

i	0	1/1000	1/500	1/400	1/300	1/200	1/120
$(H_b/d_b)_{max}$	0.55	0.55	0.55	0.56	0.58	0.64	0.78

Under irregular wave conditions, experimental studies have shown that as waves propagate from deep water to shallow water, only large waves undergo breaking. Research further indicates that the relationship between H_b of these large waves and the ratio of their corresponding deep-water wavelength L_0 to the water depth d still follows the trend shown in Fig. 5.37, but with values reduced by a factor of 0.88. That is, under irregular wave conditions, H_b is 0.88 times that under regular wave conditions.

If the wave height H_F corresponding to a certain exceedance probability for a given return period, as derived for a specific water depth in the design, is less than the limiting wave height H_b at that depth, wave breaking does not occur at that depth. When H_F equals H_b, breaking occurs at that depth. However, if H_F exceeds H_b, waves will break in deeper water before reaching that depth. For the latter two cases, i.e., when the derived wave height for a given return period and exceedance probability is greater than or equal to the shallow-water limiting wave height, the limiting wave height should be adopted as the design wave height. Please refer to ESM 5.15 to get a calculation example.

Finally, it should be noted that the above findings are only applicable when there are no structures on the beach. If the beach is built with structures such as breakwaters, the wave motion is blocked or reflected, the propagation changes significantly, and the breaking of waves in front of the breakwaters often changes and is closely related to the boundary conditions, such as the bed of the breakwaters, so it must be treated differently from the breaking of waves on the beach without breakwaters. The results of this research will be presented in the relevant specialist courses.

5.8.4 Wave Deformation Under Current Action

Waves are usually accompanied by currents, mainly tidal currents, runoff from rivers, coastal currents, and wind-blown currents, especially near tidal estuaries and bay gates. The interaction between waves and currents affects the propagation characteristics of the waves, resulting in changes in wave height, wavelength, wave speed, and wave direction. For example, near the mouth of a river, there are both river runoff and tidal currents, which together affect the waves. For example, at high tide, waves entering the estuary with the tide will increase in wavelength and decrease in wave height due to the influence of the current; while at low tide, waves entering the estuary against the current will decrease in wavelength and increase in wave height, i.e., the steepness of the waves will increase, often leading to wave breaking.

For these reasons, the actual wave data observed at sea are those influenced by local currents if they are present. If the current velocity reaches and exceeds a certain value, the deformation of the wave observations by the current is often not negligible, and therefore, when using these data, an attempt should be made to separate out the wave elements that are equivalent to those in still water.

For building design, the effect of the wave and current field on offshore buildings will also differ from the effect of a pure wave or current field.

In summary, the analysis of wave fields under the combined action of waves and currents is of great theoretical and practical importance when constructing harbor structures in marine areas with strong currents.

The wave deformation calculation methods presented below are based on the conservation of energy by wave action and assume that the energy flow of the current does not change before or after the wave and current interaction. The wave theory employed is linear, limited to regular and steady waves, while the current is assumed to be a steady, uniform flow with no consideration of energy loss along the propagation path.

Since there are many different combinations of water flow direction, wave direction, and the orientation of water depth contours, this section only introduces wave deformation under several combinations of waves, currents, and terrain. For related content, please refer to ESM 5.16.

5.8.5 Diffraction and Reflection of Wave

5.8.5.1 Wave Diffraction

When waves encounter obstacles such as island stations, headlands, or man-made structures (e.g., breakwaters) during wave propagation, some of the waves will continue to propagate around the obstacles and spread behind them, causing fluctuations in the sheltered waters. This phenomenon is called wave diffraction. The problem of wave diffraction is very important for the reasonable arrangement of port breakwaters in order to reduce waves in port waters and to ensure the safe operation of ship navigation, berthing, and loading.

Wave diffraction is a process whereby wave energy is transferred laterally along the crest line, i.e., the wave energy is redistributed from a region of high energy to a region of low energy. As a result, the wave heights along the same crest line are not equal after diffraction, but the wavelength and period remain the same. In Fig. 5.38, the incoming wave direction line passing through the head of the breakwater is called the geometric shadow line. If the waves were not bypassed, the water would be completely calm in the area covered by the breakwater to the right of the line, but in reality, due to the diffraction effect of the waves, the crest line of the incident wave extends in an arc from the geometric shadow line, centered on the head of the breakwater, towards the back of the breakwater, with the wave height decreasing the

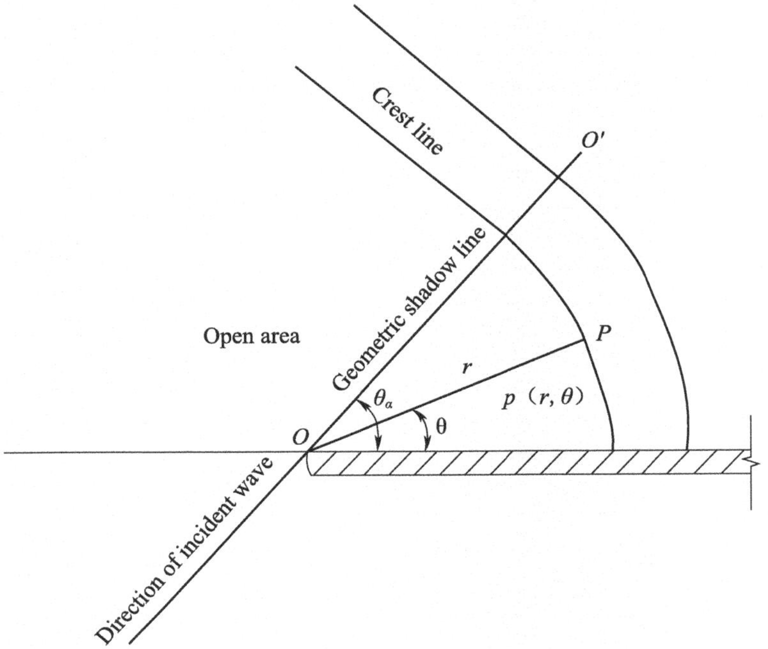

Fig. 5.38 Wave diffraction

further inwards the line extends. The incident wave to the left of the geometrically shaded line will also have a lower height as some of the energy spreads behind the dyke.

There are many research results on the wave diffraction problem. The theoretical solution of the wave diffraction problem is derived mainly by using the principle in optics, i.e., the diffraction of light waves behind a rigid thin-walled semi-infinite screen, which is the theoretical basis of the current wave diffraction calculation. The results of the theoretical approach have varying degrees of deviation from the actual situation, but there are also empirical approaches to wave diffraction calculations, based on a large number of model test results, and a set of empirical formulas or diagrams, such as the nomograph for engineering use.

The effect of wave diffraction is expressed as follows:

$$H_\mathrm{d} = K_\mathrm{d} H \tag{5.94}$$

where H_d is the bypassed wave height at a point behind the breakwater; K_d is the wave diffraction coefficient at a given point behind the breakwater, defined as the ratio of the wave height at that point to the incident wave height; H denotes the incident wave height at the breakwater gap.

The cumulative frequency of the wave train at the bypassed wave height is the same as the cumulative frequency of the incident wave height. The wave period in the diffraction zone is the same as the incident wave period at the mouth, while the wavelength is calculated based on the period and water depth [refer to Eq. (5.2)].

The latter is clearly more realistic, and the two methods recommended by our industry standard, the *Code of Hydrology for Harbour and Waterway* (JTS 145-2015)(2022) are described below.

1. Calculation of regular wave diffraction

Based on the analysis of previous theoretical and empirical calculation methods and the in-depth study of a large number of model test results at home and abroad, China has proposed an empirical formula for the calculation of wave diffraction for regular waves passing through single and double breakwater entrances.

(1) Single-projection embankments. These embankments are also divided into two categories: sloping embankments and straight wall embankments.

1) Sloping embankments.

$$\left.\begin{aligned} K_{\mathrm{d}_1} &= \frac{1}{2}\left\{\exp\left[-\frac{3}{4}\left(\frac{r}{L}\right)^{\frac{1}{3}}(\theta-\theta_0)\right]+\exp\left[-3\left(\frac{r}{L}\right)^{\frac{1}{3}}(\theta+\theta_0)\right]\right\} \\ K_{\mathrm{d}_2} &= 1-\frac{1}{2}\left\{\exp\left[-3\left(\frac{r}{L}\right)^{\frac{1}{3}}(\theta_0-\theta)\right]+\exp\left[-3\left(\frac{r}{L}\right)^{\frac{1}{3}}(\theta_0+\theta)\right]\right\} \end{aligned}\right\} \tag{5.95}$$

where θ_0 is the angle of the wave projection, i.e., the angle between the wave direction line and the axis of the jetty, in rad; r, θ are polar coordinates of the calculation point, with the origin of the coordinates at the head of the embankment, r in m, θ in rad, as shown in Fig. 5.38; K_{d_1}, K_{d_2} are wave diffraction coefficients at a point in the sheltered zone (diffraction zone, i.e., $\theta \leq \theta_0$) and the open zone (incident zone, i.e., $\theta \geq \theta_0$).

2) Upright embankments.

$$\left.\begin{aligned} K_{\mathrm{d}_1} &= \frac{1}{2}\left\{\exp\left[-\frac{1}{2}\left(\frac{r}{L}\right)^{\frac{m}{2}}(\theta_0-\theta)\right]+\exp\left[-1.9\left(\frac{r}{L}\right)^{\frac{1}{6}}(1+\theta_0-\theta)\theta_0^{\frac{1}{4}}\right]\right\} \\ K_{\mathrm{d}_2} &= 1-\frac{n}{2}\left\{1-\exp\left[-1.9\left(\frac{r}{L}\right)^{\frac{1}{6}}\theta_0^{\frac{1}{4}}\right]\right\} \end{aligned}\right\} \tag{5.96}$$

$$\left.\begin{aligned} m &= 1-\frac{7}{50}(\theta_0-\theta) \\ n &= \exp\left[-15\left(\frac{r}{L}\right)^{\frac{1}{2}}(\theta-\theta_0)^2\right] \end{aligned}\right\} \tag{5.97}$$

All of the above satisfy the following conditions:

$$\theta < \theta_0,\ r \to \infty,\ K_{d_1} \to 0$$
$$\theta > \theta_0,\ r \to \infty,\ K_{d_1} \to 1$$
$$\theta_0 = 0,\ K_{d_1} = 1$$
$$\theta = \theta_0,\ K_{d_1} = K_{d_2}$$
$$r \to 0,\ K_{d_1} = 1$$

(2) Double protruding breakwaters. The wave diffraction coefficient behind the double protruding breakwater entrance can be determined from Fig. 5.39, which is based on the Penny-Price method of electromagnetic wave diffraction analysis, improved from actual observations and model test data. The diffraction coefficient K_d is expressed as

$$K_d = \sqrt{\frac{L}{P}} f\left(\alpha, \frac{B}{L}\right) \tag{5.98}$$

where B is the mouthpiece width, when $\theta_0 \neq \frac{\pi}{2}$, the projected width of the portals in the direction perpendicular to the wave direction is used as the equivalent portal width; P is the distance between a point behind the dike and the midpoint of the gate; α is the angle between the line connecting a point behind the embankment and the midpoint of the mouthpiece gate and the wave direction line passing through the midpoint of the mouthpiece gate.

The conditions for the application of Fig. 5.1 in ESM 5.17 are $K_d \leq 1.0$, $\frac{\pi}{4} \leq \theta_0 \leq \frac{3\pi}{4}$, $B/L \leq 5.0$. If the calculated $K_d > 1$, it's still taken as $K_d = 1$. When $B/L > 5$, the calculation can be done for the left and right embankments separately as a single sudden embankment.

2. Calculation of Irregular Wave Diffraction

(1) Single protruding breakwater. It has been shown that the bypass coefficient of irregular waves is generally greater than that of regular waves in the sheltered area behind the dike and is less than that of regular waves in the open area, and this is a result confirmed by in situ observations. Therefore, according to the irregular wave diffraction model tests, combined with numerical calculations and verification of in situ observations, China has proposed a method for calculating the diffraction coefficient of irregular waves at single and double embankments, and has drawn up a set of nomographs for use in the calculation. These include 12 nomograms of K_d values for incident angles of 15°, 30°30°, 45°, 60°, 75°, 105°, 135°, and 165°, each under two irregular wave directional distribution functions.

(2) Double protruding breakwater. Figure 5.2 in ESM 5.17 shows the calculation method of irregular wave diffraction proposed by China. Firstly, the diffraction coefficients of the constituent waves of the regular waves are calculated

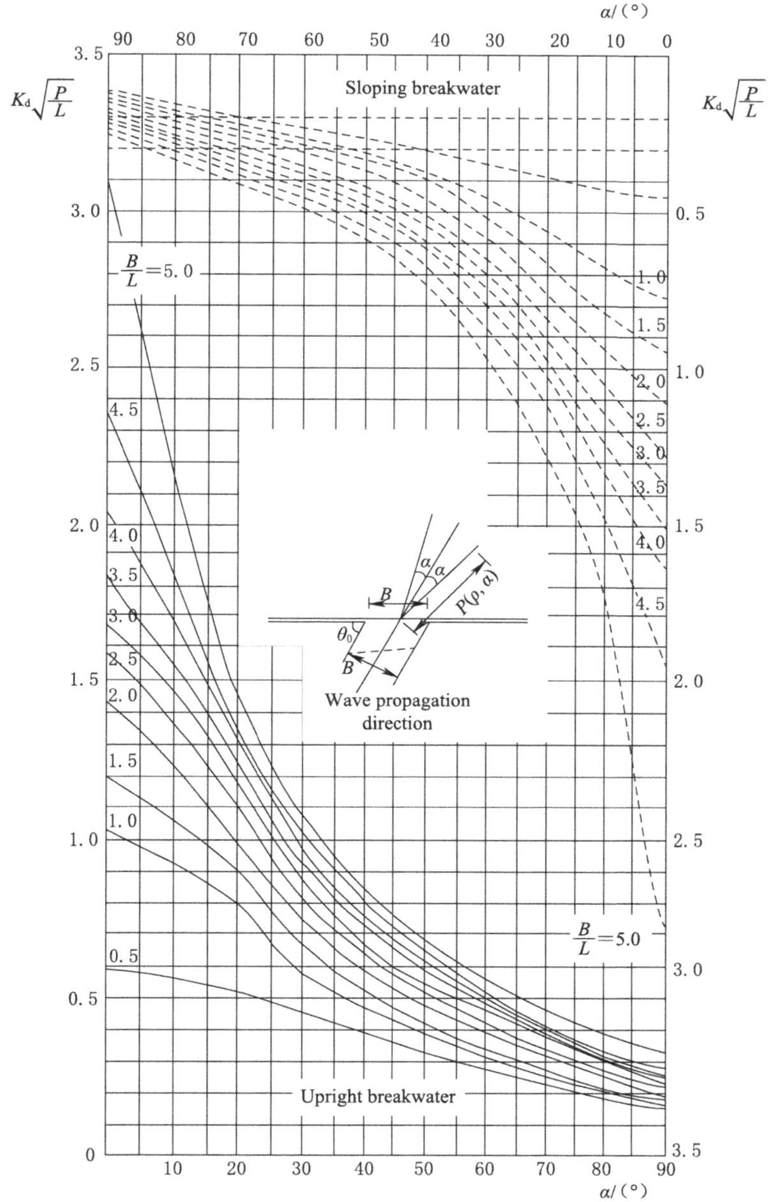

Fig. 5.39 Coefficient of regular wave diffraction behind the double protruding breakwater entrance

by the single source point method, and then, the nomogram of the irregular wave diffraction coefficient K_d is obtained by linear superposition and numerical integration of these components.

The results show that the reflection from the outer side of the embankment has little effect on the diffraction coefficient when the irregular waves are bypassed behind the double-jutting embankment; therefore, the nomograph is no longer divided into straight-walled and sloping embankments. The nomograph consists of $\theta_0 = \frac{\pi}{2}, \frac{\pi}{4}, B/L = 2.0, 4.0, 6.0, 8.0$. There are eight sheets in total, an example of which is shown in Fig. 5.1 in ESM 5.17.

The problem of regular and irregular wave diffraction from island breakwaters has also been studied in depth, but for lack of space, this will not be presented.

5.8.5.2 Wave Reflection

When a wave encounters a steep slope or an artificial structure during its propagation, all or part of its wave energy will be reflected, and a reflected wave will be formed, a phenomenon known as wave reflection. The basic principle of wave reflection is similar to that of light wave reflection, as shown in Fig. 5.40.

The reflected wave has the same wavelength and period as the incident wave, but its wave height depends on the wave energy of the reflected wave. The ratio of the reflected wave height to the incident wave height is called the reflection rate or reflection coefficient, denoted as K_R. When all wave energy is reflected, the reflected wave height equals the incident wave height, $K_R = 1.0$; when wave energy is partially reflected, the reflected wave height is less than the incident wave height, $K_R < 1.0$. The formula is

$$K_R = \frac{H_R}{H_i} \tag{5.99}$$

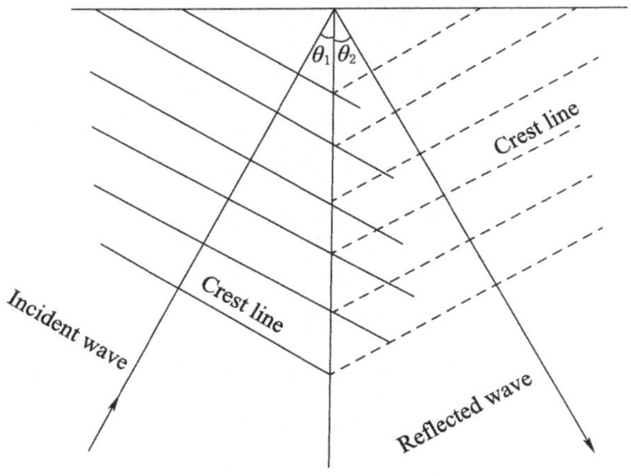

Fig. 5.40 Reflection of waves

where H_R is a reflected wave height; H_i is the incident wave height; The value of K_R ranges from 0 (no reflection) to 1.0 (total reflection). It is related to the slope, roughness, and porosity of the bank slope or artificial structure, as well as wave steepness and incident angle, and is difficult to determine accurately through calculation. For example, for an impermeable vertical-wall breakwater, the energy of normally incident waves is almost completely reflected, so $K_R \approx 1.0$.

There is both diffraction and reflection at a point in the harbor water. When the effects of both are to be considered, the wave height at this point can be approximated by the linear iterative principle of wave energy as follows:

$$H = \sqrt{H_d^2 + H_R^2} \tag{5.100}$$

where H_d is the wave height around the calculation point; H_R is the reflected wave height at the calculation point.

When there is wind, and the length of the wind area in the harbor exceeds 1 km, the wave height in the harbor can be calculated according to Eq. (5.101), taking into account the effect of local wind in the harbor.

$$H = \sqrt{H_1^2 + H_2^2} \tag{5.101}$$

where H_1, H_2 are the wave height around the calculation point and the wave height of the local wind-forming wave, and the latter is calculated according to Sect. 5.7.

When the water depth in the harbor basin varies significantly, and the wave propagation distance is long, both wave diffraction and refraction within the harbor need to be considered. Relevant research results show that within 3–4 wavelengths from the breakwater head, diffraction dominates; outside this range, the effect of wave refraction in the harbor becomes more significant. Based on this principle, the wave height at a nearshore point in the harbor can be approximately calculated using the following method.

The in-port crest line is first plotted as an approximate concentric arc in the sheltered area and as a parallel straight line in the open area, as shown in Fig. 5.40. However, for the double protruding breakwater entrance with where $B'/L \leq 1$, the wave crest lines in the harbor are approximately taken as concentric circular arcs centered at the midpoint of the entrance.

On the wave crest line, located 3–4 wavelengths. Away from the breakwater head, the diffraction coefficient K_d at each point is determined based on the regular wave diffraction principle. Beyond this crest line, a regular wave refraction diagram is plotted using the wave refraction principle. Thus, the wave height transformation coefficient at a nearshore point is

$$K' = K_d \sqrt{\frac{b_1 K_{S_2}}{b_2 K_{S_1}}} \tag{5.102}$$

where b_1, b_2 are the distances between two adjacent wave directions at the line of onset of refraction (i.e., 3–4 wavelengths from the breakwater head) and at a certain point; K_{s_1}, K_{s_2} are the shallow-water coefficients at the location.

Multiply the wave height at the breakwater gate by K' to obtain the wave height at a point in the harbor.

In fact, waves in harbors are often the result of a combination of diffraction, reflection, refraction, local wind and wave, and breakwater overtopping, and it is difficult to obtain a comprehensive and accurate result from these calculations. Therefore, for large or important projects, the most effective method is to carry out overall model tests to study the wave conditions in the harbor in a comprehensive manner and to produce results that are closer to reality.

As described in the sections of this chapter, wind waves are generated in deep water and then propagated to shallow water, where they undergo a series of changes in speed, wavelength, height, and direction due to the shallowing of water, wave refraction from the topography, and the influence of the current. Therefore, the variation of waves in shallow water of nearshore areas is extremely complex and is often the result of all these processes simultaneously. As most of the harbor structures are located in shallow waters, the analysis of nearshore waves is of great importance for harbor construction, waterway regulation, and coastal protection.

Chapter 6
The Tides

The tide of the Zhejiang River is a magnificent spectacle under heaven.
浙江之潮, 天下之伟观也。
It reaches its peak from the 16th to the 18th day of the lunar month.
自既望以至十八日为最盛。
When it first emerges at the sea gate, it appears as a silver thread.
方其远出海门, 仅如银线;
As it draws nearer, it resembles a jade wall or a snowy ridge, stretching to the sky.
既而渐近, 则玉城雪岭际天而来,
With a thunderous roar, it crashes and surges.
大声如雷霆, 震撼激射,
Swallowing the sky and drenching the sun, its might is supremely majestic.
吞天沃日, 势极雄豪。
Observing the Tidal Bore, Zhou Mi, Southern Song Dynasty
—《观潮》, 周密, 南宋
▲
This passage illustrates the magnitude of the Zhejiang River tide, from its narrow appearance at a distance to its high and turbulent front near the shore, indicating its considerable energy and strong tidal forces.

Abstract This chapter begins with tidal phenomena and their mechanisms, exploring gravitational theory, tidal constituents, and amphidromic systems. Next, tide observation and forecasting details gauge networks, satellite-based methods, and harmonic analysis. The third section covers extreme value analysis and non-stationary adjustments for climate change. Finally, storm surge fundamentals address meteorological forcing and coupling with tides.

Supplementary Information The online version contains Electronic Supplementary Material (ESM) available at https://doi.org/10.1007/978-981-95-9089-6_6.

X. Feng (ed.), *Engineering Hydrology*, https://doi.org/10.1007/978-981-95-9089-6_6

For a visual overview of the chapter's key concepts, refer to the knowledge map below.

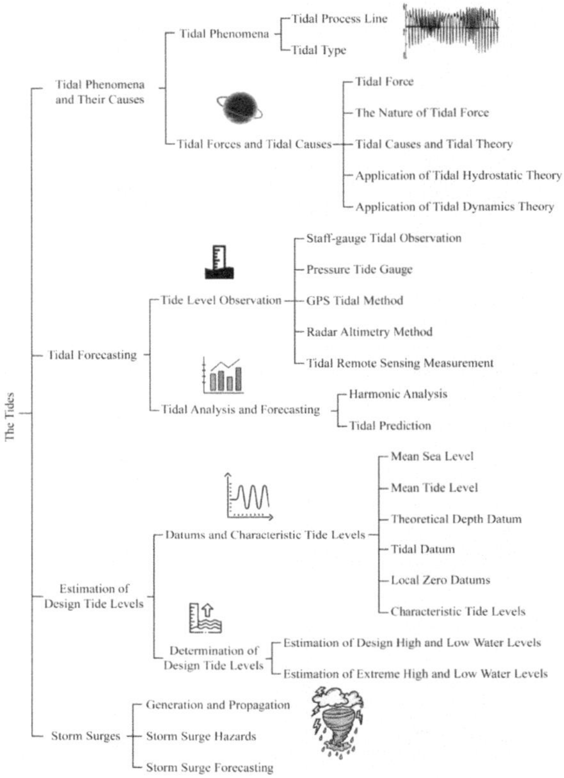

6.1 Introduction

Tides are complex oscillations of water levels caused by the combined gravitational and centrifugal forces of the Earth-Moon and Earth-Sun systems. Their periods range from 12 to 24 h, and they can lead to sea level changes from almost zero to over ten meters, depending on the location.

In 1740, D. Bernoulli proposed the equilibrium tide theory, assuming the Earth's surface was fully covered by oceans. This was followed by P. S. Laplace's dynamic theory in 1775, and further advancements in tidal theories continued, culminating in numerical solutions by J. Proudman and A. T. Doodson in the twentieth century. With the advent of electronic computing, tidal research progressed rapidly. However, nonlinear effects from complex nearshore topography, human activities, and climate change add uncertainty to tidal predictions.

Meteorological factors such as low-pressure systems, cold waves, and typhoons also influence tidal levels. These phenomena, combined with astronomical tides, can produce nonlinear effects, leading to storm surges, which are significant contributors to coastal flooding. For example, Typhoon "Yagi" in 2024 caused a storm surge of 1.5–2.3 m, flooding Hainan Province, causing widespread damage, highlighting the increasing threat posed by storm surges and extreme weather events.

The purpose of this chapter is to help readers understand the main components of the measured tidal levels, that is, the physical principles and consequences behind astronomical tides and storm surges, as well as the methodologies of tidal prediction, in order to facilitate preparation and management for coastal inundations.

6.2 Tidal Phenomena and Their Causes

The periodic upward and downward movements of the sea on the Earth, caused by the gravitational pull of the Moon, the Sun, and other celestial bodies, are called tides. The main reason for the tidal phenomenon is that the relative position of different seas on the Earth in relation to the Moon and the Sun causes different gravitational forces to be exerted on them, resulting in the relative movement of the waters on the Earth. In addition to the gravitational force of the Moon and the Sun, there is also the inertial centrifugal force generated by the Earth and the Moon and the Earth and the Sun moving around their common centre of mass. Typhoons, cold waves, and other weather systems that bring high winds or dramatic changes in air pressure can also cause abnormal sea level rise and fall, a phenomenon known as storm surge. The planning, design, construction, operation, and management of marine engineering and port projects all require an understanding of the laws of tidal changes on the sea surface. In order to correctly apply tidal level observations to derive the design tide level required for a project, this chapter firstly introduces the tidal phenomenon and its causes, the periodic characteristics of tidal phenomena, and the astronomical, meteorological, and geographical factors affecting tidal variations, so that the reader has a proper understanding of the use of tidal level information. The chapter focuses on the derivation of design tide levels, including the derivation of tide levels in the presence of long-term or a shortage of measured tide information.

6.2.1 Astronomical Knowledge Related to Tidal Phenomena

Tides are caused by the gravitational pull of celestial bodies, primarily the Moon and the Sun, as they move in relation to the Earth. The ever-changing positions of these celestial neighbors make tides rise and fall in predictable patterns. To truly understand why tides behave as they do, we need to explore some key astronomical concepts. Please refer to ESM 6.1 to discover the cosmic forces behind Earth's tides.

6.2.2 *Tidal Phenomena*

To grasp the laws of tidal change and development, one should start by understanding tidal phenomena. Under the influence of the Moon, the Sun, and other celestial bodies, the periodic movement of seawater on the sea surface is called tides. It consists of the periodic vertical rise and fall of the sea surface and the periodic horizontal inward and outward flows of the sea. It is customary to refer to the former as tides and the latter as currents.

6.2.2.1 Tidal Process Line

Figure 6.1 shows a section of tidal level information measured at a tide gauge station, i.e., the tide level as a function of time, with the horizontal coordinate being the time and the vertical coordinate being the tide level.

When the sea reaches its highest level, it is called high tide, and when it falls to its lowest level, it is called low tide. In each tidal cycle, the interval from low water to high water is termed the flood duration, and that from high water to low water is termed the ebb duration. The difference in water level between the adjacent spring and neap tides is called the ebb tide difference, and the difference in water level between the adjacent neap and spring tides is called the flood tide range.

6.2.2.2 Tide Type

The tidal curves vary with location and date and are very complex. However, in terms of long-term records, there are three broad types of tides: semidiurnal, diurnal, and mixed, as shown in Fig. 6.2.

(1) Semidiurnal tides. Two high tides and two low tides occur within one lunar day (24 h 50 min). The tide levels of the two adjacent spring and neap tides are approximately equal, the flood duration is the same as the ebb duration, and the tide curve is approximately a symmetrical cosine curve. For example, the tide pattern is typical of the semidiurnal tide pattern in Hangzhou Bay in China and the coast of Panama in Central America.
(2) Diurnal tides. Most of the days in a lunar month have a high and a low tide each day. In diurnal-tide regions, the tidal range becomes distinctly larger a few days after the Moon stands to the north or south of the equator, whereas it turns noticeably smaller—and occasionally even shows two tidal cycles per day—several days after the Moon rises due east or sets due west.
(3) Mixed tides. Mixed tides are divided into mixed semidiurnal tides and mixed diurnal tides. Mixed semidiurnal tides are those that also have two high tides and two low tides in one lunar day, but the adjacent high tides or low tides are of different heights, and the flood duration and ebb duration are also different. Mixed diurnal tides are semidiurnal tides on most days of a half month, but

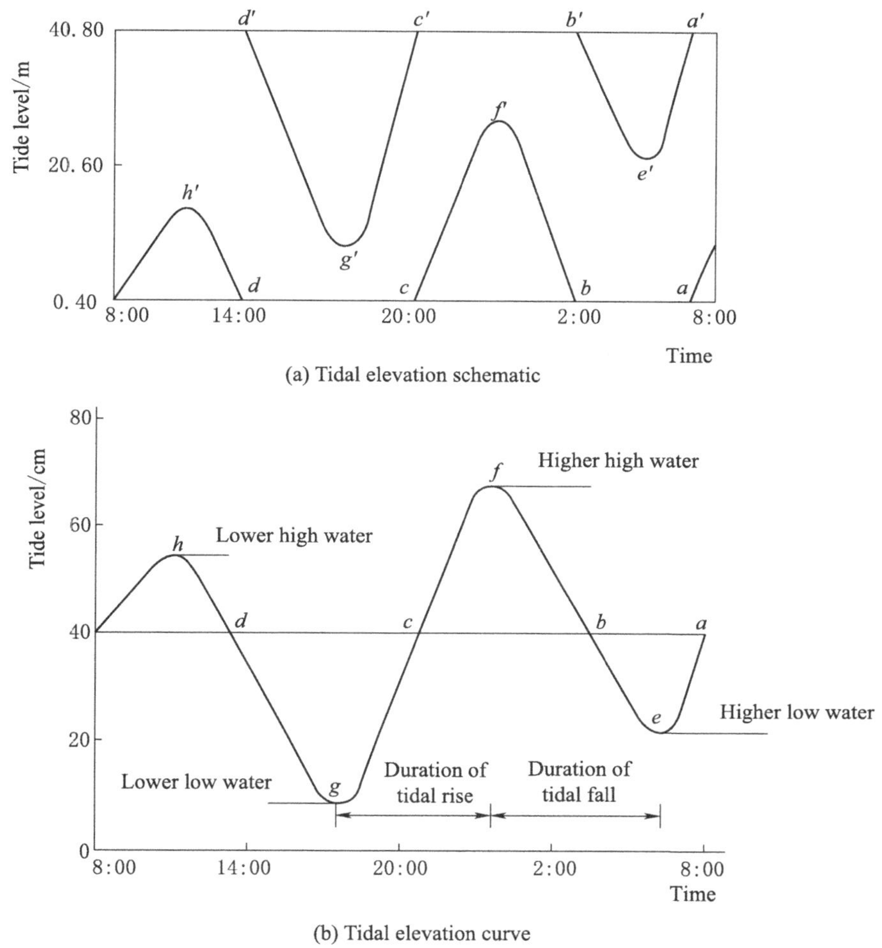

Fig. 6.1 Tidal elevation variation during a tidal day

sometimes a diurnal high tide and a diurnal low tide occur, and the number of days of diurnal tides does not exceed 7 d.

6.2.3 Tidal Forces and Tidal Causes

The understanding of tidal phenomena and their causes dates back more than two thousand years, but the principles of their mechanics were not properly explained at the time. It was not until 1687, when Newton proposed the law of gravity, that he explained the tidal phenomena occurring in the ocean due to the action of the

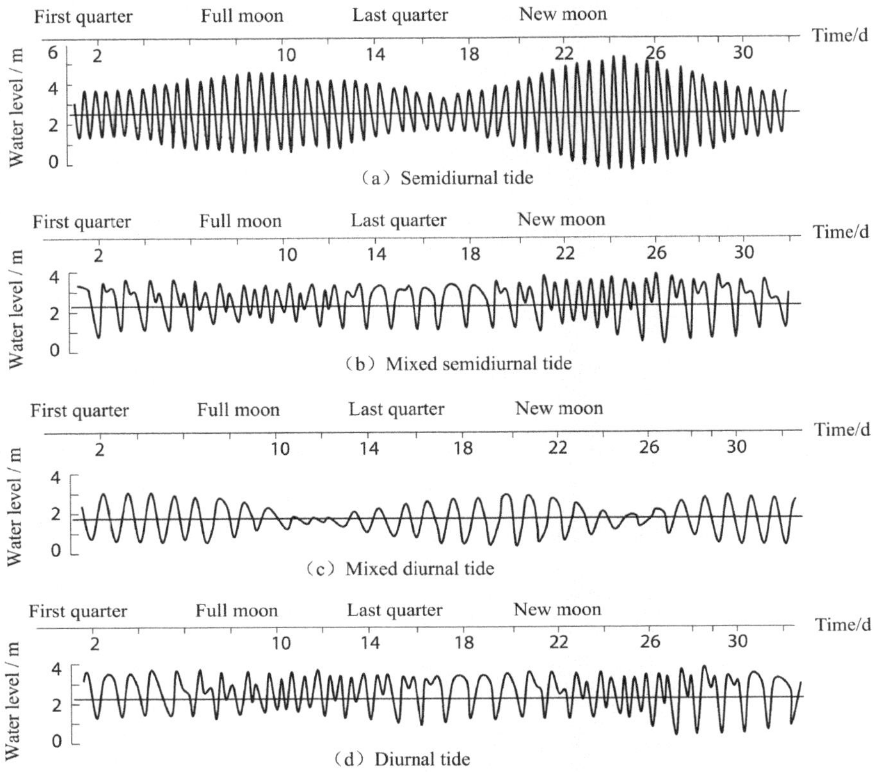

Fig. 6.2 Curves for various types of tidal monthly processes

gravitational force, and laid the foundation for modern tidal theory, which is known as the Newtonian tidal hydrostatic theory or equilibrium tidal theory. As the hydrostatic theory was unable to give a satisfactory theoretical explanation for the complex tidal phenomena, Laplace proposed a theory of tidal dynamics about 100 years after Newton proposed his theory, which suggested that periodic horizontal tidal forces caused forced vibrations with corresponding components, explaining the causes of tides from a dynamic point of view. The following is a brief introduction to the theory of tidal forcing, Newtonian static tidal theory, and Laplace's kinetic tidal theory, and the use of these tidal theories to qualitatively explain complex tidal phenomena.

6.2.3.1 Tidal Force

The expression for the tidal force is first derived from the equilibrium of the forces on a water point of unit mass on the surface of the Earth's outer water body when it is subjected to the action of the Moon, and then extended to the tidal force of celestial

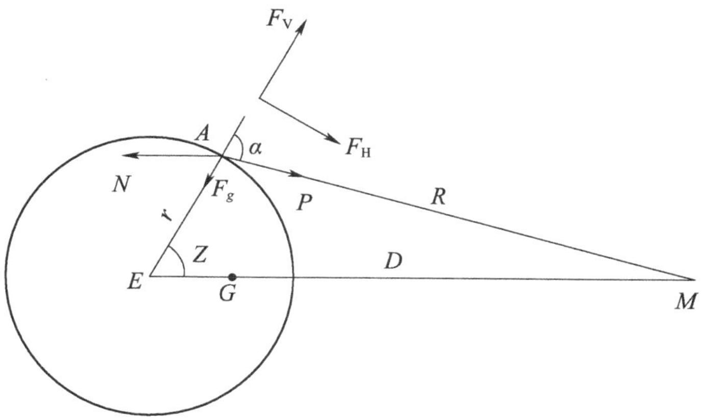

Fig. 6.3 Surface water particle stress analysis diagram

bodies in general. As shown in Fig. 6.3, a water particle of unit mass located at point A is subjected to several forces as follows.

(1) The gravitational force of the Moon. According to the law of gravitation, the total attraction between the Earth and the Moon is $P_t = K\frac{EM}{D^2}$, where K is the universal gravitational constant, E is the mass of the Earth, M is the mass of the Moon, D is the distance from the center of the Earth to the center of the Moon. Then, for a unit mass of water at point A, the gravitational force of the Moon on it is

$$P = K\frac{1 \cdot M}{R^2} \tag{6.1}$$

where R is the distance from the point of application of the force to the center of the Moon, and the force is directed towards the center of the Moon.

(2) The inertial centrifugal force. Considering the Earth and the Moon as a common system, its common center of mass G can be found within the Earth at a distance of $0.73r$ from the Earth's center, with r being the radius of the Earth. Astronomical observations have shown that the motion of the Earth around the common center of mass of the Earth and the Moon is an overall translational motion in which the trajectory of each point is circular, similar to the motion of a grinding rod when pushing a mill. Hence, at any instant, the inertial centrifugal force is uniform both in magnitude and direction over the entire Earth, identical to that at its center (Fig. 6.4).

In the relative motion of the Earth-Moon system around its common center of mass G, the Moon-point G-the Earth line (MGE) is always in a straight line. If we assume that the masses of the Earth and the Moon are concentrated at their respective centers of mass. Then the total inertial centrifugal force of the Earth should balance

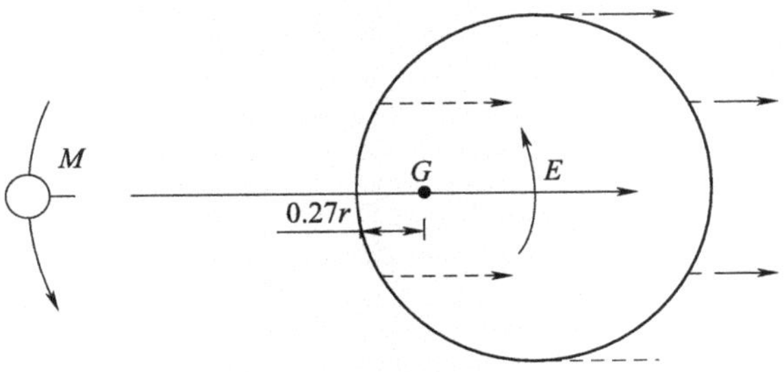

Fig. 6.4 Centrifugal forces acting at various points on the Earth

with the total attractive forces of the Earth and Moon, i.e., $EN = K\frac{ME}{D^2}$, where N is the inertial centrifugal force per unit mass of the Earth, with the other symbols as before. From this, the inertial centrifugal force per unit mass of the Earth can be found as

$$N = K\frac{M}{D^2} \tag{6.2}$$

Its direction is the same at each point as at the center of the earth, parallel to the *MGE* line and pointing away from the center of the moon.

(3) Gravity. The gravitational force per unit mass on a water particle is

$$F_g = g \tag{6.3}$$

Its direction points to the center of the earth.

Of the above three forces acting on a water particle, all are tidal forces except for gravity, so the combination of the first two forces is referred to as the tidal force.

$$\boldsymbol{F} = \boldsymbol{N} + \boldsymbol{P} \tag{6.4}$$

6.2.3.2 The Nature of Tidal Force

In order to understand the nature of the tidal force, the tidal force $\boldsymbol{F}$ is decomposed into its components along the radial (vertical) and tangential (horizontal) directions of the coordinates, as shown in Fig. 6.3. The radial component $\boldsymbol{F}_V$ and tangential component $\boldsymbol{F}_H$ are

$$F_V = P_V - N_V = K\frac{M}{R^2}\cos\alpha - K\frac{M}{D^2}\cos Z \tag{6.5}$$

$$F_H = P_H - N_H = K\frac{M}{R^2}\sin\alpha - K\frac{M}{D^2}\sin Z \tag{6.6}$$

where Z is the zenith distance of the Moon; α is the angle between the normal to the radial-force coordinate and AM, as shown in Fig. 6.3; other symbols are as before.

If you start with $R^2 = D^2 + r^2 - 2rD\cos Z$, $\cos\alpha = \frac{D\cos Z - r}{R}$ and $\sin\alpha = \frac{D\sin Z}{R}$, substituting these into Eqs. (6.5) and (6.6) and omitting $\frac{r}{D}$ of the higher-order term yields

$$F_{\text{V}} = \frac{KMr}{D^3}\left(3\cos^2 Z - 1\right) \tag{6.7}$$

$$F_{\text{H}} = \frac{3KMr}{2D^3}\sin(2Z) \tag{6.8}$$

In Eqs. (6.7) and (6.8), if the solar mass S is substituted for M, the solar zenith distance Z' is substituted for Z, and the solar-Earth distance D' is substituted for D, the vertical component of the solar gravitational force $\boldsymbol{F}'_{\text{V}}$ and the horizontal component $\boldsymbol{F}'_{\text{H}}$ can be derived, yielding the expressions for both components.

The analysis of the above equations provides an understanding of the nature and characteristics of the tidal force.

(1) As the Earth rotates around its axis, the gravitational force at each point on the surface will vary periodically, and the amount of gravitational force is proportional to the mass of the object and inversely proportional to the third power of the distance from the center of the Earth.

(2) When the Moon is in the same position as shown in Fig. 6.5, the lunar gravitational force at each point on the surface varies accordingly, as can be seen from the above equations for the vertical and horizontal components of the gravitational force. At points B ($Z = 0°$) and D ($Z = 180°$), the vertical component of the gravitational force has a maximum value $F_{\text{V max}} = \frac{2KMr}{D^3}$, and the horizontal component is zero ($F_{\text{H}} = 0$). At points A ($Z = 90°$) and C ($Z = 270°$), the vertical component of the tidal force is negative (pointing towards the center of the Earth), with $F_{\text{V}} = -\frac{KMr}{D^3}$ and the horizontal component is zero ($F_{\text{H}} = 0$). The horizontal component of the force reaches its maximum $\left(F_{\text{H max}} = \frac{3KMr}{2D^3}\right)$ only at $Z = 45°$, $135°$, $225°$, and $315°$. At points symmetrical to the line BD, the maximum value of the gravitational force is equal in the upper and lower hemispheres, and its direction can be found by means of mirror reflection.

(3) The lunar gravitational force at the center of the Earth is zero because it is equal in magnitude and opposite in direction to the inertial centrifugal force.

(4) As the Earth rotates around its axis, there are periodic variations in the gravitational forces at various points on the surface. Moreover, due to the constant movement of the celestial bodies, the tidal force has a complex periodic variation.

These are some of the basic properties of the tidal force. Figure 6.5 also shows that the maximum vertical component of the tidal force at the points B and D is $F_{\text{V max}} = \frac{2KMr}{D^3}$. Comparing F_{Vmax} with the local gravitational force per unit mass of water $F_g = \frac{KE}{r^2}$, yields a ratio of

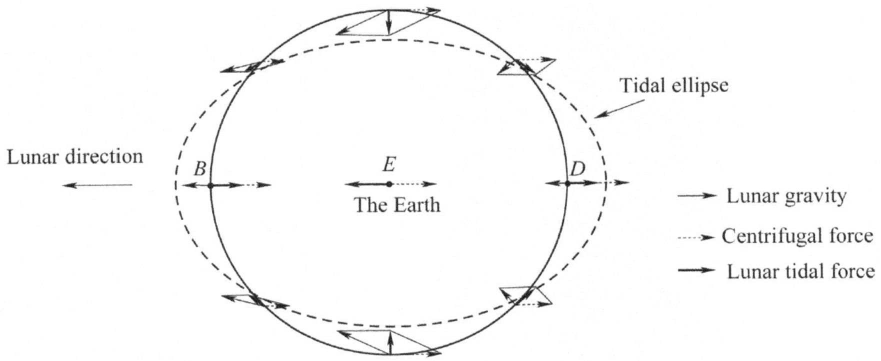

Fig. 6.5 Lunar tidal gravitational forces and tidal ellipses

$$\frac{F_{\mathrm{V\,max}}}{F_g} = \frac{\frac{2KMr}{D^3}}{\frac{KE}{r^2}} = \frac{1}{9{,}000{,}000} \tag{6.9}$$

where $D = 60.3r$ and $E = 81.5\,M$.

This shows that the maximum vertical tidal force is only one nine-millionth of the gravitational force, which means that the effect of F_{V} on tidal phenomena is very small. Although the horizontal component of the gravitational force is small, it is perpendicular to the direction of gravity and acts at all points on the Earth, causing significant tidal phenomena in the ocean. If it is assumed that the surface of the Earth is covered with seawater of equal depth, this layer of water forms a tidal ellipsoid under the action of the lunar gravitational force, as shown in Fig. 6.6.

The vertical component of the solar tidal force can be calculated in a similar way.

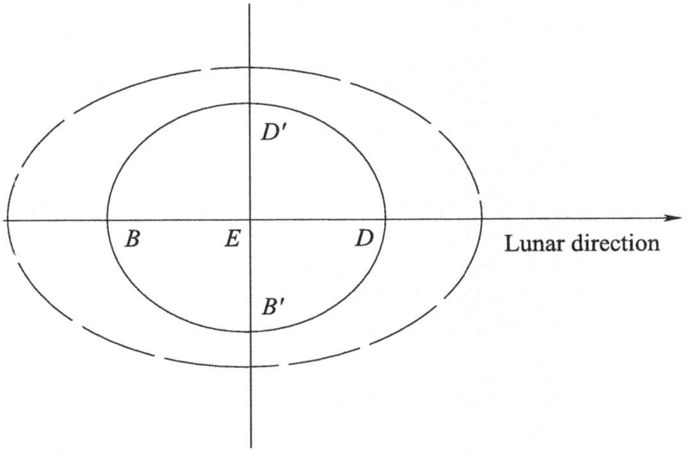

Fig. 6.6 Cross-section of the Earth and tidal ellipsoid along the equator

$$F'_V = \frac{2KSr}{D'^3} \tag{6.10}$$

$$F'_H = \frac{3KSr}{2D'^3} \tag{6.11}$$

When both the Moon and the Sun are located at the zenith at point B, the ratio of F_V to F'_V is

$$\frac{F_V}{F'_V} = \frac{2KMr}{D^3} \Big/ \frac{2KSr}{D'^3} = 2.17 \tag{6.12}$$

where $S = 333400E$, $D' = 2348\ r$.

From Eq. (6.12), it is clear that the solar gravitational force is only about 1/2 of that of the Moon.

6.2.3.3 Tidal Causes and Tidal Theory

While only tidal forces and their properties have been discussed, there are two theories of how tides are generated under the influence of tidal forces and how their causes are investigated, namely the hydrostatic theory and the dynamic theory. A brief summary is given below.

1. Tidal Hydrostatic Theory

The hydrostatic theory of tides is the study of the formation of tides through the hydrostatic equilibrium of surface water masses, and is therefore also known as the theory of equilibrium tides. The hydrostatic theory has two assumptions: ① the Earth's surface is uniformly covered with seawater of equal depth; ② the water body is influenced solely by gravitational forces, with inertia and friction neglected. The surface of the sea is formed into an ellipsoid, called a tidal ellipsoid, due to the action of the tidal forces of the Sun and Moon. This tidal ellipsoid is the result of the mutual equilibrium between the gravitational force, the gravitational force at the center of the Earth, and the pressure difference at sea level, which is the basic idea of Newton's tidal hydrostatic theory.

(1) The concept of hydrostatic tides. According to the hydrostatic tide theory, the sea surface forms a tidal ellipsoid under the action of the gravitational tidal force. Figure 6.5 shows a longitudinal section of the Earth and the tidal ellipsoid along the Earth's axis (AC) and the Moon's center M. Figure 6.6 shows a transverse section along the Earth's equatorial plane (BD) and the Moon's center M. When the Earth rotates, a point on the Earth will be in different positions at different times of the day, with high tide when the point is in positions B and D, and low tide when it is in positions B' and D', thus creating two high tides and two low tides in a day. As the Moon is also in motion, it takes 24 h 50 min for point B on the Earth (where the Moon's zenith distance $Z = 0°$) to return to the same

position, so the daily cycle of tidal change is 24 h 50 min, which is known in astronomy as a lunar day.

(2) Equation for hydrostatic tidal height. According to the theory of hydrostatics, the equilibrium tidal height can be derived by analyzing the water particle in equilibrium under the action of the hydrostatic force. Figure 6.7 shows the longitudinal section along the axis of the Earth and the plane where the center of the Moon is located. As mentioned above, assuming that the surface is covered by a uniform layer of water, when subjected to the action of the tidal force, the water body will form an ellipsoid rotating around the long axis. The water surface at point *B* rises above that at point *A* by a height *h* relative to the still-water level. The resulting neutral (equipotential) difference between points *A* and *B* is balanced by the vertical component of the tide-generating force, thus

$$\frac{KE}{r} - \frac{KE}{r+h} = \frac{3}{2}\frac{KMr^2}{D^3}\left(\cos^2 Z - \frac{1}{3}\right) \tag{6.13}$$

After Eq. (6.13) is sorted out, the formula for calculating the lunar tide height can be obtained by omitting the terms above the $(h/r)^2$:

$$h = \frac{3}{2}\frac{Mr^4}{ED^3}\left(\cos^2 Z - \frac{1}{3}\right) \tag{6.14}$$

Since the zenith distance *Z* in the tidal height formula for the tide of lunar is related to the latitude φ, lunar declination δ and time angle *T*, we can apply the spherical trigonometric formula by expanding $\cos^2 Z$ and then substituting it into Eq. (6.14) to obtain

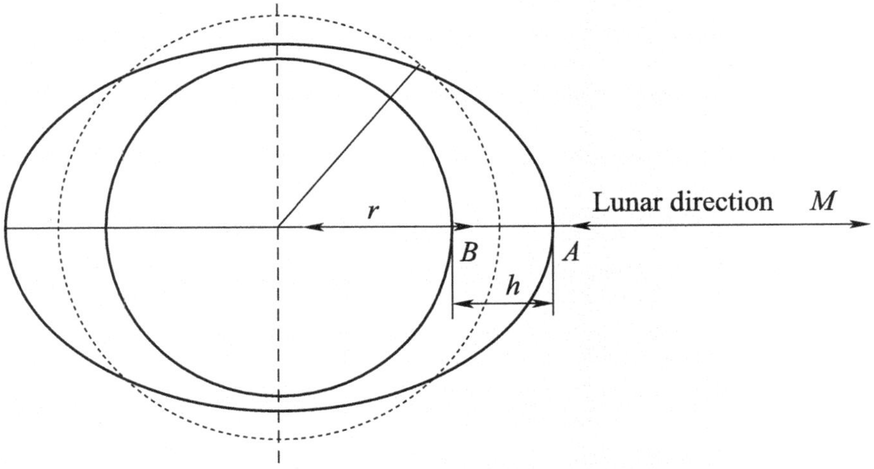

Fig. 6.7 Derivation diagram of the tidal range formula

$$h = h_1 + h_2 + h_3 \tag{6.15}$$

Of which:

$$h_1 = \frac{3}{2}(Mr^4)/(ED^3)\left[\frac{1}{2}\sin(2\varphi)\sin(2\delta)\cos T\right]$$

$$h_2 = \frac{3}{2}(Mr^4)/(ED^3)\left[\frac{1}{2}\cos^2\varphi\cos^2\delta\cos(2T)\right]$$

$$h_3 = \frac{3}{2}(Mr^4)/(ED^3)\left(\frac{1}{2} - \frac{3}{2}\sin^2\varphi\right)\left(\frac{1}{3} - \sin^2\delta\right)$$

In the above equations, T is the hour angle, whose period of variation is 24 h; δ is the lunar declination, whose period of variation is a month; φ is the local latitude, which is a fixed value for a fixed point. Thus, it can be seen that h_1 is the full-day division of the tide; h_2 is the semidiurnal division of the tide; and h_3 contains the $\sin^2\delta$ term, because $\sin^2\delta = \frac{1}{2}[1 - \cos(2\delta)]$, so that h_3 is the semidiurnal division of the tide.

Using the theory of hydrostatic tides and the same approach as above, the formula for the tidal height of a solar tide can be obtained as follows:

$$h' = \frac{3}{2}\frac{Sr^4}{ED'^3}\left(\cos^2 Z' - \frac{1}{3}\right) \tag{6.16}$$

The maximum difference between the tides caused by the above tidal formula of the tide of the Sun and the tide of the Moon is only 0.78 m, which is similar to the actual tide difference in the ocean, but is very different from the tide difference in the shallow sea. For example, the maximum tidal difference in Hangzhou Bay is 9 m, and in the Bay of Fundy, North America, the maximum tidal difference is 18–21 m. In addition, there are many discrepancies between the conclusions drawn from Newton's hydrostatic tide theory and the actual ocean tidal phenomena. For example, the hydrostatic tide theory does not take into account the movement of seawater at all, and therefore, it cannot explain the important phenomenon of tidal currents. This is due to two basic assumptions of the hydrostatic tide theory that are different from reality.

In response to these shortcomings, the tidal dynamic theory of tides has emerged from the viewpoint of seawater movement.

2. Tidal Dynamic Theory

Laplace proposed the theory of dynamics in 1775 and argued that tides are tidal wave movements under the action of the horizontal gravitational force of the Moon and the Sun, i.e., the tidal phenomenon is a fluctuation. Oceanic waters are subjected to a horizontal gravitational force field that causes tidal waves to flow, with high tides occurring when the crest of the wave is reached and low tides occurring when the trough of the wave is reached. The dynamic theory improves on the main drawback of the hydrostatic theory, namely the assumption that the surface of the ocean is at rest under the action of gravitational forces and gravity. In addition to the gravitational force, there are also the Coriolis force and the frictional force that act on the movement of the tidal waves. The latter two forces are induced once the motion has started, and disappear as soon as the motion stops. Therefore, to study the movement of tidal waves in the ocean, special consideration should be given to the Coriolis force and the frictional forces on the sea floor (especially important in shallow seas) in order to match the actual phenomenon. In his study, Laplace still assumes that the Earth is covered by an equal thickness of seawater and that tidal wave motion occurs under the action of tidal force, pressure, and the Coriolis force. In the solution, the wave motion was treated separately for three tidal species: long-period, semidiurnal, and diurnal tides. Laplace's work still suffers from the omission of frictional forces and the failure to consider the effects of topography. However, this theory treats the tide as forced vibrations. The proposed system of equations for tidal motion is still the theoretical basis for the study of tides.

Prior to the 1960s, with very few exceptions, the analytical solutions were based on idealizing the shape and depth of the actual ocean, and in this regard, Airy and Hough in the nineteenth century and Longuet-Higgins in the twentieth century were able to solve for simple-shaped oceanic tides (e.g., assuming that the globe is covered by seawater or replacing the actual ocean with a trench distributed along the meridian or a trench distributed along the latitude circle). Although the sea areas studied are idealized, the results obtained are important for gaining insight into and explaining the actual oceanic tidal movements.

Another direction is the study of tidal wave propagation in smaller scale oceans. In these studies, the Coriolis force is treated as a constant, and a right-angle coordinate system is allowed to be used. As early as 1879, Thomson (later Lord Kelvin) studied long waves in infinitely deep trenches under the action of the Earth's rotation, deriving the famous Kelvin wave, and in the 1910s, Poincaré studied a wider class of long wave motions, and a little later, Taylor derived solutions for tidal waves in bays and rectangular seas. Chen Zongbi, Fang Guohong, Wang Renshu, Hendershott, and Speranza considered the effect of friction in different ways, which played an important role in further understanding the tidal distribution in marginal seas.

Prior to the 1950s, tidal research was mainly analytical in approach. Since the advent of computers, numerical tidal solutions have flourished and have become an important tool and field in tidal research. By the late 1960s, many scholars had already calculated the tides of the actual oceans. Based on previous work, Schwiderski (1980) gave fairly good numerical results for oceanic tides, with deviations of about 10 cm compared to actual measurements. Numerical calculations of tides for marginal seas or harbors and rivers are even more numerous, and some of these results are accurate enough for practical application. The Ministry of Natural Resources of the People's Republic of China has used computers to analyze and forecast tides since 1964 and has achieved remarkable results so far.

6.2.3.4 Application of Tidal Hydrostatic Theory

According to Newtonian hydrostatic theory, the tidal ellipsoid is formed by the tidal force on the surface of the Earth, and this concept accounts for both the periodicity of tides and the diurnal inequality.

(1) Equinoctial tides and regression tides. Assuming that the Moon is over the equator, i.e., the lunar declination is zero, the tidal ellipsoid at this point is shown in Fig. 6.8, with its long axis pointing towards the Moon. If an observer is standing at point *A*, he can observe the first high tide; after a quarter of the Earth's rotational period (6 h 12.5 min), the observer can see the first low tide; after another quarter of the Earth's rotational period, he reaches point *B* and sees the second high tide; after another quarter of the Earth's rotational period, he can see the second low tide; after another quarter of the Earth's rotational period, he returns to the original point *A* and sees the high tide again. Thus, after one day, the observer sees two high tides and two low tides, which is a typical semidiurnal cycle. The two high tides (or low tides) are of equal height, and the spring and neap tides have the same duration.

When the Moon leaves the equatorial plane and moves south or north, i.e., when the declination is not zero, the lunitidal inequality occurs. This is most pronounced when the Moon reaches near the Tropic of Cancer, when the tide is called the regression tide. Figure 6.9 shows the tidal ellipsoid caused when the Moon reaches its northernmost declination. The first high tide is observed when the observer is at point *A*. Due to the rotation of the Earth, the observer can see the first low tide after a rotation of 90°. At point *B*, a second high tide is observed, but the tide height is lower than the first high tide. Thus, when the Moon's declination is not zero, the two high tides (or low tides) of the day are of different heights, and the spring and neap tides are not

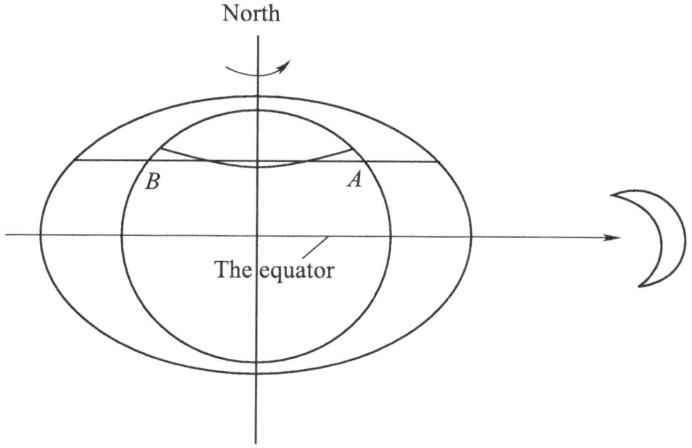

Fig. 6.8 Equinoctial tide

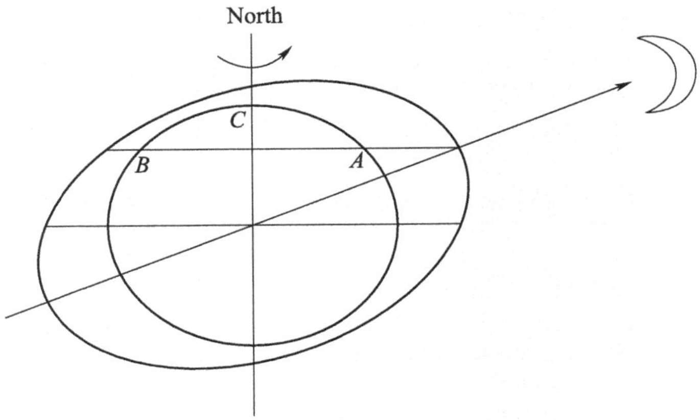

Fig. 6.9 Diurnal inequality tide

of the same duration. At high latitudes, when the lunar declination is not zero, the tide loses its semidiurnal tidal cycle and appears as a daily cycle tide. For example, at point C in Fig. 6.9, there is only one high tide and one low tide during the day.

(2) Spring and neap tides. Ocean tides are not only caused by the Moon, but the Sun can also play a role. Therefore, the periodic changes in the relative positions of the Moon and the Sun in space cause the tidal scale to change periodically. Figure 6.10 shows the relative positions of the Sun, the Moon, and the Earth during spring tides (at new and full moons) and neap tides (at the first and last quarter Moons). When the Sun and the Moon are in a straight line, the long axes of the tidal ellipsoids of the Sun and the Moon point in the same direction, and the two tides are added together to form the spring tide. When the Sun and the Moon are perpendicular to each other, the long axes of the tidal ellipsoids of the two tides are perpendicular to each other, and the two tides subtract from each other to form the neap tide. Because there is a high tide and a low tide during the semilunar cycle, the phenomenon of semilunar inequality is created.

(3) Tidal monthly, annual, and multi-annual inequalities. In addition to the tidal diurnal and semiannual inequalities mentioned above, there are also monthly, annual, and multi-annual inequalities. Because the Moon's orbit around the Earth is an ellipse with the Earth at one focus, the Moon's gravitational force pull on the Earth is 40% greater at perigee than at apogee, which results in lunar tidal inequalities. Similarly, over the annual cycle, the tide-generating force at perihelion is about 10% greater than at aphelion, giving an annual variation in tidal magnitude.

In addition, the long axis of the Moon's orbit is constantly shifting due to gravitational interactions, with the perigee constantly moving eastwards with a period of 8.85 years. The nodes of intersection between the ecliptic and the lunar orbit are

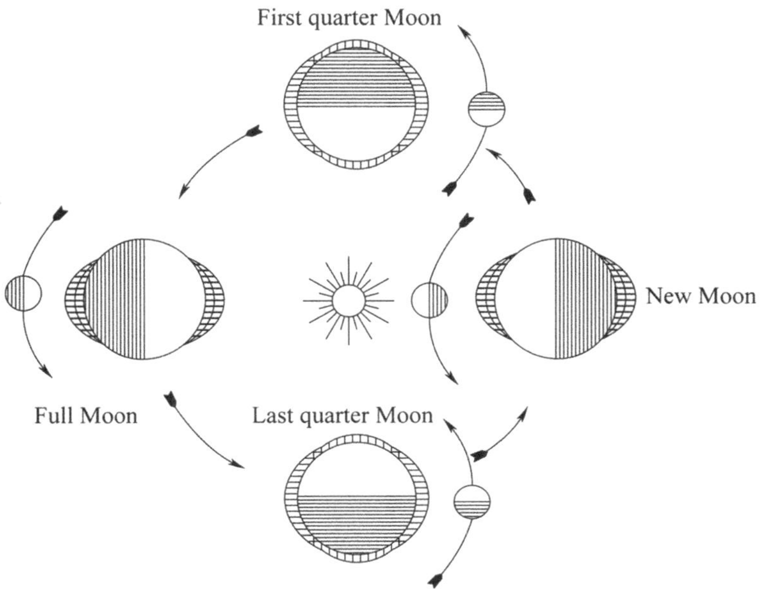

Fig. 6.10 Lunar phases and spring-neap tides

also in continuous motion, with a period of 18.61 years. The tides thus have a long period of 8.85 years and 18.61 years, which is why 9 and 19 years should be taken into account when we analyze tide level data.

6.2.3.5 Application of Tidal Dynamic Theory

The tidal cycles and tidal inequalities described above are fully consistent with the conclusions drawn from the Newtonian hydrostatic theory and the qualitative discussion of the tidal ellipsoid. However, since the two basic assumptions of the hydrostatic theory do not correspond to reality, there are many practical tidal phenomena that cannot be explained by the hydrostatic theory and can only be explained from the point of view of the motion of the sea, using the dynamic theory of the tides.

(1) Tidal lag phenomenon. According to the Newtonian theory of hydrostatic tides, if the inertia and friction of seawater were not taken into account, the high tide would occur at the midheaven moment when the Moon passes through the local meridian circle. However, in reality, due to the viscous nature of seawater, the flow of seawater is affected by the internal viscous forces of the water column and the frictional forces of the seafloor, as well as the inertia of seawater, the high tide always occurs sometime after the midheaven moment, and this time interval is called the high-water interval; the time interval from the midheaven

moment to the low tide is called the low-water interval, and both are collectively referred to as the lunitidal interval.

According to Newtonian hydrostatic tide theory, the spring tide should occur on the first and fifteenth days of every lunar month; the neap tide should occur on the first and last quarters. In practice, due to the inertia and viscous forces of the sea, the varying depths of the seabed and the underwater topography of the coast, the maximum and minimum tides are always delayed by one to three days per month, the time difference being known as the lunitidal interval or lunar epoch, which varies from place to place.

(2) The phenomenon of tidal resonance. According to the theory of hydrostatic tides, diurnal tides never occur at the equator, and semidiurnal tides predominate at lower latitudes. In reality, however, there are many equatorial and low-latitude areas where diurnal tides occur. According to tidal dynamics theory, the oceanic tidal wave has the same period as the tide-generating force and thus represents a forced oscillation. In short, it can be argued that the tidal force in a lunar day consists of two different periods: a half-day tidal force with a period of 12 h 25 min, and a diurnal tidal force with a period of 24 h 50 min. In reality, the ocean is not a layer of water of uniform thickness, but rather a body of water separated by land, such as a harbor or an ocean, which, due to its different shapes and sizes, is bound to have its own inherent vibration period. When tidal waves enter the water, if the period of a particular tidal vibration is close to the intrinsic vibration period of the sea area, a phenomenon known as resonance occurs, which intensifies the vibration of this period. For example, the Bay of Fundy in North America is one of the largest tidal bays in the world, and is a semidiurnal tidal type, because the inherent period of vibration in the bay is 11.5 h, which is closer to the semidiurnal tidal period, forming a resonance. The Gulf of Tonkin, near the equator, has an intrinsic period of 23.1 h, which is very similar to the period of the diurnal tidal constituent, and therefore, the Gulf has a diurnal tidal pattern.

(3) Amphidromic point and tidal surge phenomenon. In some semi-enclosed bays, there are often amphidromic points (regions) where the tidal range is effectively zero. Cotidal lines rotate clockwise or counterclockwise around these points, and the tidal range on opposite sides of the bay is usually unequal. Only tidal-dynamics theory can explain this: as the tidal wave propagates, it is influenced by both topography and the Coriolis force; reflections inside the bay create standing waves in the longitudinal and transverse directions. Where the nodal lines of these two vibrations intersect, the tidal range is zero, and this intersection is termed an amphidromic point.

When the tidal wave propagates landward along an estuary, the gradually decreasing water depth and narrowing channel width successively reduce its celerity. Superimposed on this are the opposing river discharge and bed friction, which jointly distort the wave profile: the leading face steepens while the trailing face flattens, so that the water level rises rapidly but falls slowly. Under these conditions, the wave may evolve into a tidal bore—a near-discontinuous surge. A classic example is the renowned bore at Haining on the Qiantang Estuary, China, where spring tidal ranges frequently exceed 8 m. A similar phenomenon occurs at the "Reversing Falls" of the Saint John River, Bay of Fundy, where the incoming tidal wave is arrested by a constricted gorge, causing an abrupt elevation of the surface and an upstream-propagating hydraulic jump resembling a waterfall in reverse.

6.3 Tidal Forecasting

6.3.1 Tidal Level Observation

The variation of nearshore tides is not only related to tidal force, but also influenced by hydro-meteorological factors, coastal morphology, and underwater topography, which vary greatly from place to place. So far, the calculations based on either the tidal hydrostatic theory or the tidal dynamic theory fail to meet the actual engineering application, especially for the edge of the sea with complex topography. Therefore, according to the *Technical Code for maintenance of Port Infrastructure* (JTS 310-2025), the construction of port projects must have more than 20 years of measured tidal level information, and new ports far away from the tide gauge station must also have at least one year of measured tidal level information, which can be used to deduce the design tidal level. Tide gauges are installed at coastal marine stations and ports in China to record tide level information systematically over a long period of time (Fig. 6.11). When designing projects in areas where tidal level information is not available, temporary tide gauges should be set up as early as possible to conduct short-term tidal level observations.

There are many means of tidal level measurement, including tide gauge, pressure tide gauge, Global Positioning System (GPS)-based tide gauge, satellite-based tidal remote sensing, and so forth.

6.3.1.1 Staff-Gauge Tidal Observation

The tide gauge (tide staff) is usually located at a coastal location that can be regularly flooded by the sea. These gauges are usually 2 m long and are mounted on special wooden stakes. If the variation in water level at the location is greater than 2 m, two or more sections are used to suit the local measurement requirements. The scale can also be located on structures such as jetties and breakwaters. After the installation of the staff gauge, the level must be used to determine the elevation of the zero point of the staff gauge according to the level point on the shore. Following the observation schedule specified in the code, the water-surface graduation on the staff is read, and the instantaneous water level is computed with the formula below; readings are recorded to the nearest centimetre.

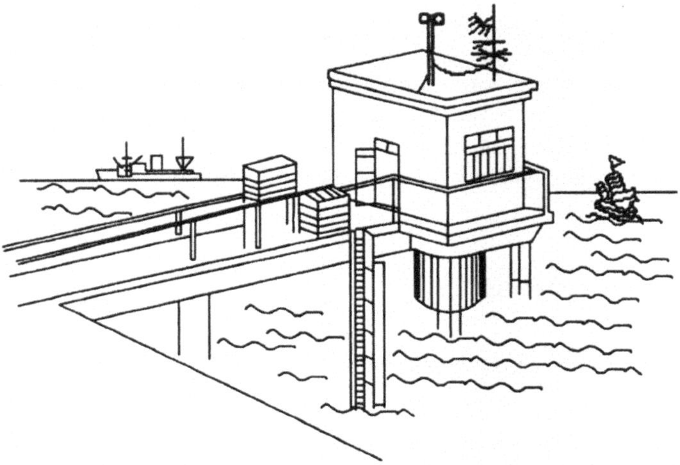

Fig. 6.11 Tidal level observation station

$$\text{Water level} = \text{tide staff reading} + \text{staff datum elevation}$$

When the water level fluctuates unsteadily, more observations must be made, the average of which must be entered into the observation log and used to calculate the value of the water level at that moment.

Tide-staff observations shall be carried out continuously over 24 h, with readings taken every hour. Within ± 30 min of predicted high and low waters, the sampling interval shall be shortened to 5–15 min, so that both the extreme levels and their times of occurrence are captured. All observed data are to be plotted immediately as a time series of tide level.

6.3.1.2 Pressure Tide Gauge

The pressure tide gauge is a commonly used type of tide gauge. It is placed in a fixed position underwater and detects changes in seawater pressure to calculate the fluctuation of the sea surface. It has a wider range of applications than the previous tide gauges, as it does not require the construction of a well or the use of the coast as a base. Therefore, the pressure tide gauge is well-suited to short-term, mobile deployments that require rapid installation, retrieval, and relocation (typically lasting one to two months).

6.3.1.3 GPS Tidal Method

GPS tidal method is a method of measuring sea surface height changes using GPS receiving devices. By setting multiple GPS receivers at different locations, researchers can obtain sea surface height data from different locations, thereby calculating the patterns of tidal changes. The GPS tidal method has the characteristics of a wide measurement range and high real-time performance, and is widely used in marine tidal measurement research.

6.3.1.4 Radar Altimetry Method

Radar altimetry is a technique that uses radar signals to measure variations in sea surface height, including ocean waves and tides. By sending radar signals and receiving reflected signals, the changes in sea surface height can be calculated. The radar altimetry method enables tidal observation in a large area of the sea, with relatively simple operation, and is widely used in ocean tidal measurement.

6.3.1.5 Tidal Remote Sensing Measurement

Tidal telemetry uses satellite-borne radar altimeters to track the rise and fall of the sea surface. Satellite altimetry provides global, particularly remote, tidal information and is characterized by its speed and economy. In principle, tidal telemetry is analogous to GNSS (GPS) tide measurements, in that both combine sensor observations with precise positioning/orbit information to estimate sea-surface height. An altimeter is mounted on a satellite, which then measures the elevation of the sea surface, from which valid elevation data is obtained to calculate valid data about the tides. This method is fast and economical and covers a wide area, but is less accurate. To obtain highly accurate and valid tidal data, it is necessary to develop highly accurate sensors to be mounted on satellites.

6.3.2 *Tidal Analysis and Forecasting*

6.3.2.1 Harmonic Analysis

Mathematical analysis has shown that the tidal curve can be approximated by the superposition of many cosine curves. Consequently, ocean tides can be considered as a superposition of tidal constituents with different amplitudes, periods, and phases, each of which is caused by a fictitious celestial body in equatorial motion on the celestial plane.

In the case of lunar gravity, for example, it is possible to assume that there are many "imaginary moons" on the celestial sphere with different velocities and different distances from the Earth, and that their orbits around the Earth are all circles. Each hypothetical body corresponds to a tidal constituent. For example, the principal lunar semidiurnal constituent (M_2) is caused by the gravitational force of a hypothetical ideal Moon, whose period around the Earth is the same as that of the Moon, but which is in a circular motion in the equatorial plane, centered on the Earth. This principle applies equally to the Sun and other celestial bodies.

Theoretically, there are many different constituents, but based on the characteristics of the factors that influence them, there are generally three types of tides.

(1) Astronomical tidal constituents. This category comprises constituents generated solely by astronomical forces. There are many different astronomical constituents, the principal ones being: the principal lunar semidiurnal constituent (M_2), the principal solar semidiurnal constituent (S_2), the luni-solar declinational diurnal constituent (K_1), the principal lunar diurnal constituent (O_1), the luni-solar declinational semidiurnal constituent (K_2), the larger lunar elliptic semidiurnal constituent (N_2), the larger lunar elliptic diurnal constituent (Q_1), and the principal solar diurnal constituent (P_1), etc.

(2) Shallow-water constituents. Each of the astronomical constituents described above is itself a sinusoidal wave pattern. However, as the tidal wave travels from deep water to the shore, the wave front becomes steeper and the trailing slope flatter as the water becomes shallower. Analogous to the deformation of waves entering shallow water, the duration of the flood tide becomes shorter than that of the ebb tide. Its period can be 1/2, 1/3, or 1/4 of the original period, forming a superposition of various overtides. Thus, when each astronomical constituent of the tide is transmitted to shallow water areas, it generates a series of shallow water constituents. Take M_2, for example, there are M_4, M_6, M_8, and so on.

(3) Meteorological constituents. Tidal cycles are also influenced by meteorological factors such as monsoons, precipitation, and atmospheric pressure. Monsoons and precipitation generally exhibit annual cycles. Other factors, such as local transient winds, have negligible periodicity in their effect on the sea surface. For example, sudden changes in atmospheric pressure may cause short-period standing wave oscillations in semi-enclosed seas, with periods ranging from a few minutes to over an hour. In harmonic analysis, the principal consideration for such changes is the solar annual constituent (S_a).

Each constituent can be represented by a cosine curve. Based on the analysis of observed data, the general expression for each constituent is

$$h = fR' \cos\left[qt + (v_0 + u) - k\right] \tag{6.17}$$

where h is the tidal height of the constituent; R' is the mean amplitude of the constituent (half the tidal range); f is the nodal correction factor for the amplitude; q is the angular speed of the constituent (Table 6.1); t is the mean local time; $v_0 + u$ is the phase angle of the constituent, referenced from 00:00 h on a given date and available in astronomical phase tables; k is the phase lag of the constituent, representing the angular delay of high tide relative to the lunar transit due to factors such as seabed friction and seawater inertia.

In Eq. (6.17), f, q, and $v_0 + u$ are known quantities. Therefore, once the harmonic constants R' and k are determined for a specific location, the tidal constituent can be fully resolved. Since R' and k are constants for a fixed location, they are collectively termed the harmonic constants of the tide.

The purpose of harmonic analysis is to derive the harmonic constants for each local constituent from the observed tidal data. The expression for each constituent is written in the standard form as follows:

$$h = R\cos(qt - \theta) \tag{6.18}$$

where R is the amplitude; θ is the phase angle.

Comparing Eq. (6.17) with Eq. (6.18), we obtain

$$R = fR',\ \theta = k - (V_0 + u)$$

Table 6.1 Angular speeds and periods of 11 principal tidal constituents

Category	Name	Symbols	q/(degrees per mean solar hour)	Cycle time /h
Semidiurnal constituents	Principal lunar semidiurnal constituent	M_2	28.98	12.42
	Larger lunar elliptic semidiurnal constituent	N_2	28.44	12.66
	Principal solar semidiurnal constituent	S_2	30.00	12.00
	Luni-solar declinational semidiurnal constituent	K_2	30.82	14.96
Diurnal constituents	Principal lunar diurnal constituent	O_1	13.94	25.82
	Larger lunar elliptic diurnal constituent	Q_1	13.40	26.87
	Principal solar diurnal constituent	P_1	14.96	24.07
	Luni-solar declinational diurnal constituent	K_1	15.04	23.93
Shallow-water constituents	Principal shallow-water constituent	M_4	57.96	6.21
	Shallow-water compound constituent	MS_4	58.98	6.10
	Shallow-water constituent	M_6	86.94	4.14

Consequently, the harmonic constants R' and k are given by:

$$R' = \frac{R}{f}, \ k = V_0 + u + \theta$$

This implies that only R and θ of the constituent are required to calculate the harmonic constants R' and k, expanding Eq. (6.18) gives

$$h = R\cos\theta\cos(qt) + R\sin\theta\sin(qt) = a\cos(qt) + b\sin(qt) \tag{6.19}$$

where $a = R\cos\theta$, $b = R\sin\theta$; then $R = \sqrt{a^2 + b^2}$, $\theta = \arctan\left|\frac{b}{a}\right|$. If a and b are known, the relationship between R and θ can then be determined.

If the observed tide height at a tide gauge station at any point in time t is expressed as $h'(t)$, the predicted tide height $h(t)$ at that time is expressed as at that time is expressed as

$$h(t) = A_0 + \sum_{j=1}^{n}\left[a_j\cos(q_j t) + b_j\sin(q_j t)\right] \tag{6.20}$$

where A_0 is the mean sea level elevation; n is the total number of constituents. Although many constituents exist in theory, most of them have negligible influence and can therefore be omitted.

In tidal prediction, calculations typically utilize standard sets of constituents (e.g., 11 or 63 constituents; see Table 6.1). Common analysis techniques include Darwin's method and the least squares method. With the rapid development of computer technology, the least squares method has become preferred for its efficiency and accuracy. Its principles are briefly described below. The method seeks to minimize the sum of squared deviations between the observed $h'(t)$ and the predicted tide level $h(t)$. Let S denote this integral:

$$S = \int_{-0}^{T}\left[h'(t) - h(t)\right]^2 \mathrm{d}t \tag{6.21}$$

The condition for the best conformity between predicted and observed values (i.e., for S to be a minimum) is that the partial derivatives with respect to each coefficient must be zero.

$$\left.\begin{aligned} \frac{\partial S}{\partial A_0} &= 0 \\ \frac{\partial S}{\partial a_i} &= 0 \\ \frac{\partial S}{\partial b_i} &= 0 \end{aligned}\right\} \tag{6.22}$$

where $i = 1, 2, \ldots, n$.

The above equation forms a system with $2n + 1$ unknowns. Solving this system of equations yields A_0, a_i, and b_i. These parameters are then converted into A_0, R_i, and θ_i, which are substituted into Eq. (6.20) to predict tidal levels.

If the tidal shape factor $F = \frac{R_{K1}+R_{O1}}{R_{M2}+R_{S2}}$ is used to classify tidal types, they can be broadly classified as:

Regular semidiurnal tide: $0 < F \leq 0.25$
Irregular semidiurnal tide: $0.25 < F \leq 1.50$
Irregular diurnal tide: $1.50 < F \leq 3.00$
Regular diurnal tide: $F > 3.00$.

In the above expression for F, R_{K1}, R_{O1}, R_{M2}, and R_{S2} represent the amplitudes of the four principal constituents K_1, O_1, M_2, and S_2, respectively. Based on calculations by relevant authorities, the general distribution of tidal types along the coast of China is characterized as follows: The Bohai Sea coast is dominated by irregular semidiurnal tides. The coasts of the Yellow Sea and the East China Sea generally feature regular semidiurnal tides. The South China Sea coast exhibits more complex patterns, comprising regular diurnal, irregular diurnal, and irregular semidiurnal tides.

6.3.2.2 Tidal Prediction

By using the Harmonic analysis method, the tides at any time can be predicted by calculating the relative positions of the Sun, the Moon, and the Earth, and using the harmonic constants at that location. Annual Tide Tables, which include tidal predictions for standard ports, are typically published in the preceding year.

(1) Contents of the tide table. The tide tables include the predicted times and heights of daily high water and low water for each standard port, together with the lunar calendar, coordinates (latitude and longitude), standard time zone, and the tidal datum. The tide table also includes a "Tidal Difference Ratio and Tidal Information Table". The table includes columns for time differences, range ratios, and mean sea level for standard ports (reference stations) and neighboring secondary ports (subordinate stations). From these data, tidal predictions for the secondary ports can be derived. The table also provides mean lunitidal intervals and spring/neap rises (referring to the height of high and low waters relative to the tidal datum; see Fig. 6.12). Instructions and auxiliary tables are also provided for calculating the tidal height at any specific time.

The tides predicted in the tide tables for coastal ports only represent the tidal conditions under normal conditions, excluding the effects of meteorological anomalies or certain special factors. Generally, a discrepancy of 20–30 min in time and within 0.3 m in height between the prediction and observation is considered consistent. However, during neap tides or periods of low sea level—especially when the diurnal inequality is significant—the difference between the predicted and actual tide times may exceed 1 h. Furthermore, when coastal areas are subject to storm

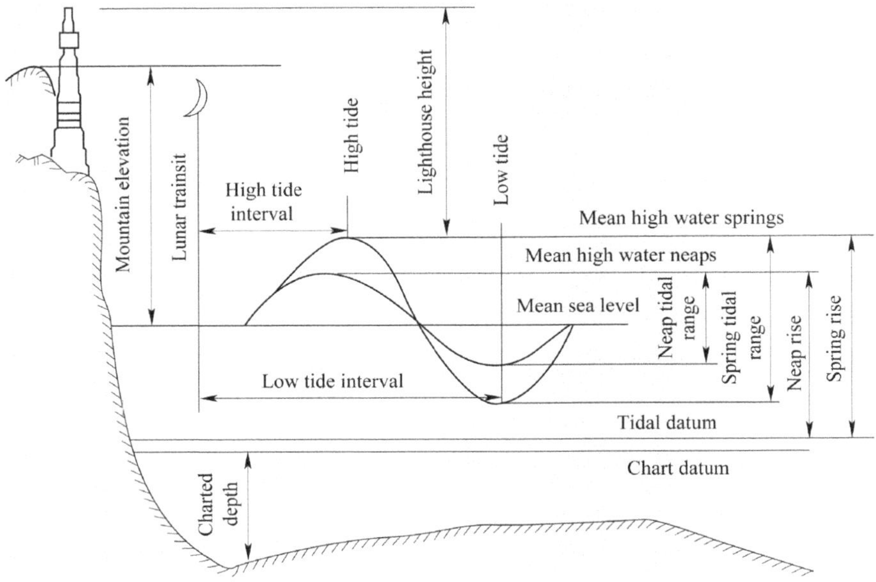

Fig. 6.12 Datum and characteristic tide levels

surges, typhoons, or cold waves, the deviations in both tidal height and time may be significantly larger in bays and estuaries.

(2) Calculation method for deriving the tidal height at any given time and the tidal time for any given height. The tidal curve is approximated as a cosine curve with a period of 360°, where the phase changes from 0 to 180° during the interval from high water to low water; that is, $\cos\theta$ varies from + 1 to − 1. Thus, to find the tidal height at any time t between high water and low water, first find the phase angle corresponding to t (starting at 0° at high water), multiply the cosine of this angle by the amplitude (semi-range), and then add the height of the mean tide level itself to find the tidal height at that time, as shown in Fig. 6.13.

The formula for finding the height of the tide at any given moment is

$$\begin{aligned} & H = Y - \tfrac{R}{2}\left[1 - \cos\left(\tfrac{t}{T} \times 180^\circ\right)\right] \\ \text{or} \quad & H = Y' + \tfrac{R}{2}\left[1 - \cos\left(\tfrac{t}{T} \times 180^\circ\right)\right] \end{aligned} \tag{6.23}$$

where H is the tidal height at any given moment; Y is the height of high water; Y' is the height of Low Water; T is the duration of flood or ebb tide; t is the time interval between the given moment and the time of high water; R is *the* tidal range.

For ease of application, the values $\frac{R}{2}\left[1 - \cos\left(\frac{t}{T} \times 180^\circ\right)\right]$ are often provided in a special table (e.g., "Table for Determining Tidal Height at Any Time") for checking. In a diurnal tide port, T is about 12 h, and the tide time is about twice that of a semidiurnal tide port.

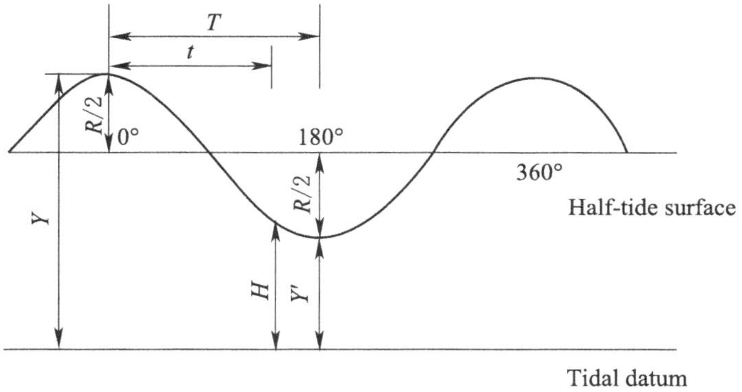

Fig. 6.13 Calculation of the tidal height at any time

6.4 Estimation of Design Tide Levels in Coastal Engineering

6.4.1 Datums and Characteristic Tide Levels

Topographic surveys are conducted prior to the construction of onshore works. For coastal engineering projects, both onshore topographic surveys and offshore bathymetric surveys are required. Measurements in both cases must be referenced to a zero point of elevation, called the datum. Vertical datums are often established at national or regional levels, and the mean sea level (MSL) of a permanent tide gauge station is usually chosen as the standard reference. For instance, in Europe, the long-term MSL at the Amsterdam tide gauge station is used as the vertical datum; in the United States, the long-term MSL at the Portland tide gauge station is used as the datum. These regional vertical datums are also known as regional geoid reference surfaces. The reference level must be clearly defined based on the MSL.

6.4.1.1 Mean Sea Level (MSL)

Mean sea level can be categorized as daily, monthly, annual, and long-term mean sea level. Analysis of the monthly mean sea level data over multiple years reveals that the mean sea level varies regularly on an annual cycle. Figure 6.14 shows the seasonal variation of the monthly mean sea level at a port over a period of one year, with the horizontal axis indicating the month. This phenomenon is attributed to the fact that atmospheric pressure in Chinese coastal waters is low in summer and autumn when southeasterly winds prevail, while the atmospheric pressure is high in winter and spring when northwesterly winds are strong. Long-term variations in mean sea level are related to astronomical factors. Since the astronomical nodal cycle is 18.6 years, to obtain an accurate long-term mean sea level, the average of the hourly tide levels

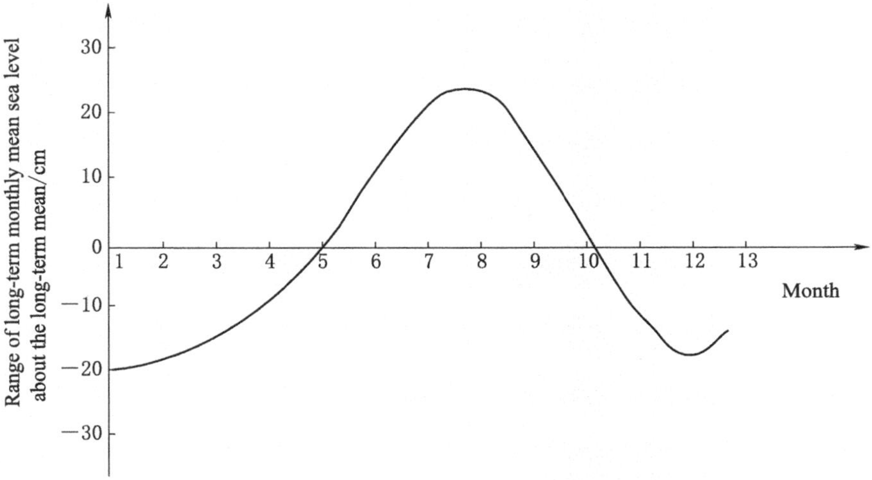

Fig. 6.14 Monthly correction for tide levels

for 19 years must be calculated. The national vertical datum in China, known as the Yellow Sea Datum, is defined by the 19-year mean sea level of the Qingdao tide gauge station.

6.4.1.2 Mean Tide Level (MTL)

The mean tide level is the arithmetic mean of high and low tide heights, also referred to as the half-tide level. Ideally, if the tidal curve follows a regular cosine function, the MTL coincides with the mean sea level (MSL). In reality, tidal curves are complex, leading to discrepancies between the two, especially in regions with significant diurnal inequality. Table 6.2 compares the MTL and MSL at a specific port in 1991. The data indicate that the difference between the two is minimal and remains relatively constant. Consequently, the MSL can be derived by applying this constant offset to the MTL.

Table 6.2 Comparison of MTL and MSL in 1991 for a port

Month	1	2	3	4	5	6	7	8	9	10	11	12	Year-round
MTL	187	186	192	194	201	226	215	221	237	220	205	174	205
MSL	183	182	188	190	199	224	214	219	234	216	202	170	202
Difference/cm	− 4	− 4	− 4	− 4	− 2	− 2	− 1	− 2	− 3	− 4	− 3	− 4	− 3

6.4.1.3 Theoretical Depth Datum (Theoretical Lowest Tide)

MSL serves as the zero point for determining land elevation. However, for hydrographic surveys, the water surface fluctuates below MSL approximately half of the time due to tidal action. If the MSL were used as the chart datum, the actual water depth would be less than the charted depth for about half of the time. To ensure navigational safety, alternative datums are adopted to ensure that the actual water depth exceeds the value indicated on the chart most of the time. Chart datum standards vary across countries, primarily including: the lowest possible tide level, the mean low water springs (MLWS), and the lowest recorded tide level. Since 1956, China has standardized the use of the theoretical depth datum as the chart datum. This datum is the theoretical lowest astronomical tide calculated from a combination of eight major tidal constituents ($M_2, S_2, N_2, K_2, K_1, O_1, P_1, Q_1$). The computational procedure is relatively complex and is detailed in the relevant chapter on tidal theory.

6.4.1.4 Tidal Datum

The zero point of the predicted tide levels in a tide table is called the tidal datum. It is situated below mean sea level and does not necessarily coincide with the chart datum. Therefore, the actual water depth at a given time and location equals the charted depth plus the predicted tide height, with an appropriate correction for the offset between the tidal datum and the chart datum. A tide level is the displacement of the sea surface relative to a specific datum; since it varies with the datum, it is a relative value. In contrast, water depth is the vertical distance from the sea surface to the seabed; it is an absolute value independent of any datum. For computational convenience in new ports or coastal engineering sites, the tidal datum and the theoretical depth datum may be set to coincide.

6.4.1.5 Local Zero Datums

In some areas of China's coast, for legacy port or coastal engineering projects, a specific zero point was historically established as the datum for all construction data, commonly referred to as the local (port construction) zero point. In these areas, prior to the adoption of a national unified depth datum, it was customary to use these local zero points, such as the Dalian (port-construction) Zero, Dagu Zero, Old Yellow River Zero, Wusong Zero, Pearl River Datum, and Yulin Datum. Therefore, when referencing historical construction data in these regions, it is essential to clarify the relationship between the local zero points and other regional datums and to convert all elevations to a unified datum. This issue is critical; any oversight can lead to significant errors in subsequent port construction and operations, as evidenced by numerous lessons from past engineering projects.

6.4.1.6 Characteristic Tide Levels

Tidal phenomena are closely related to the design and construction of port engineering. The characteristic tide levels commonly used in engineering are as follows:

(1) Maximum (minimum) water level. It refers to the maximum (minimum) tide level value ever observed in history.
(2) Mean annual maximum (minimum) tide level. The multi-year average of the maximum (minimum) tide levels recorded each year based on long-term tidal observations.
(3) Mean high (low) water springs (MHWS/MLWS). This is the multi-year average of the high (low) tide levels of spring tides per month.
(4) Mean high (low) water neaps (MHWN/MLWN). This is the multi-year average of the high (low) tide level of neap tides per month.

The interrelationships between the various datums and the different characteristic tide levels are shown in Fig. 6.12.

6.4.2 Determination of Design Tide Levels

The design tide level is an important hydrological parameter in the design and construction of port projects. It is closely related to the extent and elevation of port land, the crest elevation of hydraulic structures, and the water depth of navigation channels and harbor basins. It also influences the selection of structural forms and design calculations. The design tide levels are selected according to the scale, grade, and intended function of the port. Design tide levels usually include: design high water level (DHWL), design low water level (DLWL), extreme high water level (EHWL), extreme low water level (ELWL), etc. The following sections describe the methods for deriving each design tide level under conditions of both sufficient and insufficient measured tide data.

6.4.2.1 Estimation of Design High and Low Water Levels

DHWL and DLWL refer to the water levels specified for port hydraulic structures under normal operating conditions. For quays, the range between the DHWL and DLWL must ensure the safe berthing and cargo-handling operations of all design vessels. Furthermore, the quay structure and its foundation shall satisfy stability and strength requirements under all specified design load cases.

1. Criteria for design tide levels

Design water level standards for coastal ports vary internationally. Some countries adopt MHWS/MLWS, while others utilize tide levels with cumulative duration frequencies of 1 and 98%. The technical specification for port engineering projects in China, the *Code of Hydrology for Harbour and Waterway* (JTS 145-2015) (2022), stipulates that for coastal ports and estuarine ports with significant tidal influence, the DHWL should adopt the tide level with a 10% cumulative frequency of high tides (abbreviated as High Tide 10%). Similarly, the DLWL should adopt the tide level with a 90% cumulative frequency of low tides (abbreviated as Low Tide 90%). If duration-frequency statistics are available, the DHWL and DLWL may also be determined using the 1% and 98% cumulative duration levels, respectively. For estuarine ports where the tidal effect is not obvious during the flood season, the DHWL and DLWL should be based on the multi-year 1 and 98% duration tide levels.

2. Data record length

To determine the design high and low water levels, tidal frequency analysis should be based on multi-year observed data or, at minimum, one complete year of continuous hourly tidal observations.

3. Methods for deriving design tide levels

According to the design criteria and requirements for tide levels, two methods are commonly used for derivation: one is based on the cumulative frequency curve of hourly tide levels (all-time data), and the other uses the cumulative frequency curve of high or low tide levels. The following section provides a brief description and comparison of these two methods.

(1) Plotting of cumulative frequency curves of tide levels over time. The hourly tidal observations recorded throughout the year serve as the raw data for frequency analysis. Before plotting the curve, the reliability, consistency, and completeness of the original data must be rigorously reviewed, and any gaps or missing values should be interpolated. The steps for constructing the curve are as follows:

   1) Determine the maximum and minimum tide levels in the dataset, and group the data into 20 cm (or 10 cm) elevation bins spanning this range.
   2) Calculate the monthly occurrences of each tide level group, and then determine the annual cumulative number of occurrences for tide levels exceeding the lower limit of each group.
   3) Let the cumulative number of occurrences be m, and the total number of observations be n. The cumulative frequency P of tide levels above the lower limit of the group is calculated as: $P = \frac{m}{n} \times 100\%$.
   4) Plot the tide level on the vertical axis (y-axis), and the cumulative frequency P on the horizontal axis (x-axis). Plot the cumulative frequency values at the corresponding lower limit of each tide level and connect the points with a smooth curve to generate the tide level cumulative frequency curve. The design high tide level ($P = 1\%$) and design low tide level ($P = 98\%$) can then be determined from the curve.

   Refer to ESM 6.2 for a worked example of the cumulative frequency of tide levels.

Fig. 6.15 The cumulative frequency curves of spring and neap tides at a station

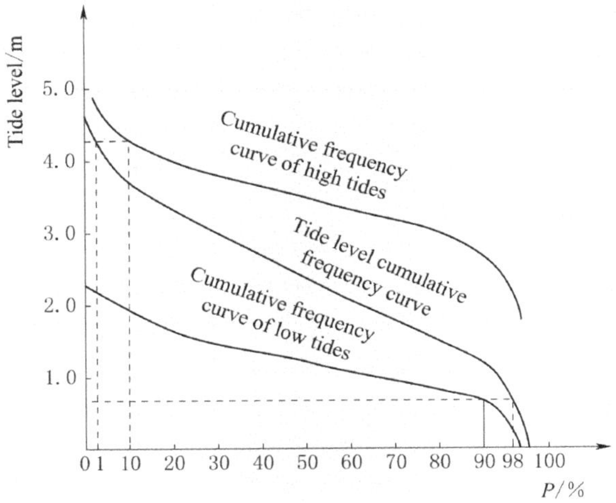

(2) Cumulative frequency curve of spring and neap tides. This method utilizes the values of the two high tides and two low tides recorded daily as the primary dataset. The curves are constructed following the same procedure described above. Figure 6.15 illustrates cumulative frequency curves for spring and neap tides at a station. According to the design codes, the tide level corresponding to a 10% cumulative high tide frequency is typically designated as the design high water level, while the level corresponding to a 90% cumulative low tide frequency is designated as the design low water level.
(3) Comparison and application of the two methods. Regarding the relationship between the tide level cumulative frequency curve (hourly data) and the cumulative frequency curves of high/low tides: an analysis of twelve coastal tide stations in China indicates that a tide level with a 1% duration-based frequency is essentially equivalent to one with a 10% high-tide cumulative frequency. The latter is generally slightly lower (by up to 15 cm), though in some cases it can be up to 5 cm higher. In the case shown in Fig. 6.15, the two are virtually identical.

It should be noted that the tidal stage-duration curve has two primary drawbacks: first, it requires a vast amount of data, making the analysis intensive and prone to errors; second, although it provides the total annual duration (in hours) that a water level exceeds a certain threshold, it does not specify the number of days over which these hours are distributed. Since 1949, coastal ports in China have consistently developed these curves, which account for long-term tidal cycles and are derived from several years of observed data. Therefore, for new engineering projects in these regions, the tidal stage-duration curve remains a reliable basis for deriving design water levels. In estuarine areas or coastal zones significantly affected by river discharge, where the fluctuations in river stage exceed the tidal range during the flood season, design water levels cannot be accurately determined using high or low tide cumulative frequency curves. In such cases, the tidal stage-duration curve must be employed. Furthermore, this curve is essential for calculating available operational windows for loading and unloading activities at a quay.

(4) The cumulative frequency curve of riding-tide water levels. To reduce the required water depth in artificial approach channels, vessels at coastal ports and shipyards frequently utilize the tidal range to enter and exit the port (a practice known as "riding the tide"). Therefore, the duration of the high tide level needs to be calculated. To ensure navigational safety, a reasonable duration, t, should be selected based on vessel traffic density and navigational requirements. The cumulative frequency, P, at which the water level reaches or exceeds this specific tide level is then determined to derive the riding-tide high-water level. The specific steps are as follows:

1) On the tidal hydrograph, measure the tide level maintained for a duration of t hours at each high-tide peak, and tally the occurrences within different water level classes.
2) The remaining steps are similar to those for constructing the high-tide (crest) cumulative frequency curve, which ultimately yields the cumulative frequency curve of riding-tide water levels for a duration of t.
3) Based on design requirements, determine the tide level corresponding to the cumulative frequency P from the constructed curve.

6.4.2.2 Estimation of Extreme High and Low Water Levels

Extreme water levels refer to the high and low tide levels that port structures may encounter under abnormal operating conditions. These levels result from a combination of meteorological surges (both positive and negative) and astronomical tides, driven by factors such as cold waves, typhoons, low pressure, earthquakes, and tsunamis. Their return periods typically span several decades. Consequently, while port structures are not required to remain fully operational at these levels, the quay must maintain a necessary factor of safety for all structural components and foundations under various loads during non-operational periods.

(1) Criteria for extreme tide levels. In the past, historical annual maximum and minimum tide levels were often adopted as the extreme high and low water levels in Chinese design practice. However, tidal records from various ports indicate that these observed extremes can vary significantly as the length of the observation period increases. Historical maximum and minimum tide levels obtained from surveys face similar consistency issues and are often less reliable. Furthermore, there have been no clear criteria for the selection or exclusion of exceptionally high and low tide levels. To address these shortcomings, the *Code of Hydrology for Harbour and Waterway* (JTS 145-2015) (2022) specifies the use of annual frequency analysis to derive tide levels with a 50-year return period as the extreme water levels. This approach provides a clear statistical basis, and other special water levels can also be determined based on specified return periods.

(2) Data record length. To determine extreme high and low water levels through frequency analysis, a minimum of 20 years of annual maximum and minimum tide observations is required. Additionally, historical occurrences of exceptional water levels must be investigated.
(3) Method of extrapolation of extreme water levels. In a semidiurnal port, over 700 spring and neap tides occur annually due to a combination of astronomical and meteorological factors. While astronomical tides are deterministic and predictable, tidal variations influenced by meteorological factors exhibit stochastic (random) characteristics. These can be represented as a series of K random variables $x_1',x_2',\ldots, x_k'$. In this series, the annual maximum tide level is denoted as x_{a}, and the annual minimum tide level as x_{b}. The probability distribution of the entire series, $G(x)$, is referred to as the parent distribution, whereas the distribution of the maximum or minimum values is the extreme value distribution, which can be derived from $G(x)$.

The *Code of Hydrology for Harbour and Waterway* (JTS 145-2015) (2022) specifies the use of the Gumbel distribution (Type I extreme value distribution) for the frequency analysis of extreme tide levels.

For a sample of n annual maximum tide levels $x_{\mathrm{a}1}, x_{\mathrm{a}2}, \ldots, x_{\mathrm{a}i}, \ldots, x_{\mathrm{a}n}$, the probability P that a value is greater than or equal to x_P according to the Gumbel distribution is

$$P(x_{\mathrm{a}} \geq x_P) = 1 - \exp\left[-\exp(-y_P)\right]$$

where the variable y_P is defined as

$$y_P = \alpha(x_P - u)$$

In the formula, α and u are the scale and location parameters of the curve, respectively, which can be obtained by Eq. (3.33). By substituting α and u, and rearranging the terms, the extreme high tide level x_P for a return period T (where $T = \frac{1}{P}$) can be expressed as

$$x_P = \overline{x} + \lambda_{P_n} S_x \tag{6.24}$$

where $\lambda_{P_n} = \frac{1}{\sigma_n}(y_P - \overline{y_n}) = \frac{1}{\sigma_n}\{-\ln[-\ln(1-P)] - \overline{y_n}\}$.

The above equation is known as the Type I extreme value distribution. For a given sample size n, the empirical frequency $P_m = \frac{m}{n+1} \times 100\%$ is used to determine y_m for each sample m. The values of σ_n and $\overline{y_n}$ are functions solely of the sample size n. Consequently, the frequency factor λ_{P_n} depends only on P and n, and its values can be found in the appended tables. The mean $\overline{x}$ and standard deviation S_n are calculated from the observed data.

Similarly, the formula for the extreme minimum tide level is derived in a form similar to Eq. (6.24), differing only in the sign of the frequency factor:

$$x_P = \overline{x} - \lambda_{P_n} S_x \tag{6.25}$$

When calculating the empirical frequency, the annual maximum series x_{ai} should be arranged in descending order, while the annual minimum series x_{bi} should be arranged in ascending order.

【Example 6.1】 Given a dataset of 22 years of annual maximum tide records (the relevant data are shown in Table 6.3), extrapolate the tide levels for 100-year and 50-year return periods.

Table 6.3 Ranked annual maximum tide levels and empirical frequencies

m	x_i/cm	P	x_i^2/cm^2	m	x_i/cm	P	x_i^2/cm^2
1	330	4.35	108,900	13	275	56.52	75,625
2	315	8.70	99,225	14	274	60.87	75,076
3	310	13.04	96,100	15	270	65.22	72,900
4	303	17.39	91,809	16	268	69.57	71,824
5	297	21.74	88,209	17	264	73.91	69,696
6	286	26.09	81,796	18	263	78.26	69,169
7	285	30.44	81,225	19	261	82.61	68,121
8	284	34.78	80,656	20	256	86.96	65,536
9	284	39.13	80,656	21	254	91.30	64,516
10	282	43.48	79,524	22	254	96.65	64,516
11	280	47.83	78,400	$\sum$	6173		1,740,763
12	278	52.17	77,284				

The solution process can be carried out as follows:

a. Arrange the 22 years of annual maximum tide data in descending order. Calculate the empirical frequency P_m for each data point using the formula: $P_m = \frac{m}{n+1} \times 100\%$. The results, including the squared values of the observations, are summarized in Table 6.3.
b. Calculate the mean tide level $\overline{x}$.

$$\overline{x} = \frac{1}{n}\sum_{i=1}^{n} x_i = \frac{6173}{22} = 280.59 \text{ cm}$$

c. Calculate the standard deviation S_x.

$$S_x = \sqrt{\frac{1}{n}\sum_{i=1}^{n} x_i^2 - \overline{x}^2} = \sqrt{\frac{1}{22} \times 1{,}740{,}763 - 280.59^2} = 19.87\,\text{cm}$$

d. Plot the theoretical frequency curve on probability paper according to Eq. (6.24). Then, plot the empirical frequency points to evaluate the goodness-of-fit (Fig. 6.16). If the initial fit is unsatisfactory, the curve can be adjusted by iteratively refining the parameters S_x and $\overline{x}$ until the theoretical curve optimally aligns with the empirical data.

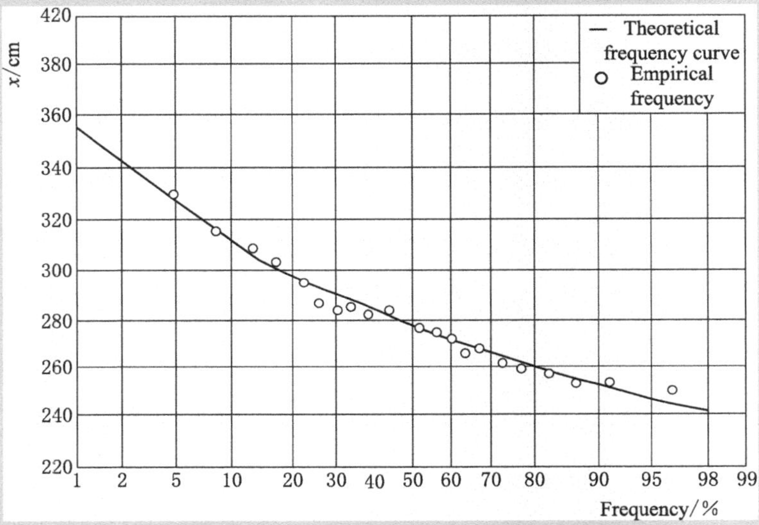

Fig. 6.16 Extrapolated line of fit diagram for extreme tide levels

e. Determine the design high tide levels. From the best-fit theoretical frequency curve, the tide levels for a 100-year return period ($x_{1\%}$) and a 50-year return period ($x_{2\%}$) are found to be 356 cm and 343 cm, respectively.

(4) Treatment of very high tide levels. If an exceptionally high or low historical tide level is identified within or beyond the period of systematic observation (n years), it should be treated as an extraordinary value (outlier) during frequency analysis. The primary task is to determine the magnitude of the extraordinary value X_N and its corresponding return period N. The following formula is used to calculate the T-year $\left(P = \frac{1}{T}\right)$ extreme tide levels:

$$X_P = \overline{X_N} \pm \lambda_{P_N} S_{XN} \tag{6.26}$$

where $\overline{X_N}$ and S_{XN} are the mean and standard deviation of the annual maximum (or minimum) tide series, adjusted to account for the extraordinary values; λ_{P_N} is the frequency factor corresponding to return period N, available in Schedule 3.

If the survey identifies only one extraordinary value X_N during a period of N years and this value lies outside the n-year systematic record, the parameters are calculated as follows:

$$\overline{X_N} = \frac{1}{N}\left(X_N + \frac{N-1}{n}\sum_{i=1}^{n} X_i\right) \tag{6.27}$$

$$S_{XN} = \sqrt{\frac{1}{N}\left(X_N^2 + \frac{N-1}{n}\sum_{i=1}^{n} X_i^2\right) - \overline{X_N}^2} \tag{6.28}$$

The empirical frequency for the extraordinary value is calculated as $P = \frac{1}{N+1} \times 100\%$, while other data points in the systematic record are determined using $P_m = \frac{m}{n+1} \times 100\%$.

In practice, the return period N for a historical value is often an estimate within a range rather than a precise figure. However, the following example demonstrates that the sensitivity of the results to the precision of N is relatively low.

In the previous dataset (Table 6.3), assume a historical high tide level $X_N = 480$ cm was identified. Investigation suggests this was caused by a typhoon and has an estimated return period of approximately 150 years. To test sensitivity, return periods of $N = 100$, 150, and 200 years are assumed. Using Eq. (6.27), we get

$$\overline{X_{100}} = \frac{1}{100}\left(480 + \frac{99}{22} \times 6173\right) = 282.6\,\text{cm}$$

$$\overline{X_{150}} = \frac{1}{150}\left(480 + \frac{149}{22} \times 6173\right) = 281.9\,\text{cm}$$

$$\overline{X_{200}} = \frac{1}{200}\left(480 + \frac{199}{22} \times 6173\right) = 281.6\,\text{cm}$$

By Eq. (6.28), we get

$$S_{X100} = \sqrt{\frac{1}{100}\left(480^2 + \frac{99}{22} \times 1740763\right) - 282.6^2} = 27.9\,\text{cm}$$

$$S_{X150} = \sqrt{\frac{1}{150}\left(480^2 + \frac{149}{22} \times 1740763\right) - 281.9^2} = 25.8\,\text{cm}$$

$$S_{X200} = \sqrt{\frac{1}{200}\left(480^2 + \frac{199}{22} \times 1740763\right) - 281.6^2} = 24.2\,\text{cm}$$

These results indicate that the mean value $\overline{X_N}$ remains stable regardless of whether N is 100 or 200 years. Although S_{XN} varies by nearly 15%, this is within the typical adjustment range (often up to 20%) used during curve fitting to achieve the best fit with empirical points. Therefore, for extraordinary value treatment, a reasonably estimated range for N is sufficient.

6.5 Storm Surges

6.5.1 Generation and Propagation of Storm Surges

Storm surges frequently pose significant hazards to coastal regions. A storm surge refers to an abnormal rise in sea level caused by intense atmospheric disturbances, including tropical cyclones (referred to as typhoons in the Western Pacific), extratropical cyclones, cold fronts, and abrupt changes in atmospheric pressure. A storm surge typically lasts from several hours to several days. The total water level resulting from the combination of the astronomical tide and the storm surge is generally referred to as the storm tide. Furthermore, wave runup—generated by wind waves and swells (typically with periods of several seconds) during a typhoon—can be superimposed on the storm tide, causing rapid sea-level elevations that lead to coastal flooding and catastrophic damage. Recent research indicates that wave setup near the shore also contributes to the rise in the mean water surface, often with an amplitude of several tens of centimeters. Conversely, negative storm surges (or storm set-down) may occur when strong offshore winds blow away from an open coast for an extended period. This causes a sharp drop in the coastal water level, exposing expansive tidal flats and severely hindering navigation and anchoring, particularly for large vessels.

Based on the characteristics of the inducing atmospheric disturbances, storm surges are generally classified into tropical cyclone (TC) surges and extratropical storm surges. Tropical cyclone surges are generated by a tropical cyclone—an intense circular storm originating over warm tropical oceans, characterized by extremely low atmospheric pressure, high winds, and heavy precipitation. These storms are known by various regional names: hurricanes in the North Atlantic and eastern North Pacific; typhoons in the Western North Pacific (impacting the Philippines, Japan, and China); and cyclones or severe tropical cyclones in the Southern Pacific Ocean and Indian Ocean. Despite these different names, they refer to the same meteorological phenomenon. Extratropical storm surges are induced by extratropical weather systems, a collective term for phenomena such as cold high-pressure systems and low-pressure systems with frontal structures, which are primarily active in mid-to-high latitudes.

The primary distinction between TC surges and extratropical surges lies in their temporal dynamics. The former is typically characterized by abrupt and rapid water level fluctuations, reflecting the fast movement and intense gradients of the cyclone's wind and pressure fields. In contrast, the latter exhibits gradual and sustained changes in water levels. In the Yellow and Bohai Seas of northern China, storm surges are also frequently triggered by cold waves or cold air outbreaks due to the vigorous interaction between polar and tropical air masses.

For a visual representation of the formation and propagation of storm surges, please refer to the example provided in ESM 6.3.

6.5.2 Storm Surge Hazards

Storm surges drive abnormal sea-level fluctuations, typically occurring in conjunction with the high winds and heavy precipitation associated with typhoons. The surge-induced rise in water level forces seawater to propagate toward the shore alongside tidal currents. This process can be analyzed as a specific form of long-wave propagation. Upon reaching the coast, storm surges can trigger severe erosion of embankments and inundate low-lying coastal areas, resulting in floods, structural collapse, and casualties. The socio-economic impacts resulting from these events are collectively defined as storm surge disasters. It is noteworthy that storm surges rarely act in isolation. The most catastrophic disasters typically arise from the synchronization of surge-induced high water levels with peak astronomical tides, further exacerbated by the destructive force of typhoon-induced waves. To visualize the dynamic coupling mechanism of tide-wave-wind interactions, refer to ESM 6.4 for a 3D simulation of storm surge propagation, illustrating the transformation of deep-ocean energy into a catastrophic flooding chain.

6.5.3 Storm Surge Forecasting

Currently, two primary methodologies are utilized globally for storm surge forecasting: the empirical-statistical method and the hydrodynamic numerical simulation method. The former predominated from the 1930s through the late 1950s and was primarily used for predicting peak storm surges at individual stations. Numerical methods gained prominence in the 1960s and 1970s, leading to the development of operational forecasting models. While initial efforts relied heavily on empirical techniques, by the late 1980s, both empirical and numerical models were employed simultaneously for public warnings, though empirical methods remained the primary tool at that time.

Today, marine agencies and disaster management organizations increasingly require forecasts that are highly precise, inundation-resolving, and applicable over large spatial scales. These demands generally necessitate numerical modeling approaches, and strengthening numerical forecasting capability has therefore become an inevitable trend. In contrast, empirical and statistical methods rely on long, high-quality historical observation records, which limit their applicability in data-scarce regions and further underscore the need for numerical techniques. Nevertheless, although numerical models are increasingly prevalent worldwide, the role of empirical methods in improving forecast accuracy should not be overlooked.

For details of these methods, refer to ESM 6.5, which presents both empirical-statistical approaches and hydrodynamic numerical simulation techniques.

Chapter 7
Offshore Currents

The imperial envoy bears the decree of heaven.
皇华使者承天敕,
Proclaiming imperial edicts to distant foreign lands.
宣布纶音往夷域。
Whale-like ships roar through the waves, sailing across the vast sea.
鲸舟吼浪泛沧溟,
Venturing far into the boundless expanse of mighty tides.
远涉洪涛渺无极。
Travelogue Poems, Ma Huan, Ming Dynasty
—《纪行诗》, 马欢, 明代

▲

This poem depicts an envoy sailing through roaring waves and boundless seas, highlighting the influence of offshore currents on navigation and the perilous yet magnificent nature of maritime voyages.

Abstract This chapter systematically introduces offshore currents, focusing on tidal currents, including their phenomena, classification into rotary and reversing types, and harmonic analysis, wind-driven currents, covering direction and velocity, as well as wave-induced currents, such as mass transport, longshore current prediction, and rip current characteristics. It also covers field observations, data collection, and analysis methods for nearshore flow systems.

For a visual overview of the chapter's key concepts, refer to the knowledge map below.

Supplementary Information The online version contains Electronic Supplementary Material (ESM) available at https://doi.org/10.1007/978-981-95-9089-6_7.

X. Feng (ed.), *Engineering Hydrology*, https://doi.org/10.1007/978-981-95-9089-6_7

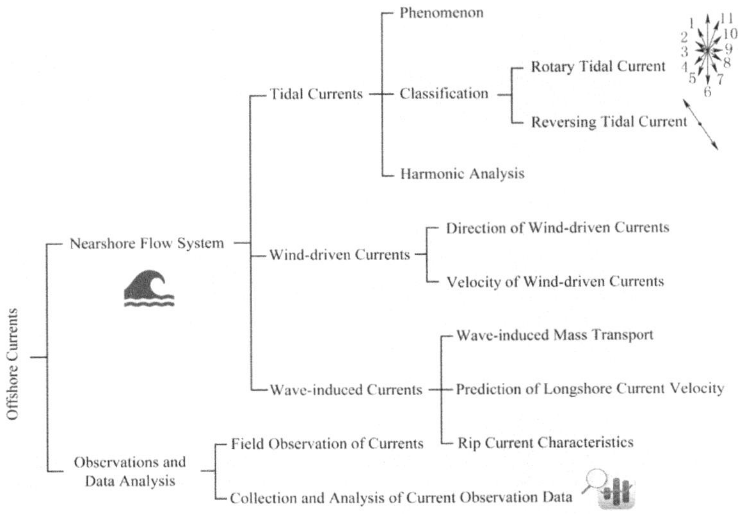

7.1 Introduction

Tides not only cause vertical water-level oscillations but also horizontal reciprocating motions known as tidal currents. Apart from tidal currents, which have definitive solutions, measured ocean currents are influenced by complex environmental factors such as weather, temperature-salinity variations, topography, and waves, making their prediction uncertain.

Historically, ocean monitoring faced significant limitations due to rudimentary technology. Early researchers used simple instruments like traditional current meters and staff gauges to manually record basic current parameters. Establishing China's first South China Sea hydrological station marked an essential milestone despite limited observational scope and accuracy, providing foundational data for future research.

Since the early twenty-first century, advancements in observational technologies have significantly improved ocean-current measurements. Instruments such as Distributed Temperature Sensors (DTS) and Acoustic Doppler Current Profilers (ADCP) enhanced measurement accuracy. High-frequency radars and Unmanned Aerial Vehicle (UAV) photogrammetry now provide detailed surface current fields suitable for monitoring near-shore rip currents. Remote sensing technologies, including satellite imagery, airborne Light Detection and Ranging (LiDAR), and infrared imaging, have deepened understanding of large-scale processes like river plume dynamics and surf-zone dispersion. Acoustic profilers further measure fluid structures, density, and sediment concentration.

Due to the limited scope of in situ observations, numerical models are crucial for understanding ocean currents at regional and global scales. Models using structured grids (orthogonal curvilinear or rectangular) and unstructured grids (triangular) follow Eulerian methodologies, with simulation quality heavily dependent on grid resolution and the complexity of grid generation. Advances in high-performance computing, parallel processing, Graphics Processing Unit (GPU) acceleration, cloud computing, and machine learning further improve modeling efficiency, data assimilation, and parameter estimation.

Simulation models often rely on parameterized processes, such as Reynolds-averaged Navier-Stokes equations and turbulent viscosity hypotheses, to model vertical turbulence. Bottom roughness and vegetation resistance parameters also contribute to modeling complexity. However, widespread ocean-current monitoring and prediction remain constrained due to high costs.

Promoting interdisciplinary cooperation among modeling communities (forward and inverse modeling, structured and unstructured grids) and fields like applied mathematics, computer science, artificial intelligence, and high-performance computing is essential. Such integration will enhance monitoring technologies, improve data-processing capabilities, and increase model generalization.

Emerging technologies, including artificial intelligence, big data analytics, and the Internet of Things, are promoting the development of ocean-current monitoring toward greater intelligence, higher resolution, and enhanced system integration. Continued advancement in these technologies is expected to provide strong technical support for marine exploration and global resource management.

7.2 The Nearshore Flow System and Its Causes

Inshore currents can generally be divided into tidal and non-tidal currents. The former is the periodic horizontal movement of seawater caused by the celestial gravitational forces of the Sun and the Moon. The latter can be divided into two categories: permanent currents are the well-known oceanic circulation, including the equatorial current, the Gulf Stream, the Kuroshio, etc.; temporary currents are mainly due to changes in meteorological factors, such as wind and wave action, which generate wind-driven currents and near-shore currents.

In coastal engineering, issues related to the selection of port sites, the layout of hydraulic structures and channels, the selection of mud disposal areas, the determination of structural forces and ship mooring forces, as well as the siltation and scouring of coastal sediments, are all related to the local current conditions.

There are many forms of offshore seawater flow, which can be classified according to the causes of their generation as follows.

(1) Tidal currents. The tide is a periodic horizontal movement of seawater caused by the gravitational forces of the Sun, the Moon, and other celestial bodies, and accompanied by tides.

(2) Wind-driven currents. Currents generated by wind stress on the sea surface are influenced by the Coriolis effect associated with Earth's rotation. As a result, the flow is deflected to the right of the wind direction in the Northern Hemisphere and to the left in the Southern Hemisphere.
(3) Pressure-gradient currents. A change in atmospheric pressure lowers water levels in high-pressure areas and increases water levels in low-pressure areas, resulting in a current known as a pressure-gradient current.
(4) Density-driven currents. Due to the uneven distribution of temperature and salinity in the water layer, the density of water masses in the sea area is also unevenly distributed, resulting in a current known as a density gradient current.
(5) Compensatory currents. The flow of water masses inevitably creates a deficit in some areas of the sea, which must be replenished by neighboring water masses, resulting in a current known as a compensating current.

In addition, there are two other currents that are unique to the nearshore.

(1) River-induced currents. These currents result from river runoff entering the sea. The horizontal flow generated in coastal waters near the river mouth is referred to as river-induced currents.
(2) Wave-induced currents. Horizontal currents generated by wave breaking in the nearshore zone are known as wave-induced currents.

The following sections focus on tidal currents, wind-driven currents, and wave-induced currents, which are closely associated with nearshore sediment transport.

7.2.1 Tidal Currents

7.2.1.1 Tidal Currents Phenomenon

Tides and tidal currents are two forms of seawater movement that occur simultaneously in response to the gravitational pull of celestial bodies such as the Sun and the Moon. For example, in the vicinity of Qinhuangdao, the tide is predominantly diurnal in character, whereas the tidal current exhibits a semidiurnal pattern. In contrast, in the offshore waters of Yantai, the tide is semidiurnal, while the tidal current is primarily diurnal. The similarity and differences of tidal and tidal current types depends on the distribution of the tidal wave system in each sea area. The tides in the offshore area are complicated by the coastal topography and the propagation of the tidal wave system.

7.2.1.2 Classification of Tidal Currents

The tidal cycle can be divided into regular and irregular semidiurnal tidal currents with a period of 6 h 12.5 min, and regular and irregular diurnal tidal currents with a period of 12 h 25 min. The criteria for distinguishing between the two are

$$K = \frac{W_{O_1} + W_{K_1}}{W_{M_2}} \tag{7.1}$$

When $K \le 0.5$, the tidal current is classified as regular semidiurnal. When $0.5 < K \le 2.0$, it is classified as irregular semidiurnal. When $2.0 < K \le 4.0$, it is classified as irregular diurnal. When $K > 4.0$, it is classified as regular diurnal. And W_{O_1}, W_{K_1} and W_{M_2} denote the lengths of the semi-major axes of the tidal current ellipses corresponding to the principal lunar diurnal constituent, the lunisolar diurnal constituent, and the principal lunar semidiurnal constituent, respectively. According to the direction of flow, tidal currents can be classified into two types: rotary tidal currents and reversing tidal currents.

(1) Rotary tidal current. Rotary tidal currents occur mainly in offshore or relatively open sea areas, as well as in coastal water seaward of river mouths. They are commonly formed by the superposition of two reversing tidal current components with different directions and phases. The flow direction and current speed vary continuously with time. In general, tidal currents in the Northern Hemisphere rotate predominantly in a clockwise direction, whereas those in the Southern Hemisphere tend to rotate counterclockwise. Taking the observation point as the origin, tidal current vectors measured at successive times over one tidal cycle can be plotted on a plane, where the vector direction represents the flow direction and the vector length represents the current speed. The curve formed by connecting the endpoints of the vectors at each moment is called the tidal ellipse. A tidal current in which the flow direction rotates continuously through 360° over one tidal cycle, while the current speed never reduces to zero, is defined as a rotary tidal current, as illustrated in Fig. 7.1.

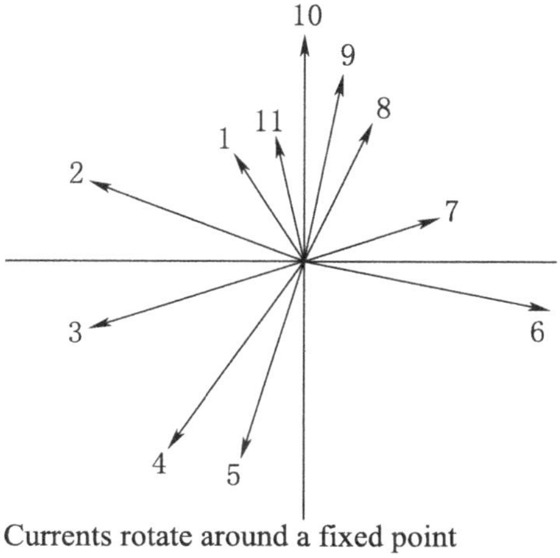

Fig. 7.1 Rotary tidal currents

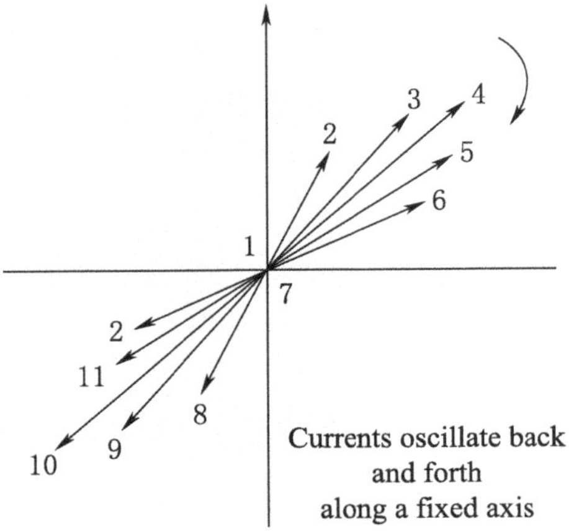

Fig. 7.2 Reversing tidal currents

(2) Reversing tidal current. Reversing tidal currents mainly occur in estuaries, straits, or narrow bays where the flow is constrained by coastal and seabed topography. Under such conditions, tidal currents oscillate back and forth along an approximately fixed direction, forming a reversing flow pattern. For a typical reversing tidal current, the variation of tidal current vectors over one tidal cycle is shown in Fig. 7.2. The flow direction is confined to a narrow range and alternates periodically between two opposite directions. The difference between the maximum and minimum current speeds is relatively large, and the current speed may approach zero during flow reversal. When the current speed becomes nearly zero, this condition is referred to as slack water.

7.2.1.3 Harmonic Analysis of Tidal Currents

Compared to the harmonic analysis of tide levels, the harmonic analysis for tidal current prediction has the following characteristics.

(1) The harmonic constants of tidal currents are distributed in two component groups. When calculating tidal current predictions at successive time steps, it is necessary first to compute the two tidal current components and then combine them to obtain the current speed and direction. For gyres, it is usually necessary to predict the current speed and direction at hourly intervals. In addition, the times at which extreme current conditions occur, together with the corresponding current speed and direction, may also be predicted.
(2) When the length of the semi-major axis of the tidal current ellipse is much greater than that of the semi-minor axis, the tidal current speed can be regarded as a reversing tidal current that varies only along two opposite directions. In this case, the current speed at successive time intervals can be calculated first, and then the extreme values can be determined. However, in addition to the extreme

values, it is also necessary to calculate the times at which the current speed becomes zero, namely the times of slack water.

(3) Because the governing conditions of tidal currents are relatively complex, harmonic constants are available at many locations for only a limited number of principal tidal constituents, derived from short-term observations through quasi-harmonic analysis. In such a case, tidal current prediction usually requires the use of empirical coefficients to represent the effects of the remaining tidal constituents.

The northward velocity component u and eastward velocity component v of the tide can be expressed by the following equations.

$$u = U_0 + \Sigma_i D_{ui} U_i \cos(\sigma_i t + v_{0i} + d_{ui} - \xi_i) \tag{7.2}$$

$$v = V_0 + \Sigma_i D_{vi} V_i \cos(\sigma_i t + v_{0i} + d_{vi} - \eta_i) \tag{7.3}$$

where, U_i, ξ_i are the harmonic constants of the northward velocity component; V_i and η_i are the harmonic constants of the eastward velocity component; U_0 and V_0 are the Eulerian residual currents; D_{ui} and D_{vi} are the amplitude correction factors, while d_{ui} and d_{vi} are the corresponding phase correction factors, and all of these correction factors are functions of time; v_{0i} and σ_i are the astronomical angular frequency and the astronomical phase of each tidal constituent, respectively.

In practical applications, Eqs. (7.2) and (7.3) are solved analytically using linearized equations, which are expressed as

$$u = U_0 + \Sigma_i D_{ui} \cos(\sigma_i t + v_{0i} + d_{ui}) U_i \cos \xi_i + \Sigma_i D_{ui} \sin(\sigma_i t + v_{0i} + d_{ui}) U_i \sin \xi_i \tag{7.4}$$

$$v = V_0 + \Sigma_i D_{vi} \cos(\sigma_i t + v_{0i} + d_{vi}) V_i \cos \eta_i + \Sigma_i D_{vi} \sin(\sigma_i t + v_{0i} + d_{vi}) V_i \sin \eta_i \tag{7.5}$$

Using the least squares method, U_0, U_i, ξ_i, V_0, $V_{i,}$ and η_i are determined. The resulting harmonic constants of the tidal currents, together with the residual currents, are then substituted into Eqs. (7.2) and (7.3), yielding u and v at the specified time.

After determining the u and v values at a specific time, we can substitute them into Eq. (7.6) to calculate the tidal flow rate ω and direction of tidal currents θ.

$$\left.\begin{aligned} \omega &= \sqrt{u^2 + v^2} \\ \theta &= \tan^{-1} \tfrac{v}{u} \end{aligned}\right\} \tag{7.6}$$

7.2.2 Wind-Driven Currents

Before discussing wind-driven currents, a brief introduction to V. W. Ekman's theory is presented, based on the following two assumptions. ① Wind-driven currents occur in the open ocean, or in the ocean far offshore. ② The ocean is of infinite depth, or at least sufficiently deep, so that a stable wind direction and force can cause constant wind-driven currents. This theory can be summarized mainly in the following conclusions.

(1) Direction of wind-driven currents. The direction of surface wind-driven currents is deflected 45° to the right of the wind in the Northern Hemisphere and 45° to the left in the Southern Hemisphere. This deflection does not vary with changes in wind speed, current speed, or latitude.
(2) Velocity of wind-driven currents. The empirical relationship between the surface velocity of wind-driven currents V_0, and wind velocity is

$$V_0 = \frac{0.0127}{\sqrt{\sin\varphi}}U \tag{7.7}$$

where U is the wind speed, m/s; φ is the latitude; 0.0127 is the wind factor.

The velocity of the wind-driven currents decreases rapidly with increasing water depth. The direction of the wind-driven currents is gradually rightward to the wind in the Northern Hemisphere as the depth increases. For an infinitely deep ocean, the complex expression for the velocity of wind-driven currents V_{wc} at a depth of Z is

$$V_{wc} = V_0 \exp\left[-\frac{\pi}{D}Z + i\left(\frac{\pi}{4} - \frac{\pi}{D}Z\right)\right] \tag{7.8}$$

where D is the friction depth, at $Z = D$, the current velocity is only 1/23 of V_0, and its direction is opposite to that at the surface. The vertical spiral variation of current velocity and direction illustrated in Fig. 7.3 is known as the Ekman spiral distribution.

D can be determined by the following equation:

$$\left.\begin{aligned} D &= \frac{7.6}{\sqrt{\sin\varphi}}U \quad (U > 6\,\text{m/s}) \\ D &= \frac{3.67}{\sqrt{\sin\varphi}}U^{3/2} \ (U < 6\,\text{m/s}) \end{aligned}\right\} \tag{7.9}$$

The friction depth of the equatorial current has been measured. In the Atlantic Ocean, the measured friction depth is approximately 150 m, whereas in the Pacific Ocean it typically ranges between 200 and 300 m. Relative to the total depth of the ocean, the wind-driven current is confined to a comparatively shallow surface layer.

Ekman pointed out that under the same wind direction and wind speed, the time taken for the formation of wind currents at each depth is highly inconsistent, varying from a few hours to as long as a month. This variability is likely due to complex interactions between wind, water, and oceanic conditions. Ekman also investigated

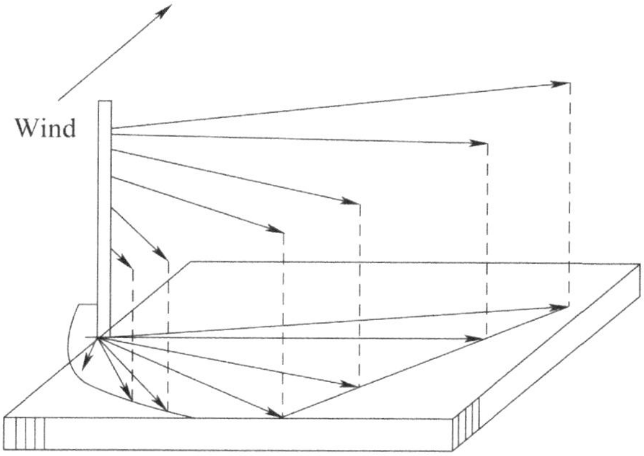

Fig. 7.3 Variation of the wind-driven current velocity with depth

Table 7.1 The relationship between relative water depth *d*/*D* and α

d/D	0	0.1	0.25	0.50	0.75	1.00	2.00	…	∞
α	0	3.7°	21.5°	45°	45.5°	45°	45°	45°	45°

the effect of finite water depth on wind-driven current direction and found that the shallower the water depth *d*, the smaller the value of the declination between surface flow direction and wind direction. The relationship between relative water depth *d*/*D* and wind direction α is shown in Table 7.1.

7.2.3 Wave-Induced Currents

In the course of wave propagation from deep water toward the shallow coast, energy is dissipated due to seabed friction, seepage through the seabed, and turbulent motions in seawater. Moreover, as the water depth decreases, wave energy becomes concentrated, and waves may break, which further dissipates energy and thus leads to a redistribution of wave energy. In near-shore waters, the near-shore current system induced by wave action is mainly composed of three parts: ① onshore mass transport of water; ② the longshore current flowing parallel to the shoreline; ③ the seaward-flowing rip current (also referred to as a rip), as shown in Fig. 7.4.

Rip currents constitute the most prominent component of the nearshore current system. They are relatively strong, narrow, surface-intensified jets that flow seaward through the surf (breaker) zone, with velocities typically exceeding 1 m/s. The narrowest section of a rip current is termed the neck, where the velocity is greatest. Farther offshore, the current may extend to about 500 m beyond the breaker line and spread over a much broader area. This expanded terminal region is called the

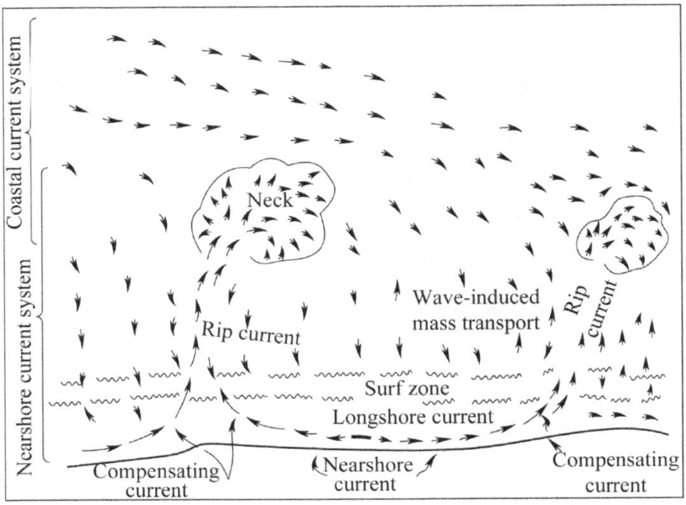

Fig. 7.4 Schematic diagram of the coastal current system

head, where the velocity is the smallest. Rip currents are maintained by the longshore current. The junction where the longshore current feeds into the rip current is referred to as the compensating current.

Longshore currents flow parallel to the shoreline, with mean speeds typically around 0.3 m/s, but occasionally exceeding 1 m/s. The velocity varies significantly in the breakers and in the coastal direction, with the maximum velocity occurring in the breakers and in the coastal direction before the flow changes to offshore. The longshore current varies markedly both across the surf zone and alongshore. It is typically strongest near the breaker line and alongshore, just upstream of where it turns seaward into a rip current. The discharge of both the longshore current and the rip current is supplied by the wave-induced mass-transport flow directed shoreward during wave propagation. Therefore, the nearshore current system plays an important role in nearshore water exchange, pollutant removal, and sediment transport.

7.2.3.1 Wave-Induced Mass Transport

During wave propagation in coastal water, the trajectory of water particles is not closed. As a result, a net transport of water exists in the direction of wave propagation; this is referred to as wave-induced mass transport. According to the second-order Stokes finite amplitude wave theory, the average mass transfer velocity expression is

$$\overline{U} = \left(\frac{\pi H}{L}\right)^2 \frac{C}{2} \frac{\cosh\left[4\pi\left(\frac{Z+d}{L}\right)\right]}{\sinh^2 \frac{2\pi d}{L}} \tag{7.10}$$

where H is the wave height; L is the wavelength; C is the wave speed; d is the water depth.

The above equation can be simplified in deep water as

$$\overline{U_0} = \left(\frac{\pi H_0}{L_0}\right)^2 C_0 \exp\left(-\frac{4\pi Z}{L_0}\right) \tag{7.11}$$

In Eq. (7.11), the subscript "0" denotes deep-water values. The Z-axis is taken with the still-water level as the origin, positive upward.

Integrating Eq. (7.11) with respect to depth Z yields the net volume of water transported forward per unit time and per unit wavelength [$cm^3/(cm\ s)$] as

$$q = \int_{-d}^{0} \overline{U_0} dZ = \frac{\pi H_0^2}{4T} \tag{7.12}$$

As waves propagate toward the shore, they cause water to pile up near the coastline, leading to a rise in the mean water level (wave set-up). This, in turn, generates a compensating return flow in the opposite direction, redistributing the water mass.

7.2.3.2 Prediction of Longshore Current Velocity

The generation of the longshore current is mainly the result of the presence of a driving component along the shoreline after wave breaking. The research methods can be divided into two categories: one applying the principle of conservation of energy and momentum to calculate the mean longshore current velocity over the entire wave breaking section, and the other applying the concept of radial stress, i.e., the residual momentum current induced by the wave, to study the lateral distribution of the longshore current velocity within the wave breaking zone.

In 1949, J. A. Putnam, W. H. Munk, and M. A. Traylor used the momentum-balance equation together with field observations to propose a method for estimating the nearshore longshore-current velocity. Later, in 1951, Inman and Quinn proposed the following formula for calculating the velocity of longshore currents based on field observations.

$$\left.\begin{aligned}
V_{\mathrm{LC}} &= \left[\left(\frac{1}{4x^2}+y\right)^2-\frac{1}{2x}\right]^2 \\
x &= \frac{646H_{\mathrm{b}}i\cos\alpha_{\mathrm{b}}}{T} \\
y &= C\sin\alpha_{\mathrm{b}} \\
C &= \sqrt{2.28gH_{\mathrm{b}}} \\
i &= \frac{d_{\mathrm{b}}}{l}
\end{aligned}\right\} \tag{7.13}$$

where V_{LC} is the longshore current velocity, m/s; H_b is the wave height at wave break-up; α_b is the angle of incidence when the wave breaks; i is the slope of the seabed; d_b is the depth of water at wave break, m; l is the distance from the breakers to the shore, m.

The V_{LC} value mentioned above only approximates the longshore current velocity near the breaker line and is not applicable within the surf zone. The calculated results for long-period waves correspond reasonably well with observed conditions.

7.2.3.3 Rip Current Characteristics

Satellite or aerial photographs of the coast show that, in addition to the longshore or shoreward transport of nearshore water, there are also concentrated, sediment-laden currents that originate at intervals along the coast and extend seaward. This is the most striking process of water exchange between the nearshore region and the offshore waters, which is a concentrated jet of water that travels through the breakers and carries the water out to sea through a narrow channel. Its occurrence is primarily associated with longshore variations in wave height and the influence of complex topography. Arthur, Bowen, and others have made theoretical discussions on rip currents and given certain models, but the flow pattern and dynamic structure of rip currents are more complex than those along the coast, and there is no flow velocity calculation formula available. The only available equations are the field observations summarized by Horikawa and co-workers.

7.3 Observations and Data Analysis of Instream Current

7.3.1 Field Observation of Currents

The foregoing has introduced several calculation methods and formulas for various types of currents. However, due to the influence of water depth, topography, and changes in hydrological and meteorological conditions in the nearshore area, the changes in the nearshore currents are very complex. Therefore, when the above-mentioned calculation methods and formulas are applied, it is essential to perform comparative analysis and verification using in situ measurements before undertaking practical calculations.

7.3.1.1 Methods of Current Observation

Observations of nearshore currents can be broadly classified into the following three methods.

(1) Single-station fixed-point continuous observation. A vessel is positioned at a fixed location to conduct continuous current measurements. The measured current speed and direction at that site are obtained in order to examine the spatial distribution and temporal variation of the current.
(2) Multi-station synchronous continuous observation. Several vessels are deployed simultaneously at designated fixed locations to conduct continuous current measurements (this method is commonly used for cross-sectional observations). The measured current information at each station is used to investigate the distribution and variability of currents.
(3) Large-area drift tracking observation. Surface drifters are released from a vessel in the nearshore area. Once the drifters have entered the designated observation area, their positions at different times are determined from shore using a theodolite or other surveying techniques, and the time-varying drifter tracks are plotted on a chart. This method provides an overall picture of the flow field, including current speed and direction, and it can also be used to identify zones (or points) of divergence and convergence.

The required duration and timing of current observations depend on the characteristics of the currents, as well as the purpose of the subsequent analysis. When short-term data are used for harmonic (quasi-harmonic) analysis, continuous observations should be conducted no fewer than three times, and should be carried out under spring-tide, intermediate-tide, and neap-tide conditions, respectively. In general tidal-current analysis, either one or two current surveys may be adopted: a single survey should be conducted during a spring tide, whereas two surveys should be conducted during spring and neap tides, respectively. The continuous duration of a single current survey should not be less than 25 h. For harmonic analysis based on long-term records, the continuous current observation period should be no less than 15 d, so as to capture the required temporal variability.

7.3.1.2 Instruments for the Observation of Currents

There are many different types of current observation instruments, which can be divided into mechanical current meters, pressure current meters, electromagnetic current meters, and acoustic current meters, depending on their principle of operation.

(1) Mechanical propeller-type current meters. The basic principle of these instruments is to determine the velocity of the current from the number of revolutions of the rotor blades driven by the current, and to determine the direction of flow using a magnetic compass (which must be corrected for magnetic differences). Depending on the characteristics of the recording method, such instruments can be grouped into types such as the Ekman current meter, the price-type current meter, the cup-type current meter, and instruments with photographic recording or magnetic recording, etc.

1) The Ekman current meter consists of five parts: the rack, the propeller, the counter, the flow box, and the rudder. The counter is a device that records the number of revolutions of the propeller, from which the current speed can be determined. The flow box is circular and has a magnetic needle with a groove in the center. The flow box is divided into 36 sectors, and for every 100 revolutions of the propeller, three small balls fall from the upper tube and roll along the groove to the cell indicated by the north pole of the magnetic needle.

During current measurement with the current meter, the counter is first set to zero (or the initial count is recorded). The meter is then lowered by winch to the specified depth. The first messenger weight is released along the wire to trigger the instrument and start the counting/recording. After the prescribed sampling time (typically 100–200 s), a second messenger weight is released to stop the counter. The meter is then recovered to the surface, and the counter reading is recorded. Next, the direction box is removed, and the distribution of the balls in the direction box is read to determine the current direction. The current speed can then be obtained from the recorded counts using the calibration relationship.

Each current meter is supplied by the manufacturer with a calibration formula and corresponding constant values, from which the measured current velocity can be calculated

$$V = K_n + C \tag{7.14}$$

where V is the velocity of currents, m/s; n is the internal rotational speed of the propeller per second; K and C are calibration constants for the current meter, different for each instrument.

The average magnetic potential direction of the current is

$$\alpha = \frac{(K_1 n_1 + K_2 n_2 + \cdots + K_n n_n) \times 10^\circ}{\sum_{i=1}^{n} n_i} \tag{7.15}$$

where $K_1, K_2, \ldots, K_n$ are sector cell numbers; $n_1, n_2, \ldots, n_n$ are the numbers of small balls falling into the corresponding sector cells.

However, this type of instrument cannot measure very low current speeds, and therefore, it is less commonly used at present.

2) The price-type current meter is a self-recording current-measurement instrument fixed to a vessel. It can automatically measure and record the mean velocity and instantaneous current direction at regular intervals at depths up to 6000 m and for a continuous period of up to two months. The recording device is spring-loaded; the time of self-recording is determined by a clock control panel, and the current velocity and direction are recorded on paper tape or photographic film.
3) The magnetic recording current meter is a fixed-point, self-recording current measurement instrument for buoys. And it works mostly by recording the measurement data in binary code on magnetic tape. The maximum depth of use is 1000–6000 m, the approximate range of current velocity is 3–400 cm/s, the accuracy of current velocity is 2%, and the accuracy of current direction is $\pm 5°$. The Geomagnetic Electrokinetograph (GEK) is an electromagnetic current meter that was widely used internationally; it has been among the most extensively used current meters worldwide.

(2) Electromagnetic current meters. An electromagnetic current meter determines current speed by applying Faraday's law of electromagnetic induction, i.e., by measuring the electromotive force induced when seawater flows through a magnetic field. Depending on the source of the magnetic field, there are two types of current meters: geomagnetic and artificial magnetic-field types.

Geomagnetic electromagnetic current meters have the advantage of autonomous recording and a simple underwater structure with good reliability. Their disadvantage is that they rely on the Earth's magnetic field and therefore are not suitable for low-latitude regions; moreover, their direction accuracy is strongly affected by the geomagnetic field intensity (approximately 0.1 A/m).

Artificial-magnetic-field electromagnetic current meters are widely used and are not limited by latitude. They are suitable for marine or anchored underwater measurements. Currently, the S4 electromagnetic current meter produced by Interocean is widely used in the world. Its main advantages are high accuracy, reliable measurement values, small size, and easy operation. The range of its measurement is from 0 to 350 cm/s, and the accuracy of the current velocity is 2%.

(3) Acoustic current meters. The focus here is on the Acoustic Doppler Profiler (ADCP), which is currently the most effective instrument for observing multi-layer current profiles in marine research projects, characterized by high accuracy, high resolution, and ease of operation. Due to the relative motion between the ultrasound source (or emitter) and the receiver (emitter), the frequency received by the receiver and the source's intrinsic frequency do not coincide. If they are in close proximity to each other, the received frequency is higher than the transmitted frequency, and vice versa. This phenomenon is known as the Doppler effect. The difference between the received frequency and the transmitted frequency is called the Doppler frequency shift. An ADCP divides the water column into multiple depth bins and determines the velocity at each bin by measuring the Doppler frequency shift of the backscattered signal. For a

particle moving at velocity v, the received frequency is

$$f' = \frac{c + v\cos\theta}{c - v\cos\theta} f_0 \tag{7.16}$$

where f' is the received signal frequency; c is the sound speed in seawater; f_0 is the transmitted frequency; v is the particle velocity (the current velocity); θ is the angle between the acoustic beam and the current direction.

Then the Doppler frequency shift is

$$f_d = f' - f_0 = \frac{2v\cos\theta}{c - v\cos\theta} f_0 \tag{7.17}$$

Since $v\cos\theta \ll c$, $f_d \approx \frac{2f_0\cos\theta}{c}$.

From the above equation, $v = \frac{cf_d}{2f_0\cos\theta}$. If $f_0\cos\theta$ is known, v can be obtained from c and f_d. c can be calculated by the following equation:

$$\begin{aligned} c = {} & 1449.2 + 4.6T - 5.5 \times 10^{-2}T^2 + 2.9 \times 10^{-4}T^3 + 1.34 \times 10^{-2}T \\ & \times (S - 35.0) \times 11.6 \times 10^{-2}D \end{aligned} \tag{7.18}$$

where T is the temperature; S is the salinity; D is the depth at the observation location.

By measuring the Doppler frequency shift of the scattered water using acoustic beams (at least three), the three-dimensional current velocity can be obtained and resolved into the components u (eastward), v (norard), and w (vertical). As sound propagates at essentially constant speed in a body of water within a certain depth range, the depth can be determined from the time difference between transmission and reception. By transmitting acoustic pulses continuously, specifying the pulse repetition interval and the sampling duration, and estimating the Doppler frequency shift as a function of range, the current-velocity field over the entire cross-section of the water column can be obtained.

7.3.2 Collection and Analysis of Current Observation Data

The following analyses can be performed for measured current data.

(1) The measured current values are used to plot the various currents from which the relevant characteristic values can be determined.
(2) Compute the cross-sectional discharge using current measurements at points along a cross-section.
(3) Prepare current-rose diagrams using data from surface current drifters.

As currents measured in the nearshore zone are typically combined currents composed of wind-driven currents and wave-induced currents, they are often broken down into periodic currents and non-periodic residual currents for ease of application. Several separation methods are available; here, the analytical method is introduced. This method assumes that, over a short time interval, the direction and velocity of the residual current remain constant. Accordingly, the hourly observed current velocity can be decomposed into eastward and northward components, calculated as follows:

$$\text{Eastward component} \quad v = U_0 \sin \theta_0 \tag{7.19}$$

$$\text{Northward component} \quad u = U_0 \cos \theta_0 \tag{7.20}$$

where U_0 is the measured current velocity; θ_0 is the current direction.

Based on the hourly eastward and northward components, the mean eastward current velocity $\overline{v}$ and northward current velocity $\overline{u}$ over a lunar day can be calculated. U and θ of the residual current are then obtained by

$$U = \sqrt{\overline{v}^2 + \overline{u}^2} \tag{7.21}$$

$$\theta = \tan^{-1} \frac{\overline{v}}{\overline{u}} \tag{7.22}$$

Calculations can be carried out in tabular form. Please refer to ESM 7.1

Schedule 1: Deviation Coefficient Φ_P for the Cumulative Frequency Curve of Pearson Type III Distribution

C_S	Cumulative frequency P/%													
	0.001	0.01	0.1	0.2	0.333	0.5	1	2	3	5	10	20	25	30
0	4.26	3.72	3.09	2.88	2.71	2.58	2.33	2.05	1.88	1.64	1.28	0.84	0.67	0.52
0.1	4.56	3.94	3.23	3.00	2.82	2.67	2.40	2.11	1.92	1.67	1.29	0.84	0.66	0.51
0.2	4.86	4.16	3.38	3.12	2.92	2.76	2.47	2.16	1.96	1.70	1.30	0.83	0.65	0.50
0.3	5.16	4.38	3.52	3.24	3.03	2.86	2.54	2.21	2.00	1.73	1.31	0.82	0.64	0.48
0.4	5.47	4.61	3.67	3.36	3.14	2.95	2.62	2.26	2.04	1.75	1.32	0.82	0.63	0.47
0.5	5.78	4.83	3.81	3.48	3.25	3.04	2.68	2.31	2.08	1.77	1.32	0.81	0.62	0.46
0.6	6.09	5.05	3.96	3.6	3.35	3.13	2.75	2.35	2.12	1.80	1.33	0.80	0.61	0.44
0.7	6.40	5.28	4.10	3.72	3.45	3.22	2.82	2.40	2.15	1.82	1.33	0.79	0.59	0.43
0.8	6.71	5.50	4.24	3.85	3.55	3.31	2.89	2.45	2.18	1.84	1.34	0.78	0.58	0.41
0.9	7.02	5.73	4.39	3.97	3.65	3.40	2.96	2.50	2.22	1.86	1.34	0.77	0.57	0.40
1.0	7.33	5.96	4.53	4.09	3.76	3.49	3.02	2.54	2.25	1.88	1.34	0.76	0.55	0.38
1.1	7.65	6.18	4.67	4.20	3.86	3.58	3.09	2.58	2.28	1.89	1.34	0.74	0.54	0.36
1.2	7.97	6.41	4.81	4.32	3.95	3.66	3.15	2.62	2.31	1.91	1.34	0.73	0.52	0.35
1.3	8.29	6.64	4.95	4.44	4.05	3.74	3.21	2.67	2.34	1.92	1.34	0.72	0.51	0.33
1.4	8.61	6.87	5.09	4.56	4.15	3.83	3.27	2.71	2.37	1.94	1.33	0.71	0.49	0.31
1.5	8.93	7.09	5.23	4.68	4.24	3.91	3.33	2.74	2.39	1.95	1.33	0.69	0.47	0.30
1.6	9.25	7.31	5.37	4.80	4.34	3.99	3.39	2.78	2.42	1.96	1.33	0.68	0.46	0.28
1.7	9.57	7.54	5.50	4.91	4.43	4.07	3.44	2.82	2.44	1.97	1.32	0.66	0.44	0.26
1.8	9.89	7.76	5.64	5.01	4.52	4.15	3.5	2.85	2.46	1.98	1.32	0.64	0.42	0.24
1.9	10.20	7.98	5.77	5.12	4.61	4.23	3.55	2.88	2.49	1.99	1.31	0.63	0.4	0.22
2.0	10.51	8.21	5.91	5.22	4.70	4.30	3.61	2.91	2.51	2.00	1.30	0.61	0.39	0.20
2.1	10.83	8.43	6.04	5.33	4.79	4.37	3.66	2.93	2.53	2.00	1.29	0.59	0.37	0.19
2.2	11.14	8.65	6.17	5.43	4.88	4.44	3.71	2.96	2.55	2.00	1.28	0.57	0.35	0.17
2.3	11.45	8.87	6.30	5.53	4.97	4.51	3.76	2.99	2.56	2.00	1.27	0.55	0.33	0.15
2.4	11.76	9.08	6.42	5.63	5.05	4.58	3.81	3.02	2.57	2.01	1.26	0.54	0.31	0.13
2.5	12.07	9.30	6.55	5.73	5.13	4.65	3.85	3.04	2.59	2.01	1.25	0.52	0.29	0.11
2.6	12.38	9.51	6.67	5.82	5.20	4.72	3.89	3.06	2.60	2.01	1.23	0.50	0.27	0.09
2.7	12.69	9.72	6.79	5.92	5.28	4.78	3.93	3.09	2.61	2.01	1.22	0.48	0.25	0.08
2.8	13.00	9.93	6.91	6.01	5.36	4.84	3.97	3.11	2.62	2.01	1.21	0.46	0.23	0.06

(continued)

X. Feng (ed.), *Engineering Hydrology*, https://doi.org/10.1007/978-981-95-9089-6

(continued)

C_s	Cumulative frequency P/%													
	0.001	0.01	0.1	0.2	0.333	0.5	1	2	3	5	10	20	25	30
2.9	13.31	10.14	7.03	6.10	5.44	4.90	4.01	3.13	2.63	2.01	1.20	0.44	0.21	0.04
3.0	13.61	10.35	7.15	6.20	5.51	4.96	4.05	3.15	2.64	2.00	1.18	0.42	0.19	0.03
3.1	13.92	10.56	7.26	6.30	5.59	5.02	4.08	3.17	2.64	2.00	1.16	0.4	0.17	0.01
3.2	14.22	10.77	7.38	6.39	5.66	5.08	4.12	3.19	2.65	2.00	1.14	0.38	0.15	−0.01
3.3	14.52	10.97	7.49	6.48	5.74	5.14	4.15	3.21	2.65	1.99	1.12	0.36	0.14	−0.02
3.4	14.81	11.17	7.60	6.56	5.80	5.20	4.18	3.22	2.65	1.98	1.11	0.34	0.12	−0.04

C_s	Cumulative frequency P/%												
	40	50	60	70	75	80	85	90	95	97	99	99.9	100
0	0.25	0	−0.32	−0.52	−0.67	−0.84	−1.04	−1.28	−1.64	−1.88	−2.33	−3.09	−∞
0.1	0.24	−0.02	−0.27	−0.53	−0.68	−0.85	−1.04	−1.27	−1.62	−1.84	−2.25	−2.95	−20.00
0.2	0.22	−0.03	−0.23	−0.55	−0.69	−0.85	−1.03	−1.26	−1.59	−1.79	−2.18	−2.81	−10.00
0.3	0.20	−0.05	−0.30	−0.56	−0.70	−0.85	−1.03	−1.24	−1.55	−1.75	−2.10	−2.67	−6.67
0.4	0.19	−0.07	−0.31	−0.57	−0.71	−0.85	−1.03	−1.23	−1.52	−1.70	−2.03	−2.54	−5.00
0.5	0.17	−0.08	−0.33	−0.58	−0.71	−0.85	−1.02	−1.20	−1.49	−1.66	−1.96	−2.40	−4.00
0.6	0.16	−0.10	−0.34	−0.59	−0.72	−0.85	−1.92	−1.18	−1.45	−1.61	−1.88	−2.27	−3.33
0.7	0.14	−0.12	−0.36	−0.60	−0.72	−0.85	−1.01	−1.17	−1.42	−1.57	−1.81	−2.14	−2.86
0.8	0.12	−0.13	−0.37	−0.60	−0.73	−0.85	−0.99	−1.15	−1.38	−1.52	−1.74	−2.02	−2.50
0.9	0.11	−0.15	−0.38	−0.61	−0.73	−0.85	−0.98	−1.13	−1.35	−1.47	−1.66	−1.90	−2.22
1.0	0.09	−0.16	−0.39	−0.62	−0.73	−0.85	−0.98	−1.13	−1.32	−1.42	−1.59	−1.79	−2.00
1.1	0.07	−0.18	−0.41	−0.62	−0.74	−0.85	−0.97	−1.10	−1.28	−1.38	−1.52	−1.68	−1.82
1.2	0.05	−0.19	−0.42	−0.63	−0.74	−0.84	−0.96	−1.08	−1.24	−1.33	−1.45	−1.58	−1.67
1.3	0.04	−0.21	−0.43	−0.63	−0.74	−0.84	−0.95	−1.06	−1.20	−1.28	−1.38	−1.48	−1.54
1.4	0.02	−0.22	−0.44	−0.64	−0.73	−0.83	−0.93	−1.04	−1.17	−1.23	−1.32	−1.39	−1.43
1.5	0	−0.24	−0.45	−0.64	−0.73	−0.82	−0.92	−1.02	−1.13	−1.19	−1.26	−1.31	−1.33
1.6	−0.02	−0.25	−0.46	−0.64	−0.73	−0.81	−0.90	−0.99	−1.10	−1.14	−1.20	−1.24	−1.25
1.7	−0.03	−0.27	−0.47	−0.64	−0.72	−0.81	−0.89	−0.97	−1.06	−1.10	−1.14	−1.17	−1.18
1.8	−0.05	−0.28	−0.48	−0.64	−0.72	−0.80	−0.87	−0.94	−1.02	−1.06	−1.09	−1.11	−1.11
1.9	−0.07	−0.29	−0.48	−0.64	−0.72	−0.79	−0.85	−0.92	−1.98	−1.01	−1.04	−1.05	−1.05
2.0	−0.08	−0.31	−0.49	−0.64	−0.71	−0.78	−0.84	−0.90	−0.95	−0.97	−0.99	−1.00	−1.00
2.1	−0.10	−0.32	−0.49	−0.64	−0.71	−0.76	−0.82	−0.87	−0.91	−0.94	−0.95	−0.95	−0.95
2.2	−0.11	−0.33	−0.50	−0.64	−0.70	−0.75	−0.80	−0.84	−0.88	−0.90	−0.91	−0.91	−0.91
2.3	−0.13	−0.34	−0.50	−0.64	−0.69	−0.74	−0.78	−0.82	−0.84	−0.87	−0.87	−0.87	−0.87
2.4	−0.15	−0.35	−0.51	−0.63	−0.68	−0.72	−0.77	−0.80	−0.82	−0.83	−0.83	−0.83	−0.83
2.5	−0.16	−0.36	−0.51	−0.63	−0.67	−0.71	−0.75	−0.77	−0.79	−0.80	−0.80	−0.80	−0.80
2.6	−0.17	−0.37	−0.51	−0.62	−0.66	−0.70	−0.73	−0.75	−0.76	−0.77	−0.77	−0.77	−0.77
2.7	−0.18	−0.37	−0.51	−0.61	−0.65	−0.68	−0.71	−0.73	−0.74	−0.74	−0.74	−0.74	−0.74
2.8	−0.20	−0.38	−0.51	−0.61	−0.64	−0.67	−0.69	−0.70	−0.71	−0.71	−0.71	−0.71	−0.71
2.9	−0.21	−0.39	−0.51	−0.60	−0.63	−0.66	−0.67	−0.68	−0.69	−0.69	−0.69	−0.69	−0.69
3.0	−0.22	−0.39	−0.51	−0.59	−0.62	−0.64	−0.65	−0.66	−0.67	−0.67	−0.67	−0.67	−0.67
3.1	−0.24	−0.40	−0.51	−0.58	−0.60	−0.62	−0.63	−0.64	−0.64	−0.65	−0.65	−0.65	−0.65
3.2	−0.25	−0.40	−0.51	−0.57	−0.59	−0.61	−0.62	−0.62	−0.63	−0.63	−0.63	−0.63	−0.63
3.3	−0.26	−0.40	−0.50	−0.56	−0.58	−0.59	−0.60	−0.60	−0.61	−0.61	−0.61	−0.61	−0.61

(continued)

(continued)

C_s	Cumulative frequency P/%												
	40	50	60	70	75	80	85	90	95	97	99	99.9	100
3.4	−0.27	−0.41	−0.50	−0.55	−0.57	−0.58	−0.58	−0.59	−0.59	−0.59	−0.59	−0.59	−0.59
C_s	Cumulative frequency P/%												
	0.001	0.01	0.1	0.2	0.333	0.5	1	2	3	5	10	20	25	30
3.5	15.11	11.37	7.72	6.65	5.86	5.25	4.22	3.23	2.65	1.97	1.09	0.32	0.10	−0.06
3.6	15.41	11.57	7.83	6.73	5.93	5.30	4.25	3.24	2.66	1.96	1.08	0.3	0.09	−0.07
3.7	15.70	11.77	7.94	6.81	5.99	5.35	4.28	3.25	2.66	1.95	1.06	0.28	0.07	−0.09
3.8	16.00	11.97	8.05	6.89	6.05	5.40	4.31	3.26	2.66	1.94	1.04	0.26	0.06	−0.10
3.9	16.29	12.16	8.15	6.97	6.11	5.45	4.34	3.27	2.66	1.93	1.02	0.24	0.04	−0.11
4.0	16.58	12.36	8.25	7.05	6.18	5.50	4.37	3.27	2.66	1.92	1.00	0.23	0.02	−0.13
4.1	16.87	12.55	8.35	7.13	6.24	5.54	4.39	3.28	2.66	1.91	0.98	0.21	0	−0.14
4.2	17.16	12.74	8.45	7.21	6.3	5.59	4.41	3.29	2.65	1.90	0.96	0.19	−0.02	−0.15
4.3	17.44	12.93	8.55	7.29	6.36	5.63	4.44	3.29	2.65	1.88	0.94	0.17	−0.03	−0.16
4.4	17.72	13.12	8.65	7.36	6.41	5.68	4.46	3.30	2.65	1.87	0.92	0.16	−0.04	−0.17
4.5	18.01	13.30	8.75	7.43	6.46	5.72	4.48	3.30	2.64	1.85	0.90	0.14	−0.05	−0.18
4.6	18.29	13.49	8.85	7.50	6.52	5.76	4.50	3.30	2.63	1.84	0.88	0.13	−0.06	−0.18
4.7	18.57	13.67	8.95	7.56	6.57	5.80	4.52	3.30	2.62	1.82	0.86	0.11	−0.07	−0.19
4.8	18.85	13.85	9.04	7.63	6.63	5.84	4.54	3.30	2.61	1.80	0.84	0.09	−0.08	−0.20
4.9	19.13	14.04	9.13	7.70	6.68	5.88	4.55	3.30	2.60	1.78	0.82	0.08	−0.10	−0.21
5.0	19.41	14.22	9.22	7.77	6.73	5.92	4.57	3.30	2.60	1.77	0.80	0.06	−0.11	−0.22
5.1	19.68	14.4	9.31	7.84	6.78	5.95	4.58	3.30	2.59	1.75	0.78	0.05	−0.12	−0.22
5.2	19.95	14.57	9.40	7.90	6.83	5.99	4.59	3.30	2.58	1.73	0.76	0.03	−0.13	−0.22
5.3	20.22	14.75	9.49	7.96	6.87	6.02	4.60	3.30	2.57	1.72	0.74	0.02	−0.14	−0.22
5.4	20.46	14.92	9.57	8.02	6.91	6.05	4.62	3.28	2.56	1.70	0.72	0	−0.14	−0.23
5.5	20.76	15.10	9.66	8.08	6.96	6.08	4.63	3.28	2.55	1.68	0.70	−0.01	−0.15	−0.23
5.6	21.03	15.27	9.74	8.14	7.00	6.11	4.64	3.28	2.53	1.66	0.67	−0.03	−0.16	−0.24
5.7	21.31	15.45	9.82	8.21	7.04	6.14	4.65	3.27	2.52	1.65	0.65	−0.04	−0.17	−0.24
5.8	21.58	15.62	9.91	8.27	7.08	6.17	4.67	3.27	2.51	1.63	0.63	−0.05	−0.18	−0.25
5.9	21.84	15.78	9.99	8.32	7.12	6.20	4.68	3.26	2.49	1.61	0.61	−0.06	−0.18	−0.25
6.0	22.10	15.94	10.07	8.38	7.15	6.23	4.68	3.25	2.48	1.59	0.59	−0.07	−0.19	−0.25
6.1	22.37	16.11	10.15	8.43	7.19	6.26	4.69	3.24	2.46	1.57	0.57	−0.08	−0.19	−0.26
6.2	22.63	16.28	10.22	8.49	7.23	6.28	4.70	3.23	2.45	1.55	0.55	−0.09	−0.20	−0.26
6.3	22.89	16.45	10.30	8.54	7.26	6.30	4.70	3.22	2.43	1.53	0.53	−0.10	−0.20	−0.26
6.4	23.15	16.61	10.38	8.60	7.30	6.32	4.71	3.21	2.41	1.51	0.51	−0.11	−0.21	−0.26
C_s	Cumulative frequency P/%												
	40	50	60	70	75	80	85	90	95	97	99	99.9	100
3.5	−0.28	−0.41	−0.50	−0.54	−0.55	−0.56	−0.56	−0.57	−0.57	−0.57	−0.57	−0.57	−0.57
3.6	−0.29	−0.41	−0.49	−0.53	−0.54	−0.55	−0.55	−0.56	−0.56	−0.56	−0.56	−0.56	−0.56
3.7	−0.29	−0.42	−0.48	−0.52	−0.53	−0.54	−0.54	−0.54	−0.54	−0.54	−0.54	−0.54	−0.54
3.8	−0.30	−0.42	−0.48	−0.51	−0.52	−0.52	−0.52	−0.53	−0.53	−0.53	−0.53	−0.53	−0.53
3.9	−0.30	−0.41	−0.47	−0.50	−0.51	−0.51	−0.51	−0.51	−0.51	−0.51	−0.51	−0.51	−0.51
4.0	−0.31	−0.41	−0.46	−0.49	−0.50	−0.50	−0.50	−0.50	−0.50	−0.50	−0.50	−0.50	−0.50
4.1	−0.32	−0.41	−0.46	−0.48	−0.48	−0.49	−0.49	−0.49	−0.49	−0.49	−0.49	−0.49	−0.49

(continued)

(continued)

C_s	Cumulative frequency P/%												
	40	50	60	70	75	80	85	90	95	97	99	99.9	100
4.2	−0.32	−0.41	−0.45	−0.47	−0.47	−0.48	−0.48	−0.48	−0.48	−0.48	−0.48	−0.48	−0.48
4.3	−0.33	−0.41	−0.44	−0.46	−0.46	−0.46	−0.46	−0.47	−0.47	−0.47	−0.47	−0.47	−0.47
4.4	−0.33	−0.40	−0.44	−0.45	−0.45	−0.45	−0.45	−0.46	−0.46	−0.46	−0.46	−0.46	−0.46
4.5	−0.33	−0.40	−0.43	−0.44	−0.44	−0.44	−0.44	−0.44	−0.44	−0.44	−0.44	−0.44	−0.44
4.6	−0.33	−0.40	−0.42	−0.43	−0.44	−0.44	−0.44	−0.44	−0.44	−0.44	−0.44	−0.44	−0.44
4.7	−0.33	−0.39	−0.42	−0.42	−0.43	−0.43	−0.43	−0.43	−0.43	−0.43	−0.43	−0.43	−0.43
4.8	−0.33	−0.39	−0.41	−0.41	−0.42	−0.42	−0.42	−0.42	−0.42	−0.42	−0.42	−0.42	−0.42
4.9	−0.33	−0.38	−0.40	−0.40	−0.41	−0.41	−0.41	−0.41	−0.41	−0.41	−0.41	−0.41	−0.41
5.0	−0.33	−0.38	−0.40	−0.40	−0.40	−0.40	−0.40	−0.40	−0.40	−0.40	−0.40	−0.40	−0.40
5.1	−0.32	−0.37	−0.39	−0.39	−0.39	−0.39	−0.39	−0.39	−0.39	−0.39	−0.39	−0.39	−0.39
5.2	−0.32	−0.37	−0.38	−0.38	−0.39	−0.39	−0.39	−0.39	−0.39	−0.39	−0.39	−0.39	−0.39
5.3	−0.32	−0.36	−0.37	−0.38	−0.38	−0.38	−0.38	−0.38	−0.38	−0.38	−0.38	−0.38	−0.38
5.4	−0.32	−0.36	−0.37	−0.37	−0.37	−0.37	−0.37	−0.37	−0.37	−0.37	−0.37	−0.37	−0.37
5.5	−0.32	−0.35	−0.36	−0.36	−0.36	−0.36	−0.36	−0.36	−0.36	−0.36	−0.36	−0.36	−0.36
5.6	−0.32	−0.35	−0.36	−0.36	−0.36	−0.36	−0.36	−0.36	−0.36	−0.36	−0.36	−0.36	−0.36
5.7	−0.32	−0.34	−0.35	−0.35	−0.35	−0.35	−0.35	−0.35	−0.35	−0.35	−0.35	−0.35	−0.35
5.8	−0.32	−0.34	−0.34	−0.35	−0.35	−0.35	−0.35	−0.35	−0.35	−0.35	−0.35	−0.35	−0.35
5.9	−0.31	−0.33	−0.34	−0.34	−0.34	−0.34	−0.34	−0.34	−0.34	−0.34	−0.34	−0.34	−0.34
6.0	−0.31	−0.33	−0.33	−0.33	−0.33	−0.33	−0.33	−0.33	−0.33	−0.33	−0.33	−0.33	−0.33
6.1	−0.31	−0.33	−0.33	−0.33	−0.33	−0.33	−0.33	−0.33	−0.33	−0.33	−0.33	−0.33	−0.33
6.2	−0.30	−0.32	−0.32	−0.32	−0.32	−0.32	−0.32	−0.32	−0.32	−0.32	−0.32	−0.32	−0.32
6.3	−0.30	−0.32	−0.32	−0.32	−0.32	−0.32	−0.32	−0.32	−0.32	−0.32	−0.32	−0.32	−0.32
6.4	−0.30	−0.31	−0.31	−0.31	−0.31	−0.31	−0.31	−0.31	−0.31	−0.31	−0.31	−0.31	−0.31

Schedule 2: Modulus Ratio Coefficient k_P for the Cumulative Frequency Curve of Pearson Type III Distribution

C_v	Cumulative frequency $P/\%$														
	0.01	0.1	0.2	0.33	0.5	1	2	5	10	20	50	75	90	95	99
(1) $C_s = 1.0C_v$															
0.05	1.19	1.16	1.15	1.14	1.13	1.12	1.11	1.00	1.07	1.04	1.00	0.97	0.94	0.92	0.89
0.10	1.39	1.32	1.30	1.28	1.27	1.24	1.21	1.17	1.13	1.08	1.00	0.93	0.87	0.84	0.78
0.15	1.61	1.50	1.46	1.43	1.41	1.37	1.32	1.26	1.20	1.13	1.00	0.90	0.81	0.77	0.67
0.20	1.83	1.68	1.62	1.58	1.55	1.49	1.43	1.34	1.26	1.17	0.99	0.86	0.75	0.68	0.56
0.25	2.07	1.86	1.80	1.74	1.70	1.63	1.55	1.43	1.33	1.21	0.99	0.83	0.69	0.61	0.47
0.30	2.31	2.06	1.97	1.91	1.86	1.76	1.66	1.52	1.39	1.25	0.98	0.79	0.63	0.54	0.37
0.35	2.57	2.26	2.16	2.08	2.02	1.91	1.78	1.61	1.46	1.29	0.98	0.76	0.57	0.47	0.28
0.40	2.84	2.47	2.34	2.26	2.18	2.05	1.90	1.70	1.53	1.33	0.97	0.72	0.51	0.39	0.19
0.45	3.13	2.69	2.54	2.44	2.35	2.19	2.03	1.79	1.60	1.37	0.97	0.69	0.45	0.33	0.10
0.50	3.42	2.91	2.74	2.63	2.52	2.34	2.16	1.89	1.66	1.40	0.96	0.65	0.39	0.26	0.02
0.55	3.72	3.14	2.95	2.82	2.70	2.49	2.29	1.98	1.73	1.44	0.95	0.61	0.34	0.20	−0.06
0.60	4.03	3.38	3.16	3.01	2.88	2.65	2.41	2.08	1.80	1.48	0.94	0.57	0.28	0.13	−0.13
0.65	4.36	3.62	3.38	3.21	3.07	2.81	2.55	2.18	1.87	1.52	0.93	0.53	0.23	0.07	−0.20
0.70	4.70	3.87	3.60	3.42	3.25	2.97	2.68	2.27	1.93	1.55	0.92	0.50	0.17	0.01	−0.27
0.75	5.05	4.13	3.84	3.63	3.45	3.14	2.82	2.37	2.00	1.59	0.91	0.46	0.12	−0.05	−0.33
0.80	5.40	4.39	4.08	3.84	3.65	3.31	2.96	2.47	2.07	1.62	0.90	0.42	0.06	−0.10	−0.39
0.85	5.78	4.67	4.33	4.07	3.86	3.49	3.11	2.57	2.14	1.66	0.88	0.37	0.01	−0.16	−0.44
0.90	6.16	4.95	4.57	4.29	4.06	3.66	3.25	2.67	2.21	1.69	0.86	0.34	−0.04	−0.22	−0.49
0.95	6.56	5.24	4.83	4.53	4.28	3.84	3.40	2.78	2.28	1.73	0.85	0.31	−0.09	−0.27	−0.55
1.00	6.96	5.53	5.09	4.76	4.49	4.02	3.54	2.86	2.34	1.76	0.84	0.27	−0.13	−0.32	−0.59
1.05	7.38	5.83	5.35	5.01	4.72	4.21	3.69	2.98	2.41	1.78	0.82	0.22	−0.17	−0.37	−0.63
1.10	7.80	6.14	5.62	5.25	4.94	4.40	3.84	3.08	2.47	1.81	0.80	0.19	−0.21	−0.41	−0.67
1.15	8.24	6.45	5.90	5.50	5.17	4.59	3.99	3.19	2.54	1.85	0.79	0.14	−0.26	−0.45	−0.71
1.20	8.69	6.77	6.18	5.74	5.39	4.78	4.14	3.29	2.61	1.88	0.77	0.11	−0.30	−0.49	−0.74
1.25	9.16	7.10	6.48	6.01	5.63	4.98	4.31	3.40	2.68	1.91	0.75	0.07	−0.34	−0.53	−0.77
1.30	9.63	7.44	6.77	6.27	5.86	5.17	4.47	3.50	2.74	1.94	0.73	0.04	−0.38	−0.56	−0.79
1.35	10.12	7.78	7.08	6.54	6.11	5.38	4.63	3.61	2.81	1.97	0.71	0.01	−0.42	−0.60	−0.82

(continued)

X. Feng (ed.), *Engineering Hydrology*, https://doi.org/10.1007/978-981-95-9089-6

(continued)

C_v	Cumulative frequency $P/\%$														
	0.01	0.1	0.2	0.33	0.5	1	2	5	10	20	50	75	90	95	99
1.40	10.62	8.13	7.38	6.81	6.36	5.58	4.79	3.72	2.88	1.99	0.69	−0.02	−0.46	−0.64	−0.85
1.45	11.12	8.48	7.70	7.09	6.62	5.79	4.95	3.82	2.94	2.02	0.66	−0.06	−0.50	−0.67	−0.87
1.50	11.64	8.85	8.02	7.36	6.87	6.00	5.11	3.92	3.00	2.04	0.64	−0.10	−0.53	−0.70	−0.89
(2) $C_s = 1.5C_v$															
0.05	1.19	1.16	1.15	1.14	1.13	1.12	1.10	1.08	1.06	1.04	1.00	0.97	0.94	0.92	0.89
0.10	1.40	1.33	1.31	1.29	1.27	1.24	1.21	1.17	1.13	1.08	1.00	0.93	0.87	0.84	0.78
0.15	1.63	1.51	1.47	1.44	1.42	1.37	1.32	1.26	1.19	1.12	1.00	0.90	0.81	0.77	0.68
0.20	1.88	1.70	1.65	1.60	1.57	1.51	1.44	1.35	1.26	1.16	1.00	0.86	0.75	0.69	0.58
0.25	2.14	1.91	1.83	1.78	1.73	1.65	1.56	1.44	1.33	1.20	0.99	0.83	0.69	0.62	0.49
0.30	2.42	2.12	2.03	1.96	1.90	1.80	1.68	1.53	1.40	1.25	0.98	0.79	0.63	0.55	0.40
0.35	2.71	2.35	2.23	2.15	2.07	1.95	1.81	1.62	1.46	1.28	0.97	0.75	0.58	0.49	0.33
0.40	3.02	2.58	2.44	2.34	2.25	2.10	1.94	1.72	1.53	1.32	0.96	0.71	0.52	0.42	0.25
0.45	3.35	2.83	2.66	2.54	2.44	2.26	2.07	1.82	1.60	1.35	0.95	0.68	0.47	0.36	0.18
0.50	3.70	3.08	2.89	2.75	2.64	2.43	2.21	1.92	1.67	1.39	0.94	0.64	0.41	0.30	0.11
0.55	4.06	3.35	3.13	2.97	2.84	2.60	2.35	2.02	1.73	1.42	0.93	0.60	0.36	0.25	0.06
0.60	4.44	3.63	3.38	3.19	3.04	2.78	2.50	2.12	1.80	1.46	0.91	0.56	0.31	0.19	0
0.65	4.84	3.92	3.64	3.42	3.25	2.95	2.64	2.22	1.87	1.49	0.90	0.52	0.27	0.14	−0.04
0.70	5.25	4.22	3.90	3.67	3.48	3.12	2.79	2.32	1.94	1.52	0.88	0.48	0.22	0.09	−0.08
0.75	5.68	4.53	4.17	3.91	3.70	3.32	2.87	2.42	2.00	1.55	0.87	0.45	0.18	0.05	−0.12
0.80	6.13	4.85	4.46	4.16	3.93	3.52	2.96	2.53	2.07	1.58	0.85	0.41	0.14	0.01	−0.16
0.85	6.60	5.18	4.75	4.42	4.16	3.72	3.19	2.63	2.19	1.61	0.83	0.37	0.10	−0.02	−0.19
0.90	7.09	5.52	5.05	4.69	4.40	3.92	3.42	2.74	2.21	1.65	0.80	0.33	0.06	−0.06	−0.22
0.95	7.58	5.87	5.37	4.96	4.50	4.12	3.58	2.84	2.27	1.67	0.78	0.30	0.02	−0.09	−0.24
1.00	8.09	6.23	5.68	5.24	4.91	4.33	3.74	2.95	2.33	1.69	0.76	0.27	−0.02	−0.13	−0.26
1.05	8.62	6.60	6.01	5.53	5.17	4.54	3.91	3.05	2.39	1.71	0.74	0.24	−0.05	−0.16	−0.27
1.10	9.16	6.98	6.34	5.82	5.43	4.76	4.08	3.16	2.45	1.74	0.71	0.21	−0.08	−0.19	−0.29
1.15	9.73	7.37	6.68	6.12	5.70	4.97	4.25	3.27	2.51	1.75	0.69	0.18	−0.10	−0.20	−0.30
1.20	10.31	7.77	7.01	6.42	5.98	5.20	4.42	3.38	2.58	1.77	0.66	0.14	−0.13	−0.22	−0.31
1.25	10.91	8.17	7.37	6.72	6.26	5.32	4.59	3.48	2.64	1.79	0.63	0.10	−0.15	−0.23	−0.31
1.30	11.52	8.59	7.72	7.05	6.54	5.65	4.76	3.59	2.70	1.81	0.61	0.07	−0.17	−0.25	−0.32
1.35	12.16	9.02	8.09	7.38	6.83	5.88	4.93	3.69	2.75	1.82	0.59	0.04	−0.19	−0.26	−0.32
1.40	12.80	9.46	8.46	7.70	7.12	6.12	5.10	3.80	2.81	1.83	0.55	0.01	−0.22	−0.28	−0.32
1.45	13.46	9.90	8.84	8.04	7.42	6.36	5.28	3.90	2.85	1.83	0.52	0	−0.23	−0.29	−0.33
1.50	14.14	10.36	9.22	8.39	7.72	6.60	5.47	4.00	2.90	1.84	0.49	−0.05	−0.25	−0.30	−0.33
(3) $C_s = 2.0C_v$															
0.05	1.20	1.16	1.15	1.14	1.13	1.12	1.11	1.08	1.06	1.04	1.00	0.97	0.94	0.92	0.89
0.10	1.42	1.34	1.31	1.29	1.27	1.25	1.21	1.17	1.13	1.08	1.00	0.93	0.87	0.84	0.78
0.15	1.67	1.54	1.48	1.46	1.43	1.38	1.33	1.26	1.20	1.12	0.99	0.90	0.81	0.77	0.69
0.20	1.92	1.73	1.67	1.63	1.59	1.52	1.45	1.35	1.26	1.16	0.99	0.86	0.75	0.70	0.59
0.25	2.22	1.96	1.87	1.81	1.77	1.67	1.58	1.45	1.33	1.20	0.98	0.82	0.70	0.63	0.52
0.30	2.52	2.19	2.08	2.01	1.94	1.83	1.71	1.54	1.40	1.24	0.97	0.78	0.64	0.56	0.44
0.35	2.86	2.44	2.31	2.22	2.13	2.00	1.84	1.64	1.47	1.28	0.96	0.75	0.59	0.51	0.37

(continued)

(continued)

C_v	Cumulative frequency $P/\%$														
	0.01	0.1	0.2	0.33	0.5	1	2	5	10	20	50	75	90	95	99
0.40	3.20	2.70	2.54	2.42	2.32	2.16	1.98	1.74	1.54	1.31	0.95	0.71	0.53	0.45	0.30
0.45	3.59	2.98	2.80	2.65	2.53	2.33	2.13	1.84	1.60	1.35	0.93	0.67	0.48	0.40	0.26
0.50	3.98	3.27	3.05	2.88	2.74	2.51	2.27	1.94	1.67	1.38	0.92	0.64	0.44	0.34	0.21
0.55	4.42	3.58	3.32	3.12	2.97	2.70	2.42	2.04	1.74	1.41	0.90	0.59	0.40	0.30	0.16
0.60	4.85	3.89	3.59	3.37	3.20	2.89	2.57	2.15	1.80	1.44	0.89	0.56	0.35	0.26	0.13
0.65	5.33	4.22	3.89	3.64	3.44	3.09	2.74	2.25	1.87	1.47	0.87	0.52	0.31	0.22	0.10
0.70	5.81	4.56	4.19	3.91	3.68	3.29	2.90	2.36	1.94	1.50	0.85	0.49	0.27	0.18	0.08
0.75	6.33	4.93	4.52	4.19	3.93	3.50	3.06	2.46	2.00	1.52	0.82	0.45	0.24	0.15	0.06
0.80	6.85	5.30	4.84	4.47	4.19	3.71	3.22	2.57	2.06	1.54	0.80	0.42	0.21	0.12	0.04
0.85	7.41	5.69	5.17	4.77	4.46	3.93	3.39	2.68	2.12	1.56	0.77	0.39	0.18	0.10	0.03
0.90	7.98	6.08	5.51	5.07	4.74	4.15	3.56	2.78	2.19	1.58	0.75	0.35	0.15	0.08	0.02
0.95	8.59	6.48	5.86	5.38	5.02	4.38	3.74	2.89	2.25	1.60	0.72	0.31	0.13	0.07	0.01
1.00	9.21	6.91	6.22	5.70	5.30	4.61	3.91	3.00	2.30	1.61	0.69	0.29	0.11	0.05	0.01
1.05	9.86	7.35	6.59	6.03	5.59	4.84	4.08	3.10	2.35	1.62	0.66	0.26	0.09	0.04	0.01
1.10	10.52	7.79	6.97	6.37	5.88	5.08	4.26	3.20	2.41	1.63	0.64	0.23	0.07	0.03	0
1.15	11.21	8.24	7.36	6.71	6.19	5.32	4.44	3.30	2.46	1.64	0.61	0.21	0.06	0.02	0
1.20	11.90	8.70	7.76	7.06	6.50	5.57	4.62	3.41	2.51	1.65	0.58	0.18	0.05	0.02	0
1.25	11.63	9.18	8.16	7.41	6.82	5.81	4.80	3.51	2.56	1.65	0.55	0.16	0.04	0.01	0
1.30	13.36	9.67	8.57	7.76	7.14	6.06	4.98	3.61	2.60	1.65	0.52	0.14	0.03	0.01	0
1.35	14.13	10.17	8.99	8.13	7.46	6.31	5.16	3.71	2.65	1.65	0.50	0.12	0.02	0.01	0
1.40	14.90	10.67	9.41	8.50	7.78	6.56	5.35	3.81	2.69	1.64	0.47	0.10	0.02	0.01	0
1.45	15.71	11.20	9.85	8.89	8.11	6.82	5.54	3.91	2.73	1.64	0.44	0.09	0.01	0	0
1.50	16.53	11.73	10.30	9.27	8.44	7.08	5.73	4.00	2.77	1.63	0.42	0.07	0.01	0	0

(4) $C_s = 2.5C_v$

0.05	1.20	1.16	1.15	1.14	1.14	1.12	1.11	1.08	1.07	1.04	1.00	0.97	0.94	0.92	0.89
0.10	1.43	1.35	1.31	1.29	1.28	1.25	1.22	1.17	1.13	1.08	1.00	0.93	0.88	0.84	0.79
0.15	1.70	1.55	1.50	1.47	1.44	1.39	1.34	1.26	1.20	1.12	0.99	0.89	0.82	0.77	0.70
0.20	1.97	1.76	1.70	1.65	1.61	1.54	1.46	1.35	1.26	1.16	0.98	0.86	0.76	0.70	0.61
0.25	2.29	2.00	1.92	1.85	1.79	1.70	1.60	1.45	1.33	1.20	0.97	0.82	0.70	0.64	0.54
0.30	2.62	2.25	2.14	2.05	1.98	1.86	1.73	1.55	1.40	1.24	0.96	0.78	0.65	0.58	0.47
0.35	3.00	2.53	2.39	2.27	2.19	2.03	1.87	1.65	1.47	1.27	0.95	0.75	0.60	0.53	0.41
0.40	3.38	2.81	2.64	2.50	2.40	2.21	2.02	1.75	1.54	1.30	0.94	0.71	0.55	0.47	0.36
0.45	3.82	3.12	2.91	2.75	2.62	2.40	2.17	1.85	1.60	1.33	0.92	0.67	0.51	0.43	0.32
0.50	4.26	3.44	3.19	3.00	2.85	2.59	2.32	1.96	1.67	1.36	0.90	0.63	0.47	0.39	0.29
0.55	4.75	3.79	3.50	3.27	3.10	2.79	2.48	2.07	1.73	1.39	0.88	0.60	0.43	0.35	0.26
0.60	5.25	4.14	3.81	3.54	3.35	3.00	2.64	2.17	1.80	1.42	0.86	0.56	0.39	0.32	0.24
0.65	5.80	4.52	4.14	3.83	3.61	3.21	2.81	2.27	1.86	1.44	0.83	0.53	0.36	0.30	0.23
0.70	6.36	4.90	4.47	4.13	3.88	3.43	2.98	2.39	1.92	1.46	0.81	0.50	0.33	0.27	0.22
0.75	6.96	5.31	4.82	4.44	4.16	3.66	3.15	2.49	1.98	1.47	0.78	0.46	0.31	0.26	0.21
0.80	7.57	5.73	5.18	4.76	4.44	3.89	3.33	2.60	2.04	1.49	0.75	0.43	0.28	0.24	0.21
0.85	8.22	6.17	5.55	5.09	4.73	4.12	3.50	2.70	2.10	1.50	0.72	0.40	0.27	0.23	0.21
0.90	8.88	6.61	5.93	5.43	5.03	4.36	3.68	2.80	2.15	1.50	0.70	0.37	0.25	0.22	0.20

(continued)

(continued)

C_v	Cumulative frequency P/%														
	0.01	0.1	0.2	0.33	0.5	1	2	5	10	20	50	75	90	95	99
0.95	9.59	7.09	6.33	5.78	5.34	4.60	3.86	2.90	2.20	1.51	0.67	0.35	0.24	0.21	0.20
1.00	10.30	7.55	6.73	6.13	5.65	4.85	4.04	3.01	2.25	1.52	0.64	0.33	0.23	0.21	0.20
1.05	11.05	8.04	7.14	6.49	5.97	5.10	4.22	3.11	2.29	1.52	0.61	0.31	0.22	0.20	0.20
1.10	11.80	8.54	7.56	6.85	6.29	5.35	4.41	3.21	2.34	1.52	0.58	0.29	0.21	0.20	0.20
1.15	12.61	9.06	8.00	7.23	6.62	5.60	4.59	3.30	2.38	1.51	0.55	0.27	0.21	0.20	0.20
1.20	13.42	9.58	8.44	7.61	6.95	5.86	4.78	3.40	2.42	1.50	0.53	0.26	0.21	0.20	0.20
1.25	14.27	10.12	8.90	8.01	7.29	6.12	4.97	3.50	2.44	1.49	0.50	0.25	0.21	0.20	0.20
1.30	15.13	10.67	9.37	8.41	7.64	6.38	5.16	3.60	2.47	1.48	0.48	0.24	0.20	0.20	0.20
1.35	16.02	11.24	9.84	8.80	8.00	6.64	5.34	3.68	2.50	1.46	0.45	0.23	0.20	0.20	0.20
1.40	16.92	11.81	10.31	9.20	8.35	6.91	5.52	3.76	2.53	1.45	0.43	0.23	0.20	0.20	0.20
1.45	17.86	12.40	10.79	9.61	8.70	7.17	5.70	3.83	2.56	1.43	0.40	0.22	0.20	0.20	0.20
1.50	18.81	12.99	11.28	10.03	9.06	7.44	5.88	3.91	2.58	1.41	0.37	0.22	0.20	0.20	0.20
(5) $C_s = 3.0C_v$															
0.05	1.20	1.17	1.15	1.14	1.14	1.12	1.11	1.08	1.07	1.04	1.00	0.97	0.94	0.92	0.89
0.10	1.44	1.35	1.32	1.30	1.29	1.25	1.22	1.17	1.13	1.08	0.99	0.93	0.88	0.85	0.79
0.15	1.71	1.56	1.51	1.48	1.45	1.40	1.35	1.26	1.20	1.12	0.99	0.89	0.82	0.78	0.70
0.20	2.02	1.79	1.72	1.67	1.63	1.55	1.47	1.35	1.27	1.16	0.98	0.86	0.76	0.71	0.62
0.25	2.35	2.05	1.95	1.88	1.82	1.72	1.61	1.46	1.34	1.20	0.97	0.82	9.71	0.65	0.56
0.30	2.72	2.32	2.19	2.10	2.02	1.89	1.75	1.56	1.40	1.23	0.96	0.78	0.66	0.60	0.50
0.35	3.12	2.61	2.46	2.33	2.24	2.07	1.90	1.66	1.47	1.26	0.94	0.74	0.61	0.55	0.46
0.40	3.56	2.92	2.73	2.58	2.46	2.26	2.05	1.76	1.54	1.28	0.92	0.70	0.57	0.50	0.42
0.45	4.04	3.26	3.03	2.85	2.70	2.46	2.21	1.87	1.60	1.32	0.90	0.67	0.53	0.47	0.39
0.50	4.55	3.62	3.34	3.12	2.96	2.67	2.37	1.98	1.67	1.35	0.88	0.64	0.49	0.44	0.37
0.55	5.09	3.99	3.66	3.42	3.21	2.88	2.54	2.08	1.73	1.36	0.86	0.60	0.46	0.41	0.36
0.60	5.66	4.38	4.01	3.71	3.49	3.10	2.71	2.19	1.79	1.38	0.83	0.57	0.44	0.39	0.35
0.65	6.26	4.81	4.36	4.03	3.77	3.33	3.88	2.29	1.85	1.40	0.80	0.53	0.41	0.37	0.34
0.70	6.90	5.23	4.73	4.35	4.06	3.56	3.05	2.40	1.90	1.41	0.78	0.50	0.39	0.36	0.34
0.75	7.57	5.68	5.12	4.69	4.36	3.80	3.24	2.50	1.96	1.42	0.76	0.48	0.38	0.35	0.34
0.80	8.26	6.14	5.50	5.04	4.66	4.05	3.42	2.61	2.01	1.43	0.72	0.46	0.36	0.34	0.34
0.85	9.00	6.62	5.92	5.40	4.98	4.29	3.59	2.71	2.06	1.43	0.69	0.44	0.35	0.34	0.34
0.90	9.75	7.11	6.33	5.75	5.30	4.54	3.78	2.81	2.10	1.43	0.67	0.42	0.35	0.34	0.33
0.95	10.54	7.62	6.76	6.13	5.62	4.80	3.96	2.91	2.14	1.43	0.64	0.39	0.34	0.34	0.33
1.00	11.35	8.15	7.20	6.51	5.96	5.05	4.15	3.00	2.18	1.42	0.61	0.38	0.34	0.34	0.33
1.05	12.20	8.68	7.66	6.90	6.31	5.32	4.34	3.10	2.21	1.41	0.58	0.37	0.34	0.33	0.33
1.10	13.07	9.24	8.13	7.31	6.65	5.57	4.50	3.19	2.23	1.40	0.56	0.36	0.34	0.33	0.33
1.15	13.96	9.81	8.59	7.70	7.00	5.83	4.70	3.26	2.26	1.38	0.54	0.35	0.34	0.33	0.33
1.20	14.88	10.40	9.08	8.12	7.36	6.10	4.89	3.35	2.30	1.36	0.51	0.35	0.33	0.33	0.33
1.25	15.84	11.00	9.57	8.53	7.72	6.36	5.07	3.43	2.31	1.34	0.49	0.35	0.33	0.33	0.33
1.30	16.81	11.60	10.06	8.94	8.09	6.64	5.25	3.51	2.33	1.31	0.47	0.34	0.33	0.33	0.33
1.35	17.80	12.21	10.57	9.38	8.45	6.91	5.42	3.59	2.34	1.30	0.45	0.34	0.33	0.33	0.33
1.40	18.84	12.83	11.09	9.82	8.88	7.17	5.61	3.66	2.34	1.27	0.43	0.34	0.33	0.33	0.33
1.45	19.88	13.47	11.62	10.26	9.20	7.45	5.77	3.72	2.35	1.23	0.42	0.34	0.33	0.33	0.33

(continued)

(continued)

C_v	Cumulative frequency P/%														
	0.01	0.1	0.2	0.33	0.5	1	2	5	10	20	50	75	90	95	99
1.50	20.95	14.13	12.15	10.69	9.58	7.72	5.95	3.78	2.35	1.21	0.40	0.33	0.33	0.33	0.33
(6) $C_s = 3.5C_v$															
0.05	1.20	1.17	1.16	1.15	1.14	1.12	1.11	1.09	1.07	1.04	1.00	0.97	0.94	0.92	0.89
0.10	1.45	1.36	1.33	1.31	1.29	1.26	1.22	1.17	1.13	1.08	0.99	0.93	0.88	0.85	0.79
0.15	1.73	1.58	1.52	1.49	1.46	1.41	1.35	1.27	1.20	1.12	0.99	0.89	0.82	0.78	0.71
0.20	2.06	1.82	1.74	1.69	1.64	1.56	1.48	1.36	1.27	1.16	0.98	0.86	0.76	0.72	0.64
0.25	2.42	2.09	1.99	1.91	1.85	1.74	1.62	1.46	1.34	1.19	0.96	0.82	0.71	0.66	0.58
0.30	2.82	2.38	2.24	2.14	2.06	1.92	1.77	1.57	1.40	1.22	0.95	0.78	0.67	0.61	0.53
0.35	3.26	2.70	2.52	2.39	2.29	2.11	1.92	1.67	1.47	1.26	0.93	0.74	0.62	0.57	0.50
0.40	3.75	3.04	2.82	2.66	2.53	2.31	2.08	1.78	1.53	1.28	0.91	0.71	0.58	0.53	0.47
0.45	4.27	3.40	3.14	2.94	2.79	2.52	2.25	1.88	1.60	1.31	0.89	0.67	0.55	0.50	0.45
0.50	4.82	3.78	3.48	3.24	3.06	2.74	2.42	1.99	1.66	1.33	0.86	0.64	0.52	0.48	0.44
0.55	5.41	4.20	3.83	3.55	3.34	2.96	2.58	2.10	1.72	1.34	0.84	0.60	0.50	0.46	0.44
0.60	6.06	4.62	4.20	3.87	3.62	3.20	2.76	2.20	1.77	1.35	0.81	0.57	0.48	0.45	0.43
0.65	6.73	5.08	4.58	4.22	3.92	3.44	2.94	2.30	1.83	1.36	0.78	0.55	0.46	0.44	0.43
0.70	7.43	5.54	4.98	4.56	4.23	3.68	3.12	2.41	1.88	1.37	0.75	0.53	0.45	0.44	0.43
0.75	8.16	6.02	5.38	4.92	4.55	3.92	3.30	2.51	1.92	1.38	0.72	0.50	0.44	0.43	0.43
0.80	8.94	6.53	5.81	5.29	4.87	4.18	3.49	2.61	1.97	1.37	0.70	0.49	0.44	0.43	0.43
0.85	9.75	7.05	6.25	5.67	5.20	4.43	3.67	2.70	2.00	1.36	0.67	0.47	0.44	0.43	0.43
0.90	10.60	7.59	6.71	6.06	5.54	4.69	3.86	2.80	2.04	1.35	0.64	0.46	0.43	0.43	0.43
0.95	11.46	8.15	7.18	6.47	5.89	4.95	4.05	2.89	2.06	1.34	0.61	0.45	0.43	0.43	0.43
1.00	12.37	8.72	7.65	6.86	6.25	5.22	4.23	2.97	2.09	1.32	0.59	0.45	0.43	0.43	0.43
1.05	13.31	9.31	8.13	7.27	6.60	5.49	4.41	3.05	2.11	1.29	0.56	0.44	0.43	0.43	0.43
1.10	14.28	9.91	8.62	7.69	6.97	5.76	4.59	3.13	2.13	1.28	0.54	0.44	0.43	0.43	0.43
1.15	15.26	10.51	9.13	8.12	7.33	6.03	4.76	3.20	2.14	1.26	0.53	0.43	0.43	0.43	0.43
1.20	16.29	11.14	9.65	8.56	7.71	6.29	4.95	3.28	2.15	1.23	0.51	0.43	0.43	0.43	0.43
1.25	17.33	11.78	10.18	8.99	8.10	6.56	5.12	3.34	2.16	1.20	0.50	0.43	0.43	0.43	0.43
1.30	18.41	12.44	10.70	9.44	8.46	6.84	5.29	3.40	2.16	1.18	0.48	0.43	0.43	0.43	0.43
1.35	19.50	13.11	11.24	9.89	8.84	7.11	5.45	3.44	2.16	1.14	0.47	0.43	0.43	0.43	0.43
1.40	20.66	13.78	11.78	10.35	9.23	7.37	5.62	3.49	2.15	1.11	0.47	0.43	0.43	0.43	0.43
1.45	21.80	14.46	12.34	10.81	9.61	7.64	5.73	3.55	2.14	1.07	0.46	0.43	0.43	0.43	0.43
1.50	23.00	15.17	12.90	11.28	10.01	7.89	5.93	3.59	2.12	1.04	0.45	0.43	0.43	0.43	0.43
(7) $C_s = 3.5C_v$															
0.05	1.20	1.17	1.16	1.15	1.14	1.12	1.11	1.09	1.07	1.04	1.00	0.97	0.94	0.92	0.89
0.10	1.45	1.36	1.33	1.31	1.29	1.26	1.22	1.17	1.13	1.08	0.99	0.93	0.88	0.85	0.79
0.15	1.73	1.58	1.52	1.49	1.46	1.41	1.35	1.27	1.20	1.12	0.99	0.89	0.82	0.78	0.71
0.20	2.06	1.82	1.74	1.69	1.64	1.56	1.48	1.36	1.27	1.16	0.98	0.86	0.76	0.72	0.64
0.25	2.42	2.09	1.99	1.91	1.85	1.74	1.62	1.46	1.34	1.19	0.96	0.82	0.71	0.66	0.58
0.30	2.82	2.38	2.24	2.14	2.06	1.92	1.77	1.57	1.40	1.22	0.95	0.78	0.67	0.61	0.53
0.35	3.26	2.70	2.52	2.39	2.29	2.11	1.92	1.67	1.47	1.26	0.93	0.74	0.62	0.57	0.50
0.40	3.75	3.04	2.82	2.66	2.53	2.31	2.08	1.78	1.53	1.28	0.91	0.71	0.58	0.53	0.47
0.45	4.27	3.40	3.14	2.94	2.79	2.52	2.25	1.88	1.60	1.31	0.89	0.67	0.55	0.50	0.45

(continued)

(continued)

C_v	Cumulative frequency P/%														
	0.01	0.1	0.2	0.33	0.5	1	2	5	10	20	50	75	90	95	99
0.50	4.82	3.78	3.48	3.24	3.06	2.74	2.42	1.99	1.66	1.33	0.86	0.64	0.52	0.48	0.44
0.55	5.41	4.20	3.83	3.55	3.34	2.96	2.58	2.10	1.72	1.34	0.84	0.60	0.50	0.46	0.44
0.60	6.06	4.62	4.20	3.87	3.62	3.20	2.76	2.20	1.77	1.35	0.81	0.57	0.48	0.45	0.43
0.65	6.73	5.08	4.58	4.22	3.92	3.44	2.94	2.30	1.83	1.36	0.78	0.55	0.46	0.44	0.43
0.70	7.43	5.54	4.98	4.56	4.23	3.68	3.12	2.41	1.88	1.37	0.75	0.53	0.45	0.44	0.43
0.75	8.16	6.02	5.38	4.92	4.55	3.92	3.30	2.51	1.92	1.38	0.72	0.50	0.44	0.43	0.43
0.80	8.94	6.53	5.81	5.29	4.87	4.18	3.49	2.61	1.97	1.37	0.70	0.49	0.44	0.43	0.43
0.85	9.75	7.05	6.25	5.67	5.20	4.43	3.67	2.70	2.00	1.36	0.67	0.47	0.44	0.43	0.43
0.90	10.60	7.59	6.71	6.06	5.54	4.69	3.86	2.80	2.04	1.35	0.64	0.46	0.43	0.43	0.43
0.95	11.46	8.15	7.18	6.47	5.89	4.95	4.05	2.89	2.06	1.34	0.61	0.45	0.43	0.43	0.43
1.00	12.37	8.72	7.65	6.86	6.25	5.22	4.23	2.97	2.09	1.32	0.59	0.45	0.43	0.43	0.43
1.05	13.31	9.31	8.13	7.27	6.60	5.49	4.41	3.05	2.11	1.29	0.56	0.44	0.43	0.43	0.43
1.10	14.28	9.91	8.62	7.69	6.97	5.76	4.59	3.13	2.13	1.28	0.54	0.44	0.43	0.43	0.43
1.15	15.26	10.51	9.13	8.12	7.33	6.03	4.76	3.20	2.14	1.26	0.53	0.43	0.43	0.43	0.43
1.20	16.29	11.14	9.65	8.56	7.71	6.29	4.95	3.28	2.15	1.23	0.51	0.43	0.43	0.43	0.43
1.25	17.33	11.78	10.18	8.99	8.10	6.56	5.12	3.34	2.16	1.20	0.50	0.43	0.43	0.43	0.43
1.30	18.41	12.44	10.70	9.44	8.46	6.84	5.29	3.40	2.16	1.18	0.48	0.43	0.43	0.43	0.43
1.35	19.50	13.11	11.24	9.89	8.84	7.11	5.45	3.44	2.16	1.14	0.47	0.43	0.43	0.43	0.43
1.40	20.66	13.78	11.78	10.35	9.23	7.37	5.62	3.49	2.15	1.11	0.47	0.43	0.43	0.43	0.43
1.45	21.80	14.46	12.34	10.81	9.61	7.64	5.73	3.55	2.14	1.07	0.46	0.43	0.43	0.43	0.43
1.50	23.00	15.17	12.90	11.28	10.01	7.89	5.93	3.59	2.12	1.04	0.45	0.43	0.43	0.43	0.43
(8) $C_s = 4.0C_v$															
0.05	1.21	1.17	1.16	1:15	1.14	1.12	1.11	1.08	1.06	0.04	1.00	0.97	0.94	0.92	0.89
0.10	1.46	1.37	1.34	1.31	1.30	1.26	1.23	1.18	1.13	1.08	0.99	0.93	0.88	0.85	0.80
0.15	1.76	1.59	1.54	1.50	1.47	1.41	1.35	1.27	1.20	1.12	0.98	0.89	0.82	0.78	0.72
0.20	2.10	1.85	1.77	1.71	1.66	1.58	1.49	1.37	1.27	1.16	0.97	0.85	0.77	0.72	0.65
0.25	2.49	2.13	2.02	1.94	1.87	1.76	1.64	1.47	1.34	1.19	0.96	0.82	0.72	0.67	0.60
0.30	2.92	2.44	2.30	2.18	2.10	1.94	1.79	1.57	1.40	1.22	0.94	0.78	0.68	0.63	0.56
0.35	3.40	2.78	2.60	2.45	2.34	2.14	1.95	1.68	1.47	1.25	0.92	0.74	0.64	0.59	0.54
0.40	3.92	3.15	2.92	2.74	2.60	2.36	2.11	1.78	1.53	1.27	0.90	0.71	0.60	0.56	0.52
0.45	4.49	3.54	3.25	3.03	2.87	2.58	2.28	1.89	1.59	1.29	0.87	0.68	0.58	0.54	0.51
0.50	5.10	3.96	3.61	3.35	3.15	2.80	2.46	2.00	1.65	1.30	0.84	0.64	0.55	0.53	0.51
0.55	5.76	4.39	3.99	3.68	3.44	3.04	2.63	2.10	1.70	1.31	0.82	0.62	0.54	0.52	0.50
0.60	6.45	4.35	4.38	4.03	3.75	3.29	2.81	2.21	1.76	1.32	0.79	0.59	0.52	0.51	0.50
0.65	7.18	5.34	4.78	4.38	4.07	3.53	2.99	2.51	1.80	1.32	0.76	0.57	0.51	0.50	0.50
0.70	7.95	5.84	5.21	4.75	4.39	3.78	3.18	2.41	1.85	1.32	0.73	0.55	0.51	0.50	0.50
0.75	8.76	6.36	5.65	5.13	4.72	4.04	3.36	2.50	1.88	1.32	0.71	0.54	0.51	0.50	0.50
0.80	9.62	6.90	6.11	5.53	5.06	4.30	3.55	2.60	1.91	1.30	0.68	0.53	0.50	0.50	0.50
0.85	10.49	7.46	6.58	5.93	5.42	4.55	3.74	2.68	1.94	1.29	0.65	0.52	0.50	0.50	0.50
0.90	11.41	8.05	7.06	6.34	5.77	4.82	3.92	2.76	1.97	1.27	0.63	0.51	0.50	0.50	0.50
0.95	12.37	8.65	7.55	6.75	6.13	5.09	4.10	2.84	1.99	1.25	0.60	0.51	0.50	0.50	0.50
1.00	13.36	9.25	8.05	7.18	6.50	5.37	4.27	2.92	2.00	1.23	0.59	0.50	0.50	0.50	0.50

(continued)

(continued)

C_v	Cumulative frequency $P/\%$														
	0.01	0.1	0.2	0.33	0.5	1	2	5	10	20	50	75	90	95	99
1.05	14.38	9.87	8.57	7.62	6.87	5.63	4.46	3.00	2.01	1.20	0.57	0.50	0.50	0.50	0.50
1.10	15.43	10.52	9.10	8.05	7.25	5.91	4.63	3.06	2.01	1.18	0.56	0.50	0.50	0.50	0.50
1.15	16.51	11.18	9.62	8.50	7.62	6.18	4.80	3.12	2.01	1.15	0.54	0.50	0.50	0.50	0.50
1.20	17.62	11.85	10.17	8.96	8.01	6.45	4.96	3.16	2.01	1.11	0.53	0.50	0.50	0.50	0.50
1.25	18.78	12.52	10.71	9.41	8.40	6.71	5.12	3.21	2.00	1.07	0.53	0.50	0.50	0.50	0.50
1.30	19.94	13.22	11.27	9.88	8.79	6.96	5.29	3.25	1.99	1.04	0.52	0.50	0.50	0.50	0.50
1.35	21.14	13.92	11.83	10.33	9.17	7.24	5.44	3.30	1.97	1.00	0.52	0.50	0.50	0.50	0.50
1.40	22.38	14.64	12.40	10.80	9.55	7.50	5.59	3.32	1.94	0.96	0.51	0.50	0.50	0.50	0.50
1.45	23.65	15.37	12.09	11.27	9.95	7.77	5.74	3.36	1.91	0.93	0.51	0.50	0.50	0.50	0.50
1.50	24.91	16.10	13.57	11.72	10.34	8.02	5.88	3.38	1.88	0.90	0.51	0.50	0.50	0.50	0.50
(9) $C_s = 5.0C_v$															
0.05	1.21	1.17	1.16	1.15	1.14	1.13	1.11	1.09	1.07	1.04	1.00	0.97	0.94	0.92	0.89
0.10	1.48	1.38	1.35	1.33	1.30	1.27	1.23	1.18	1.13	1.08	0.99	0.93	0.88	0.85	0.80
0.15	1.81	1.63	1.57	1.53	1.49	1.43	1.36	1.27	1.20	1.12	0.98	0.89	0.82	0.79	0.73
0.20	2.19	1.91	1.82	1.75	1.70	1.60	1.51	1.38	1.27	1.15	0.97	0.85	0.77	0.74	0.68
0.25	2.63	2.22	2.10	2.00	1.93	1.80	1.66	1.48	1.34	1.18	0.95	0.81	0.74	0.69	0.65
0.30	3.13	2.67	2.40	2.27	2.17	2.00	1.82	1.58	1.40	1.21	0.93	0.78	0.69	0.66	0.62
0.35	3.68	2.95	2.74	2.57	2.44	2.21	1.99	1.69	1.46	1.23	0.90	0.75	0.67	0.64	0.61
0.40	4.28	3.36	3.09	2.88	2.72	2.44	2.16	1.80	1.52	1.24	0.88	0.72	0.64	0.62	0.60
0.45	4.94	3.81	3.47	3.22	3.01	2.68	2.34	1.90	1.56	1.25	0.85	0.69	0.63	0.61	0.60
0.50	5.65	4.28	3.87	3.57	3.32	2.92	2.52	2.00	1.62	1.26	0.82	0.67	0.61	0.60	0.60
0.55	6.40	4.77	4.28	3.93	3.65	3.17	2.71	2.11	1.67	1.26	0.79	0.65	0.61	0.60	0.60
0.60	7.21	5.29	4.72	4.31	3.98	3.43	2.88	2.20	1.71	1.25	0.77	0.63	0.61	0.60	0.60
0.65	8.07	5.83	5.18	4.71	4.32	3.69	3.08	2.30	1.73	1.24	0.74	0.62	0.60	0.60	0.60
0.70	8.96	6.40	5.66	5.10	4.68	3.95	3.26	2.38	1.76	1.22	0.71	0.62	0.60	0.60	0.60
0.75	9.90	7.00	6.14	5.52	5.03	4.22	3.44	2.46	1.79	1.20	0.68	0.61	0.60	0.60	0.60
0.80	10.89	7.60	6.64	5.94	5.40	4.50	3.61	2.54	1.80	1.18	0.67	0.61	0.60	0.60	0.60
0.85	11.91	8.23	7.16	6.48	5.77	4.76	3.80	2.61	1.81	1.15	0.65	0.60	0.60	0.60	0.60
0.90	12.97	8.88	7.69	6.81	6.15	5.03	3.97	2.66	1.81	1.13	0.64	0.60	0.60	0.60	0.60
0.95	14.07	9.55	8.22	7.27	6.53	5.30	4.14	2.72	1.81	1.10	0.63	0.60	0.60	0.60	0.60
1.00	15.22	10.20	8.77	7.73	6.92	5.57	4.30	2.77	1.80	1.06	0.62	0.60	0.60	0.60	0.60
1.05	16.39	10.92	9.33	8.19	7.31	5.82	4.47	2.81	1.79	1.03	0.62	0.60	0.60	0.60	0.60
1.10	17.61	11.63	9.89	8.66	7.69	6.09	4.61	2.85	1.77	0.99	0.61	0.60	0.60	0.60	0.60
1.15	18.87	12.34	10.48	9.12	8.08	6.36	4.76	2.89	1.74	0.95	0.61	0.60	0.60	0.60	0.60
1.20	20.13	13.08	11.06	9.58	8.46	6.62	4.90	2.91	1.71	0.92	0.61	0.60	0.60	0.60	0.60
1.25	21.46	13.83	11.64	10.06	8.86	6.88	5.03	2.93	1.68	0.88	0.60	0.60	0.60	0.60	0.60
(10) $C_s = 6.0C_v$															
0.05	1.22	1.18	1.16	1.15	1.14	1.13	1.11	1.09	1.06	1.04	1.00	0.97	0.94	0.93	0.91
0.10	1.51	1.40	1.36	1.34	1.31	1.28	1.24	1.18	1.13	1.08	0.99	0.93	0.88	0.86	0.81
0.15	1.86	1.66	1.60	1.55	1.51	1.45	1.38	1.28	1.20	1.12	0.98	0.89	0.83	0.81	0.76
0.20	2.28	1.96	1.86	1.79	1.73	1.63	1.52	1.38	1.27	1.15	0.96	0.85	0.78	0.75	0.71
0.25	2.77	2.31	2.16	2.06	1.98	1.83	1.69	1.48	1.33	1.17	0.94	0.82	0.75	0.72	0.69

(continued)

(continued)

C_v	Cumulative frequency P/%														
	0.01	0.1	0.2	0.33	0.5	1	2	5	10	20	50	75	90	95	99
0.30	3.33	2.69	2.50	2.36	2.24	2.05	1.86	1.59	1.40	1.19	0.92	0.78	0.72	0.69	0.67
0.35	3.95	3.11	2.87	2.68	2.53	2.28	2.03	1.69	1.45	1.21	0.89	0.76	0.70	0.68	0.67
0.40	4.63	3.57	3.25	3.02	2.83	2.52	2.21	1.80	1.50	1.22	0.86	0.73	0.68	0.67	0.67
0.45	5.39	4.06	3.66	3.38	3.15	2.77	2.39	1.90	1.54	1.22	0.83	0.71	0.68	0.67	0.67
0.50	6.10	4.58	4.10	3.76	3.48	3.02	2.58	2.00	1.59	1.21	0.80	0.69	0.67	0.67	0.67
0.55	7.03	5.12	4.50	4.16	3.83	3.28	2.76	2.09	1.62	1.20	0.78	0.69	0.67	0.67	0.67
0.60	7.94	5.70	5.04	4.56	4.18	3.55	2.94	2.18	1.65	1.18	0.75	0.68	0.67	0.67	0.67
0.65	8.90	6.30	5.53	4.97	4.54	3.82	3.12	2.25	1.66	1.16	0.73	0.68	0.67	0.67	0.67
0.70	9.92	6.92	6.05	5.41	4.91	4.09	3.30	2.33	1.67	1.13	0.71	0.67	0.67	0.67	0.67
0.75	10.98	7.56	6.57	5.85	5.29	4.36	3.47	2.39	1.68	1.10	0.70	0.67	0.67	0.67	0.67
0.80	12.08	8.23	7.11	6.30	5.67	4.63	3.64	2.44	1.67	1.07	0.69	0.67	0.67	0.67	0.67
0.85	13.24	8.91	7.66	6.76	6.06	4.89	3.80	2.49	1.66	1.08	0.68	0.67	0.67	0.67	0.67
0.90	14.43	9.61	8.22	7.22	6.45	5.16	3.96	2.53	1.65	1.00	0.68	0.67	0.67	0.67	0.67
0.95	15.68	10.33	8.80	7.68	6.83	5.42	4.10	2.56	1.62	0.96	0.67	0.67	0.67	0.67	0.67
1.00	16.94	11.07	9.38	8.15	7.22	5.68	4.25	2.59	1.59	0.93	0.67	0.67	0.67	0.67	0.67
1.05	18.27	11.82	9.97	8.62	7.62	5.94	4.38	2.61	1.56	0.89	0.67	0.67	0.67	0.67	0.67

Schedule 3: λ_{Pn} for Type I Extreme Value Distribution

n	Cumulative frequency P/%							
	0.1	0.2	0.5	1	2	4	5	10
8	7.103	6.336	5.321	4.551	3.779	3.001	2.749	1.953
9	6.909	6.162	5.174	4.425	3.673	2.916	2.670	1.895
10	6.752	6.021	5.055	4.322	3.587	2.847	2.606	1.848
11	6.622	5.905	4.957	4.238	3.516	2.789	2.553	1.809
12	6.513	5.807	4.874	4.166	3.456	2.741	2.509	1.777
13	6.418	5.723	4.802	4.105	3.405	2.699	2.470	1.748
14	6.337	5.650	4.741	4.052	3.360	2.663	2.437	1.724
15	6.265	5.586	4.687	4.005	3.321	2.632	2.408	1.703
16	6.196	5.523	4.634	3.959	3.283	2.601	2.379	1.682
17	6.137	5.471	4.589	3.921	3.250	2.575	2.355	1.664
18	6.087	5.426	4.551	3.888	3.223	2.552	2.335	1.649
19	6.043	5.387	4.518	3.860	3.199	2.533	2.317	1.636
20	6.006	5.354	4.490	3.836	3.179	2.517	2.302	1.625
22	5.933	5.288	4.435	3.788	3.139	2.484	2.272	1.603
24	5.870	5.232	4.387	3.747	3.104	2.457	2.246	1.584
26	5.816	5.183	4.346	3.711	3.074	2.433	2.224	1.568
28	5.769	5.141	4.310	3.680	3.048	2.412	2.205	1.553
30	5.727	5.104	4.279	3.653	3.026	2.393	2.188	1.541
35	5.642	5.027	4.214	3.598	2.979	2.356	2.153	1.515
40	5.576	4.968	4.164	3.554	2.942	2.326	2.126	1.495
45	5.522	4.920	4.123	3.519	2.913	2.303	2.104	1.479
50	5.479	4.881	4.090	3.491	2.889	2.283	2.087	1.466
60	5.410	4.820	4.038	3.446	2.852	2.253	2.059	1.446
70	5.359	4.774	4.000	3.413	2.824	2.230	2.038	1.430

(continued)

X. Feng (ed.), *Engineering Hydrology*, https://doi.org/10.1007/978-981-95-9089-6

(continued)

n	Cumulative frequency P/%							
	0.1	0.2	0.5	1	2	4	5	10
80	5.319	4.738	3.970	3.387	2.802	2.213	2.022	1.419
90	5.287	4.710	3.945	3.366	2.784	2.199	2.008	1.409
100	5.261	4.686	3.925	3.349	2.770	2.187	1.998	1.401
200	5.130	4.568	3.826	3.263	2.698	2.129	1.944	1.362
500	5.032	4.481	3.752	3.200	2.645	2.086	1.905	1.333
1000	4.992	4.445	3.722	3.174	2.623	2.069	1.889	1.321
∞	4.936	4.395	3.679	3.137	2.592	2.044	1.886	1.305

n	Cumulative frequency P/%							
	25	50	75	90	95	97	99	99.9
8	0.842	−0.130	−0.897	−1.458	−1.749	−1.923	−2.224	−2.673
9	0.814	−0.133	−0.879	−1.426	−1.709	−1.879	−2.172	−2.609
10	0.790	−0.136	−0.865	−1.400	−1.677	−1.843	−2.129	−2.556
11	0.771	−0.138	−0.854	−1.378	−1.650	−1.813	−2.095	−2.514
12	0.755	−0.139	−0.844	−1.360	−1.628	−1.788	−2.065	−2.478
13	0.741	−0.141	−0.836	−1.345	−1.609	−1.769	−2.040	−2.447
14	0.729	−0.142	−0.829	−1.331	−1.592	−1.748	−2.018	−2.420
15	0.718	−0.143	−0.823	−1.320	−1.578	−1.732	−1.999	−2.396
16	0.708	−0.145	−0.817	−1.308	−1.564	−1.716	−1.980	−2.373
17	0.699	−0.146	−0.811	−1.299	−1.552	−1.703	−1.965	−2.354
18	0.692	−0.146	−0.807	−1.291	−1.541	−1.691	−1.951	−2.338
19	0.685	−0.147	−0.803	−1.283	−1.532	−1.681	−1.939	−2.323
20	0.680	−0.148	−0.800	−1.277	−1.525	−1.673	−1.930	−2.311
22	0.669	−0.149	−0.794	−1.265	−1.510	−1.657	−1.910	−2.287
24	0.659	−0.150	−0.788	−1.255	−1.497	−1.642	−1.893	−2.266
26	0.651	−0.151	−0.783	−1.246	−1.486	−1.630	−1.879	−2.249
28	0.644	−0.152	−0.779	−1.239	−1.477	−1.619	−1.866	−2.233
30	0.638	−0.153	−0.776	−1.232	−1.468	−1.610	−1.855	−2.219
35	0.625	−0.154	−0.768	−1.218	−1.451	−1.591	−1.832	−2.191
40	0.615	−0.155	−0.762	−1.208	−1.438	−1.576	−1.814	−2.170
45	0.607	−0.156	−0.758	−1.198	−1.427	−1.564	−1.800	−2.152
50	0.601	−0.157	−0.754	−1.191	−1.418	−1.553	−1.788	−2.138
60	0.591	−0.158	−0.748	−1.180	−1.404	−1.538	−1.770	−2.115
70	0.583	−0.159	−0.744	−1.172	−1.394	−1.526	−1.756	−2.098
80	0.577	−0.159	−0.740	−1.165	−1.386	−1.517	−1.746	−2.085
90	0.572	−0.160	−0.737	−1.160	−1.379	−1.510	−1.737	−2.075
100	0.568	−0.160	−0.735	−1.155	−1.374	−1.504	−1.720	−2.066

(continued)

(continued)

n	Cumulative frequency $P/\%$							
	25	50	75	90	95	97	99	99.9
200	0.549	−0.162	−0.723	−1.134	−1.347	−1.474	−1.694	−2.023
500	0.535	−0.164	−0.714	−1.117	−1.326	−1.451	−1.668	−1.990
1000	0.529	−0.164	−0.710	−1.110	−1.318	−1.442	−1.657	−1.976
∞	0.520	−0.164	−0.705	−1.110	−1.306	−1.428	−1.641	−1.957

Schedule 4: β for Poisson-Gumbel Compound Extreme Value Distribution

Return period/years	*n*	*N*						
		10	15	20	25	30	40	50
100	8	4.298	3.846	3.524	3.274	3.069	2.743	2.489
	10	4.317	3.887	3.582	3.344	3.149	2.841	2.601
	12	4.347	3.933	3.638	3.409	3.221	2.925	2.693
	14	4.381	3.978	3.691	3.468	3.286	2.997	2.773
	16	4.412	4.017	3.737	3.519	3.341	3.059	2.839
	18	4.446	4.058	3.783	3.568	3.393	3.117	2.901
	20	4.485	4.103	3.831	3.620	3.447	3.174	2.962
	22	4.519	4.141	3.872	3.664	3.493	3.223	3.014
	24	4.551	4.177	3.911	3.704	3.536	3.269	3.062
	26	4.581	4.211	3.947	3.743	3.576	3.311	3.106
	28	4.611	4.243	3.982	3.779	3.613	3.351	3.147
	30	4.639	4.274	4.015	3.813	3.649	3.389	3.187
	35	4.706	4.346	4.091	3.893	3.730	3.474	3.275
	40	4.768	4.412	4.159	3.963	3.803	3.550	3.353
	45	4.824	4.472	4.221	4.027	3.868	3.618	3.423
	50	4.876	4.527	4.278	4.086	3.928	3.679	3.486
	60	4.971	4.625	4.380	4.190	4.034	3.788	3.598
	70	5.054	4.711	4.468	4.280	4.126	3.882	3.693
	80	5.128	4.788	4.547	4.360	4.207	3.965	3.778
	90	5.195	4.858	4.618	4.432	4.280	4.040	3.853
	100	5.257	4.920	4.682	4.497	4.345	4.106	3.921
50	8	3.518	3.063	2.738	2.483	2.274	1.941	1.679
	10	3.576	3.144	2.836	2.595	2.398	2.083	1.837
	12	3.633	3.216	2.919	2.688	2.498	2.196	1.960

(continued)

X. Feng (ed.), *Engineering Hydrology*, https://doi.org/10.1007/978-981-95-9089-6

(continued)

Return period/years	n	N						
		10	15	20	25	30	40	50
	14	3.686	3.281	2.992	2.768	2.583	2.291	2.062
	16	3.732	3.336	3.054	2.834	2.655	2.369	2.147
	18	3.778	3.389	3.112	2.896	2.720	2.440	2.222
	20	3.826	3.442	3.169	2.957	2.783	2.507	2.292
	22	3.867	3.488	3.219	3.009	2.837	2.565	2.353
	24	3.906	3.531	3.264	3.057	2.887	2.618	2.409
	26	3.943	3.571	3.307	3.101	2.933	2.667	2.460
	28	3.977	3.609	3.346	3.143	2.976	2.712	2.507
	30	4.010	3.644	3.384	3.182	3.016	2.755	2.551
	35	4.086	3.726	3.470	3.271	3.108	2.850	2.650
	40	4.155	3.799	3.545	3.349	3.188	2.934	2.736
	45	4.217	3.864	3.613	3.419	3.259	3.008	2.812
	50	4.274	3.924	3.675	3.482	3.324	3.074	2.880
	60	4.376	4.030	3.784	3.593	3.437	3.191	3.001
	70	4.464	4.121	3.878	3.689	3.535	3.291	3.101
	80	4.543	4.203	3.961	3.774	3.620	3.378	3.190
	90	4.614	4.275	4.035	3.849	3.697	3.456	3.269
	100	4.678	4.341	4.102	3.917	3.765	3.526	3.340
25	8	2.726	2.263	1.929	1.667	1.449	1.098	0.816
	10	2.825	2.386	2.072	1.826	1.622	1.294	1.034
	12	2.909	2.487	2.186	1.950	1.755	1.443	1.196
	14	2.982	2.573	2.280	2.052	1.863	1.563	1.325
	16	3.044	2.644	2.359	2.136	1.953	1.661	1.431
	18	3.102	2.710	2.430	2.212	2.032	1.747	1.522
	20	3.159	2.773	2.497	2.282	2.106	1.825	1.605
	22	3.209	2.827	2.556	2.344	2.169	1.893	1.676
	24	3.255	2.877	2.609	2.399	2.227	1.954	1.740
	26	3.297	2.924	2.658	2.450	2.280	2.010	1.799
	28	3.337	2.967	2.703	2.498	2.329	2.062	1.853
	30	3.375	3.007	2.745	2.542	2.375	2.110	1.903
	35	3.460	3.099	2.841	2.641	2.476	2.216	2.013
	40	3.536	3.179	2.925	2.727	2.565	2.308	2.108
	45	3.604	3.250	2.999	2.803	2.643	2. 389	2. 191
	50	3.666	3.315	3.065	2.871	2.713	2.461	2.265
	60	3.775	3.429	3.182	2.991	2.834	2.586	2.393
	70	3.869	3.526	3.282	3.093	2.938	2.692	2.502

(continued)

(continued)

Return period/years	n	N						
		10	15	20	25	30	40	50
	80	3.953	3.612	3.370	3.182	3.028	2.785	2.596
	90	4.027	3.688	3.448	3.261	3.108	2.866	2.679
	100	4.094	3.757	3.518	3.332	3.180	2.940	2.753
10	8	1.629	1.138	0.775	0.479	0.225	−0.215	−0.615
	10	1.790	1.332	0.996	0.726	0.498	0.113	−0.216
	12	1.916	1.479	1.160	0.907	0.693	0.340	0.045
	14	2.019	1.597	1.291	1.048	0.845	0.512	0.238
	16	2.104	1.694	1.398	1.163	0.967	0.649	0.390
	18	2.180	1.779	1.490	1.261	1.072	0.764	0.515
	20	2.252	1.857	1.573	1.349	1.163	0.863	0.622
	22	2.313	1.924	1.645	1.425	1.243	0.950	0.715
	24	2.369	1.985	1.710	1.493	1.314	1.026	0.797
	26	2.421	2.041	1.769	1.555	1.379	1.095	0.870
	28	2.468	2.092	1.823	1.612	1.437	1.158	0.936
	30	2.512	2. 140	1.873	1.664	1.491	1.215	0.996
	35	2.612	2.246	1.984	1.779	1.610	1.341	1.128
	40	2.698	2.337	2.079	1.878	1.712	1.447	1.239
	45	2.775	2.418	2.163	1.964	1.800	1.539	1.334
	50	2.844	2.490	2.237	2.040	1.878	1.621	1.419
	60	2.963	2.614	2.366	2.172	2.013	I. 760	1.562
	70	3.066	2.720	2.474	2.283	2.126	1.876	1.681
	80	3.155	2.812	2.569	2.379	2.223	1.977	1.784
	90	3.234	2.894	2.652	2.463	2.309	2.064	1.874
	100	3.305	2.967	2.726	2.539	2.386	2.143	1.953

Schedule 5: Normal Probability Integral Table $\left[P(t) = \int_{t}^{\infty} \frac{1}{\sqrt{2\pi}} \exp\left(-\frac{t^2}{2}\right) dt\right]$

t	0.00	0.01	0.02	0.03	0.04	0.05	0.06	0.07	0.08	0.09
0.0	0.5000	0.4960	0.4920	0.4880	0.4840	0.4801	0.4761	0.4721	0.4681	0.4641
0.1	0.4602	0.4562	0.4522	0.4483	0.4443	0.4404	0.4364	0.4325	0.4286	0.4247
0.2	0.4207	0.4168	0.4129	0.4090	0.4052	0.4013	0.3974	0.3936	0.3897	0.3859
0.3	0.3821	0.3783	0.3745	0.3707	0.3669	0.3632	0.3594	0.3557	0.3520	0.3483
0.4	0.3446	0.3409	0.3372	0.3336	0.3300	0.3264	0.3228	0.3192	0.3156	0.3121
0.5	0.3085	0.3050	0.3015	0.2981	0.2946	0.2912	0.2877	0.2843	0.2810	0.2776
0.6	0.2743	0.2709	0.2676	0.2643	0.2611	0.2578	0.2546	0.2514	0.2483	0.2451
0.7	0.2420	0.2389	0.2358	0.2327	0.2297	0.2266	0.2236	0.2206	0.2177	0.2148
0.8	0.2119	0.2090	0.2061	0.2033	0.2005	0.1977	0.1949	0.1922	0.1894	0.1867
0.9	0.1841	0.1814	0.1788	0.1762	0.1736	0.1711	0.1685	0.1660	0.1635	0.1611
1.0	0.1587	0.1562	0.1539	0.1515	0.1492	0.1469	0.1446	0.1423	0.1401	0.1379
1.1	0.1357	0.1335	0.1314	0.1292	0.1271	0.1251	0.1230	0.1210	0.1190	0.1170
1.2	0.1151	0.1131	0.1112	0.1093	0.1075	0.1056	0.1038	0.1020	0.1003	0.0985
1.3	0.0968	0.0951	0.0934	0.0918	0.0901	0.0885	0.0869	0.0853	0.0838	0.0823
1.4	0.0808	0.0793	0.0778	0.0764	0.0749	0.0736	0.0722	0.0708	0.0694	0.0681
1.5	0.0668	0.0655	0.0643	0.0630	0.0618	0.0608	0.0594	0.0582	0.0571	0.0559
1.6	0.0548	0.0537	0.0526	0.0516	0.0505	0.0495	0.0485	0.0475	0.0465	0.0455
1.7	0.0446	0.0436	0.0427	0.0418	0.0409	0.0401	0.0392	0.0384	0.0375	0.0367
1.8	0.0359	0.0352	0.0344	0.0336	0.0329	0.0322	0.0314	0.0307	0.0301	0.0294
1.9	0.0287	0.0281	0.0274	0.0268	0.0262	0.0256	0.0250	0.0244	0.0239	0.0233
2.0	0.0228	0.0222	0.0217	0.0212	0.0207	0.0202	0.0197	0.0192	0.0188	0.0183
2.1	0.0179	0.0174	0.0170	0.0166	0.0162	0.0158	0.0154	0.0150	0.0146	0.0143
2.2	0.0139	0.0136	0.0132	0.0129	0.0126	0.0122	0.0119	0.0116	0.0113	0.0110
2.3	0.0107	0.0104	0.0102	0.0099	0.0096	0.0094	0.0091	0.0089	0.0087	0.0084
2.4	0.0082	0.0080	0.0078	0.0075	0.0073	0.0071	0.0069	0.0068	0.0066	0.0064

(continued)

X. Feng (ed.), *Engineering Hydrology*, https://doi.org/10.1007/978-981-95-9089-6

(continued)

t	0.00	0.01	0.02	0.03	0.04	0.05	0.06	0.07	0.08	0.09
–0.0	0.5000	0.5040	0.5080	0.5120	0.5160	0.5199	0.5239	0.5276	0.5319	0.5359
−0.1	0.5398	0.5438	0.5478	0.5517	0.5557	0.5596	0.5636	0.5675	0.5714	0.5753
−0.2	0.5793	0.5832	0.5871	0.5910	0.5948	0.5987	0.6026	0.6064	0.6103	0.6141
−0.3	0.6179	0.6217	0.6255	0.6293	0.6331	0.6368	0.6406	0.6443	0.6480	0.6517
−0.4	0.6554	0.6591	0.6628	0.6664	0.6700	0.6736	0.6772	0.6808	0.6844	0.6879
−0.5	0.6915	0.6950	0.6985	0.7019	0.7054	0.7088	0.7123	0.7157	0.7190	0.7224
−0.6	0.7257	0.7291	0.7324	0.7357	0.7389	0.7422	0.7454	0.7486	0.7517	0.7549
−0.7	0.7580	0.7611	0.7642	0.7673	0.7703	0.7734	0.7764	0.7794	0.7823	0.7852
−0.8	0.7881	0.7910	0.7939	0.7967	0.7995	0.8023	0.8051	0.8078	0.8106	0.8133
−0.9	0.8159	0.8186	0.8212	0.8238	0.8264	0.8289	0.8315	0.8340	0.8365	0.8389
−1.0	0.8413	0.8438	0.8461	0.8485	0.8508	0.8531	0.8554	0.8577	0.8599	0.8621
−1.1	0.8643	0.8665	0.8686	0.8708	0.8729	0.8749	0.8770	0.8790	0.8810	0.8830
−1.2	0.8849	0.8869	0.8888	0.8970	0.8925	0.8944	0.8962	0.8980	0.8997	0.9015
−1.3	0.9032	0.9049	0.9066	0.9082	0.9099	0.9115	0.9131	0.9147	0.9162	0.9177
−1.4	0.9192	0.9207	0.9222	0.9236	0.9251	0.9265	0.9278	0.9292	0.9306	0.9319
−1.5	0.9332	0.9345	0.9357	0.9370	0.9382	0.9394	0.9406	0.9418	0.9429	0.9441
−1.6	0.9452	0.9463	0.9474	0.9485	0.9495	0.9505	0.9515	0.9525	0.9535	0.9545
−1.7	0.9554	0.9564	0.9573	0.9582	0.9591	0.9599	0.9608	0.9616	0.9625	0.9633
−1.8	0.9641	0.9649	0.9656	0.9664	0.9671	0.9678	0.9686	0.9693	0.9699	0.9706
−1.9	0.9713	0.9719	0.9726	0.9732	0.9738	0.9744	0.9750	0.9756	0.9761	0.9767
−2.0	0.9772	0.9778	0.9783	0.9788	0.9793	0.9798	0.9803	0.9808	0.9812	0.9817
−2.1	0.9821	0.9826	0.9830	0.9834	0.9838	0.9842	0.9846	0.9850	0.9854	0.9857
−2.2	0.9861	0.9864	0.9868	0.9871	0.9875	0.9878	0.9881	0.9884	0.9887	0.9890
−2.3	0.9893	0.9896	0.9898	0.9901	0.9904	0.9906	0.9909	0.9911	0.9913	0.9916
−2.4	0.9918	0.9920	0.9922	0.9925	0.9927	0.9929	0.9931	0.9932	0.9934	0.9936

Schedule 6: Relationship Table for Shallow Water Wave Height, Wave Velocity, Wavelength, and Relative Water Depth

$\frac{d}{L_0}$	$\frac{d}{L}$	$\frac{C}{C_0}$ and $\frac{L}{L_0}$	$\frac{H}{H_0}$	$\frac{d}{L_0}$	$\frac{d}{L}$	$\frac{C}{C_0}$ and $\frac{L}{L_0}$	$\frac{H}{H_0}$
0.0020	0.0179	0.1119	2.1190	0.0800	0.1232	0.6493	0.9548
0.0025	0.0200	0.1250	2.0050	0.0850	0.1277	0.6655	0.9481
0.0030	0.0219	0.1369	1.9170	0.0900	0.1322	0.6808	0.9422
0.0035	0.0237	0.1477	1.8470	0.0950	0.1366	0.6953	0.9371
0.0040	0.0253	0.1579	1.7880	0.1000	0.1410	0.7093	0.9327
0.0045	0.0269	0.1674	1.7370	0.1050	0.1453	0.7226	0.9290
0.0050	0.0284	0.1764	1.6920	0.1100	0.1496	0.7352	0.9257
0.0055	0.0298	0.1848	1.6540	0.1150	0.1539	0.7474	0.9228
0.0060	0.0311	0.1929	1.6200	0.1200	0.1581	0.7589	0.9204
0.0065	0.0324	0.2007	1.5890	0.1250	0.1624	0.7700	0.9186
0.0070	0.0336	0.2082	1.5610	0.1300	0.1665	0.7804	0.9169
0.0075	0.0348	0.2154	1.5360	0.1350	0.1708	0.7905	0.9156
0.0080	0.0360	0.2223	1.5120	0.1400	0.1749	0.8002	0.9146
0.0085	0.0371	0.2290	1.4910	0.1450	0.1791	0.8094	0.9139
0.0090	0.0382	0.2356	1.4710	0.1500	0.1833	0.8183	0.9133
0.0095	0.0393	0.2419	1.4520	0.1550	0.1875	0.8267	0.9131
0.0100	0.0403	0.2480	1.4350	0.1600	0.1917	0.8349	0.9130
0.0150	0.0496	0.3022	1.3070	0.1650	0.1958	0.8427	0.9131
0.0200	0.0576	0.3470	1.2260	0.1700	0.2000	0.8501	0.9134
0.0250	0.0648	0.3860	1.1680	0.1750	0.2042	0.8572	0.9139
0.0300	0.0714	0.4205	1.1250	0.1800	0.2083	0.8640	0.9145
0.0350	0.0775	0.4517	1.0920	0.1850	0.2125	0.8706	0.9152
0.0400	0.0833	0.4802	1.0640	0.1900	0.2167	0.8767	0.9161
0.0450	0.0888	0.5066	1.0420	0.1950	0.2209	0.8827	0.9170
0.0500	0.0942	0.5310	1.0230	0.2000	0.2251	0.8884	0.9181

(continued)

X. Feng (ed.), *Engineering Hydrology*, https://doi.org/10.1007/978-981-95-9089-6

(continued)

$\frac{d}{L_0}$	$\frac{d}{L}$	$\frac{C}{C_0}$ and $\frac{L}{L_0}$	$\frac{H}{H_0}$	$\frac{d}{L_0}$	$\frac{d}{L}$	$\frac{C}{C_0}$ and $\frac{L}{L_0}$	$\frac{H}{H_0}$
0.0550	0.0993	0.5538	1.0070	0.2050	0.2293	0.8939	0.9193
0.0600	0.1043	0.5753	0.9932	0.2100	0.2336	0.8991	0.9205
0.0650	0.1092	0.5954	0.9815	0.2150	0.2378	0.9041	0.9218
0.0700	0.1139	0.6144	0.9713	0.2200	0.2421	0.9088	0.9231
0.0750	0.1186	0.6324	0.9624	0.2250	0.2463	0.9134	0.9245
0.2300	0.2506	0.9178	0.9261	0.3800	0.3860	0.9845	0.9717
0.2350	0.2549	0.9219	0.9276	0.3850	0.3907	0.9854	0.9728
0.2400	0.2592	0.9259	0.9291	0.3900	0.3955	0.9862	0.9739
0.2450	0.2635	0.9296	0.9307	0.3950	0.4002	0.9870	0.9750
0.2500	0.2679	0.9332	0.9323	0.4000	0.4050	0.9877	0.9761
0.2550	0.2722	0.9367	0.9340	0.4050	0.4098	0.9885	0.9770
0.2600	0.2766	0.9400	0.9356	0.4100	0.4145	0.9891	0.9780
0.2650	0.2810	0.9431	0.9373	0.4150	0.4193	0.9898	0.9790
0.2700	0.2854	0.9461	0.9390	0.4200	0.4241	0.9904	0.9798
0.2750	0.2898	0.9490	0.9406	0.4250	0.4289	0.9909	0.9808
0.2800	0.2942	0.9516	0.9423	0.4300	0.4337	0.9914	0.9816
0.2850	0.2987	0.9542	0.9440	0.4350	0.4385	0.9919	0.9824
0.2900	0.3031	0.9567	0.9456	0.4400	0.4434	0.9924	0.9832
0.2950	0.3076	0.9590	0.9473	0.4450	0.4482	0.9929	0.9839
0.3000	0.3121	0.9611	0.9490	0.4500	0.4531	0.9933	0.9847
0.3050	0.3166	0.9633	0.9505	0.4550	0.4579	0.9937	0.9852
0.3100	0.3211	0.9653	0.9522	0.4600	0.4628	0.9941	0.9860
0.3150	0.3257	0.9672	0.9538	0.4650	0.4676	0.9944	0.9867
0.3200	0.3302	0.9690	0.9553	0.4700	0.4725	0.9947	0.9873
0.3250	0.3349	0.9707	0.9568	0.4750	0.4774	0.9951	0.9878
0.3300	0.3394	0.9723	0.9583	0.4800	0.4822	0.9953	0.9885
0.3350	0.3440	0.9738	0.9598	0.4850	0.4871	0.9956	0.9890
0.3400	0.3468	0.9753	0.9613	0.4900	0.4920	0.9959	0.9896
0.3450	0.3532	0.9767	0.9626	0.4950	0.4969	0.9961	0.9900
0.3500	0.3579	0.9780	0.9640	0.5000	0.5018	0.9964	0.9905
0.3550	0.3625	0.9792	0.9654	0.6000	0.6006	0.9990	0.9965
0.3600	0.3672	0.9804	0.9667	0.7000	0.7002	0.9997	0.9988
0.3650	0.3719	0.9815	0.9680	0.8000	0.8001	0.9999	0.9996
0.3700	0.3766	0.9825	0.9693	0.9000	0.9000	1.0000	0.9999
0.3750	0.3813	0.9835	0.9705	1.0000	1.0000	1.0000	1.0000

The manufacturer's authorised representative in the EU is Springer Nature Customer Service Centre GmbH, Europaplatz 3, 69115 Heidelberg, Germany. If you have any concerns regarding our products, please contact ProductSafety@springernature.com

Printed and bound by CPI Group (UK) Ltd, Croydon, CR0 4YY
07/10/2026
02229373-0002